Implementing Industry 4.0 in SMEs

Implementing Industry 4.0 in SMEs

Dominik T. Matt · Vladimír Modrák ·
Helmut Zsifkovits
Editors

Implementing Industry 4.0 in SMEs

Concepts, Examples and Applications

Editors
Dominik T. Matt
Faculty of Science and Technology
Free University of Bozen-Bolzano
Bolzano, Italy

Vladimír Modrák
Department of Manufacturing
Technologies
Technical University of Kosice
Presov, Slovakia

Helmut Zsifkovits
Chair of Industrial Logistics
Montanuniversität Leoben
Leoben, Austria

ISBN 978-3-030-70515-2   ISBN 978-3-030-70516-9   (eBook)
https://doi.org/10.1007/978-3-030-70516-9

This Palgrave Macmillan imprint is published by the registered company Springer Nature Switzerland AG
The registered company address is: Gewerbestrasse 11, 6330 Cham, Switzerland

Preface

The term Industry 4.0 describes the ongoing revolution of manufacturing industry around the world. Large companies in particular have rapidly embraced the challenges of Industry 4.0 and are currently working intensively on the introduction of the corresponding enabling technologies. Small- and medium-sized enterprises (SMEs) face the hurdle of possessing neither human nor financial resources to systematically investigate the potential and risks for introducing Industry 4.0. However, in most of the countries SMEs form the backbone of the economy, they account for the largest share of the gross domestic product and are also important employers. In this respect, concepts, examples and applications of Industry 4.0 have to be provided specifically for SMEs, thus paving the way for a successful implementation of Industry 4.0.

The central question in this book is: Which concepts, methods and tools can SMEs use to successfully implement Industry 4.0 in manufacturing, logistics and to digitalize the company organization? What practical examples of applications are there to give SMEs an insight into the experiences of other companies that have already dealt with the introduction of Industry 4.0?

With this book, the research consortium of the H2020 MSCA RISE project "SME 4.0—Industry 4.0 for SMEs" (grant agreement No 734713) encourages other researchers to conduct research in the field of Industry 4.0 specifically for SMEs and thus expanding the community in SME research. In addition, practical methods, instruments and best practice case studies should inspire practitioners from SMEs in introducing Industry 4.0 in their companies.

After a first book on challenges, opportunities and requirements for Industry 4.0 in SMEs, this second book focuses on the implementation of Industry 4.0 in SMEs. The editors and contributors provide not only helpful methods, instruments and examples but also valuable experiences from the collaboration with SMEs to implement and test different kinds of Industry 4.0 technologies and concepts.

We would like to thank all authors for their refreshing ideas and interesting contributions to this topic.

Bolzano, Italy   Dominik T. Matt
Košice, Slovakia   Vladimír Modrák
Leoben, Austria   Helmut Zsifkovits
January 2021

Acknowledgments

This project has received funding from the European Union's Horizon 2020 research and innovation program under the Marie Skłodowska-Curie grant agreement No 734713 (Project title: SME 4.0—Industry 4.0 for SMEs).

About This Book/Project

This book summarizes the research results of the second phase of the project "SME 4.0—Industry 4.0 for SMEs: Smart Manufacturing and Logistics for SMEs in an X-to-order and Mass Customization Environment" in the period from 2019 to 2021. The project started in January 2017 and is funded by the European Union's Horizon 2020 research and innovation program under the Marie Skłodowska-Curie grant agreement No 734713. In a previous book published with Palgrave Macmillan (book title: Industry 4.0 for SMEs: Challenges, Opportunities and Requirements), which is summarizing the first phase of the project (2017–2018), the main challenges, requirements as well as opportunities of Industry 4.0 were addressed in order to prepare SMEs for Industry 4.0. This second book is focusing on the implementation of Industry 4.0 in SMEs providing not only helpful theoretical concepts but also practical tools and instruments as well as examples of applications in SMEs.

A great opportunity for the future lies in the transfer of Industry 4.0 technologies in small- and medium-sized enterprises (SMEs). The above-mentioned research project aims to close and overcome the gap in this transfer through the establishment of an international and interdisciplinary research network for this topic. This network has the objectives of identifying the requirements, the challenges and the opportunities for a smart and intelligent SME factory, creating adapted concepts, instruments and technical solutions for manufacturing and logistics systems in SMEs and developing suitable organization and management models. The practical applicability of the results is guaranteed through a close collaboration of the network with small- and medium-sized enterprises from Europe, USA, Thailand and India.

The book is structured in three parts with a total of 12 chapters:

Part I—Implementing Industry 4.0 for Smart Manufacturing in SMEs

In the first part, the focus lies on manufacturing in SMEs. Many small- and medium-sized companies are currently planning the introduction of Industry 4.0 technologies in manufacturing. In the production area, manufacturing systems have to become smart and highly adaptable orchestrating the single intelligent elements in form of a fully connected cyber-physical production systems and to enable the cooperation of human and machines by means of flexible automation. In this part, we present on the one hand concepts for the implementation of smart and adaptable manufacturing systems as well as artificial intelligence and machine learning in manufacturing. This part further provides examples of practical applications for the introduction of human-machine interaction and worker assistance systems.

Part II—Implementing Industry 4.0 for Smart Logistics in SMEs

The second part concentrates on Industry 4.0 in logistics. Industry 4.0 and the accompanying digitalization have, in addition to the change in production, a major impact on logistics in companies. In this part, we use practical examples and studies to present the potential of Industry 4.0 for logistics and how this can be implemented in practice. For example, an industrial case study confirms the advantages of real-time data in planning and control or in the determination and monitoring of logistics KPIs. In addition, a readiness model is presented which SMEs can use in the context of the introduction of Industry 4.0 for self-assessment in logistics. Furthermore, this part investigates also the impact of human factors in smart logistics with a case study research.

Part III—Organizational and Management Models for Smart SMEs

The third part deals with organization and management models for smart SMEs. The introduction of Industry 4.0 in companies entails not only a technological change but also a change in the organization. In this part, therefore organizational and business models as well as tools to support SMEs in the phase of introducing Industry 4.0 are presented. This includes business model concepts, insights of a study about the impact of Industry 4.0 technologies on business models, a survey on Industry 4.0 awareness and case study-based research on appropriate implementation strategies.

Contents

Notes on Contributors

Assistant Professor Dr. Tanyanuparb Anantana earned both bachelor's and master's degree in Industrial Engineering with First Class Honors from Chiang Mai University (CMU) and a doctoral degree from Tokyo Institute of Technology, Japan with the research topic on innovation strategy and new product development approach of Japanese companies (more than 350 companies) in 14 major Japanese industries. He is the founder of Science and Technology Park, CMU and the chairman of 14 universities networking in Northern Science Park (NSP). He has expertise and over-15 years working experience in Innovation Management, Startup, Technology Transfer, Innovation Policy and Strategic Planning for various government agencies and private sectors.

Rungchat Chompu-inwai is currently an Associate Professor in Industrial Engineering at Chiang Mai University, Thailand. She received her B.Eng. degree in Industrial Engineering from Chiang Mai University, Thailand in 1997, M.Eng.Sc. degree in Manufacturing Engineering from The University of New South Wales, Sydney, Australia in 2000, and Ph.D. in Industrial and Manufacturing Engineering from Oregon State University, USA in 2006. Her teaching and research interests include lean manufacturing, process and quality improvement, application of Industrial Engineering (IE) techniques for productivity improvement, and economic analysis.

Patrick Dallasega received the degree(s) from the Free University of Bolzano, Bolzano, Italy, the Polytechnic University of Turin, Turin, Italy, and the Ph.D. degree from the University of Stuttgart, Stuttgart, Germany. He is an Assistant Professor of project management and industrial plants design with the Faculty of Science and Technology, the Free University of Bolzano. He was a Visiting Scholar with the Excellence Center in Logistics and Supply Chain Management Chiang Mai University, Chiang Mai, Thailand, and the Worcester Polytechnic Institute in Massachusetts, Worcester, MA, USA. His main research interests are supply chain management, Industry 4.0, lean construction, lean manufacturing and production planning, and control in MTO and ETO enterprises.

Darya Dancaková is an Assistant Professor in the Department of Banking and Investment at the Faculty of Economics, Technical University of Kosice. She completed her M.Sc. and Ph.D. in Finance at the same university. Her research interests lie in the area of intangible assets, intellectual capital, human resources management, Internet of Things and corporate finance. Dr. Dancaková has collaborated with other researchers in several projects, particularly focused on the Industry 4.0 implementation within SMEs and investments in intellectual capital.

Laura Davi is a Student in the degree course of Industrial and Mechanical Engineering at the Free University of Bolzano. Her research interest is on Industry 4.0 and the impact on business model development in small- and medium-sized enterprises.

Alexander Hošovský works as an Associate Professor at Faculty of Manufacturing Technologies with seat in Prešov, Technical University of Košice, Slovakia. His main research interests are soft actuators and soft robotics as well as computational intelligence (neural networks, fuzzy systems, evolutionary computing, PSO, artificial immune systems) and automatic control, modeling and identification of nonlinear systems and time series analysis. He is the author and co-author of more than 70 research papers in peer-reviewed journals and conference proceedings. In addition, he also authored or co-authored more than 30 utility models and/or national patents. In 2016, he was awarded Research and Technology Award for Young Researchers (under 35 years age) under Ministry of Education, Science, Research and Sport of the Slovak Republic. He served as a reviewers for many respected journal (e.g., IEEE Transaction on Industrial Electronics, Neural Computing and Applications, Materials and Design, IEEE Access and many more) and was (or is) a guest editors of three special issues in different journals (Advances in Mechanical Engineering, International Journal of Advanced Robotic Systems and Actuators).

Johannes A. Kapeller received the Dipl.-Ing. Degree in Industrial Logistics from the Montanuniversitaet Leoben, Austria. In 2018, he completed his Ph.D in Industrial Engineering with special focus on the sequential combination of production control strategies within the line production area. Since 2018, he worked as a management consultant at the Boston Consulting Group and currently as Vice President for Operational Excellence at All4Labels. Prior he has been a Postdoc at the Montanuniversitaet in Leoben and a visiting researcher within the research project "SME 4.0—Industry 4.0 for SMEs" at the University of Chiang Mai, Thailand. Dr. Kapeller published his work in high-quality peer-reviewed journals.

Benedikt G. Mark is a Ph.D. Candidate and Research Fellow in the Smart Mini Factory for Industry 4.0 at the Free University of Bozen-Bolzano in the research group Industrial Engineering and Automation (IEA). Previously, he studied food and packaging technology at the University of applied sciences of Kempten (Germany) and the Università Carlo Cattaneo (Italy), as well as industrial and mechanical engineering at the Free University of Bozen-Bolzano (Italy). His research area includes smart manufacturing and worker assistance systems in production. He is involved in the EU project "SME 4.0—Industry 4.0 for SMEs" as well as in industry projects on worker assistance and workplace design. In 2019, he was visiting researcher at the Worcester Polytechnic Institute (WPI).

Dominik T. Matt holds the Chair for Production Systems and Technologies and heads the research department "Industrial Engineering and Automation (IEA)" at the Faculty of Science and Technology at the Free University of Bozen-Bolzano. Moreover, Prof. Matt is the Director of the Research Center Fraunhofer Italia in Bolzano. Prof. Matt coordinates as Principal Investigator the Horizon 2020 research project SME 4.0 as Lead Partner. The research of Prof. Matt focuses primarily on the areas of Industry 4.0 and Smart Factory, Lean and Agile Production, Artificial Intelligence in manufacturing, on the planning and optimization of assembly processes and systems, as well as on organizational and technical aspects of in-house logistics. He has authored more than 200 scientific and technical papers in journals and conference proceedings and is member of numerous national and international scientific organizations and committees (e.g., AITeM—Associazione Italiana di Tecnologia Meccanica|Acatech—National Academy of Science and Engineering—Germany|WGAB—Academic Society for Work and Industrial Organization|EVI—European Virtual Institute on Innovation in Industrial Supply Chains and Logistic Networks).

Vladimír Modrák is a Full Professor of Manufacturing Technology at Faculty of Manufacturing Technologies of Technical University of Kosice. He obtained his Ph.D. degree at the same University in 1989. His research interests include cellular manufacturing systems design, mass customized manufacturing and planning/scheduling optimization. Since 2015, he is a Fellow of the European Academy for Industrial Management (AIM). He was the leading editor of three international books, Operations Management Research and Cellular Manufacturing Systems, Handbook of Research on Design and Management of Lean Production Systems, and Mass Customized Manufacturing: Theoretical Concepts and Practical Approaches. He is also active as editorial board member in several scientific journals and committee member of many international conferences.

Guido Orzes received his M.Sc. in Management Engineering from the University of Udine (Italy) in 2011 with summa cum laude. In 2015, he obtained his Ph.D. degree in Industrial and Information Engineering (topic: operations management) from the University of Udine

(Italy). Currently, he is an Assistant Professor in Management Engineering at the Free University of Bozen-Bolzano (Italy). He is also Honorary Research Fellow at the University of Exeter Business School (UK) and Visiting Scholar at the Worcester Polytechnic Institute (USA). His research focuses on Industry 4.0 and sustainability and international sourcing and manufacturing and their social and environmental implications. He has published over 70 scientific works on these topics in leading operations management and international business journals as well as in conference proceedings and books. He is involved in various research projects on global operations management and Industry 4.0. He is also Associate Editor of the Electronic Journal of Business Research Methods and member of the board of the European division of the Decision Science Institute.

Corina Pacher is an Education Project Manager at the Resources Innovation Center (RIC) of the Montanuniversitaet Leoben, Austria. She studied pedagogical and educational science at the University of Klagenfurt with a specialization on social and inclusive education as well as on professional education. During and after her studies, she gained work experience, e.g., as the head of educational programs and in different social public service enterprises as social education worker. Currently, she is mainly focusing on raising the awareness for a resource-orientated handling of raw materials by connecting research, education, and society, e.g., by building and expanding skills and competencies of students through training initiatives.

Ján Piteľ is a Full Professor in automation at the Technical University of Košice (Slovakia) and he currently works at the Faculty of Manufacturing Technologies with a seat in Prešov as vice-dean and head of the Institute of Production Control. His research activities include modeling, simulation, automatic control and monitoring of machines and processes. He is author of more than 60 papers registered in databases WoS and Scopus with more than 200 SCI citations. He is inventor and co-inventor more than 50 patents and utility models. He has been leader many national projects (e.g., EU Structural Funds projects), and currently he participates in 2 EC-funded projects (H2020, Erasmus+). As vice-dean for

xxii   Notes on Contributors

external relationships, he is responsible for mobility programs under the framework of projects ERASMUS and CEEPUS.

Chiara Raith is currently working as University Assistant at the Chair of Industrial Logistics at the University of Leoben. In 2019, she obtained her Master's degree in Industrial Logistics with focus on Automation and Logistics System Engineering from the Montanuniversitaet Leoben. In the same year, she received her Certificate of the Delta Academy & Management Certificate of the University of St.Gallen (CAS HSG). In 2020 she started her Ph.D.-studies at Montanuniversitaet Leoben. In her research, she is dealing with the modeling and simulation of material flows. The focus lies on developing a metamodel for the analysis of discrete manufacturing processes. She has conducted research projects with voestalpine AG, MAGNA International Europe GmbH, Österreichische Post AG, Jungheinrich Systemlösungen GmbH, etc. In her current position, she educates students in the field of discrete-event simulation and basics and concepts of logistics and also supervises several bachelor theses in this field.

Associate Professor Dr. Sakgasem Ramingwong graduated from Chiang Mai University with First Class Honors in B.Eng. Mechanical Engineering in 2000. In 2004, he completed his Ph.D. in Advanced Manufacturing System Engineering at Royal Melbourne Institute of Technology (RMIT) of Australia. He is currently a Full-time Lecturer at the Industrial Engineering Department, Faculty of Engineering, Chiang Mai University. He is also a Deputy Head of Excellence Center in Logistics and Supply Chain Management (E-LSCM).

Sakgasit Ramingwong is currently an Associate Professor at Department of Computer Engineering, Faculty of Engineering, Chiang Mai University, Chiang Mai, Thailand. He received his Ph.D. from the University of New England, Australia, in 2009. His main research focuses on software project management, risk management, software process improvement and gamification of software engineering aspects.

Lachana Ramingwong is an Assistant Professor, teaches at the Department of Computer Engineering, Faculty of Engineering, Chiang Mai

University, Chiang Mai, Thailand, where she grows her interests in software process models, software requirements analysis, software testing, and Human-Computer Interaction.

Erwin Rauch holds a M.Sc. in Mechanical Engineering from the Technical University Munich (TUM) and a M.Sc. in Business Administration from the TUM Business School and obtained his Ph.D. degree in Mechanical Engineering from the University of Stuttgart with summa cum laude. Currently, he is an Assistant Professor for Manufacturing Technology and Systems at the Free University of Bolzano, where he is the Head of the Smart Mini Factory laboratory for Industry 4.0. His current research is on Industry 4.0, Social Sustainability in Production, Smart and Sustainable Production Systems, Smart Shopfloor Management and Engineer/Make to Order. He has 10 years of experience as Consultant and later Associate Partner in an industrial consultancy firm operating in production and logistics. He is Project Manager of the EU-funded H2020 research project "SME 4.0—Industry 4.0 for SMEs" in an international partner consortium. Further, he is author and co-author of more than 150 scientific and non-scientific books, chapters of books, articles and other contributions and received several awards for scientific contributions.

Manuel A. Ruiz Garcia received his Master's degree in Control Engineering in 2013 and his Ph.D. in Engineering in Computer Science in 2018, both from Sapienza University of Rome, Rome, Italy. He is a postdoc research fellow in the Smart Mini-Factory Laboratory of the Free University of Bozen-Bolzano, Bolzano, Italy. His research interests include reactive control of mobile manipulators, robotics perception, collaborative robotics and human-robot cooperation. He had been involved in the EU Projects "NIFTi—Natural human-robot cooperation in dynamic environments" and "TRADR—Long-term human-robot teaming for disaster response" and is currently involved in the EU Project "SME 4.0—Industry 4.0 for SMEs" as well as in industry projects on collaborative robotics and robotics perception.

Leoš Šafár graduated B.Sc., M.Sc. in Finance, Banking and Investments and Ph.D. in Finance from Technical University of Košice, Slovakia (2019), where he currently works as the Assistant Professor. During his studies and current working position, he obtained additional experience also from Belgrade Banking Institute (Serbia), SACS MAVMM Engineering College in Madurai (India) and Chiang Mai University (Thailand). His research focuses on financial markets, quantitative easing, and Industry 4.0. He has participated in several national (VEGA) and international projects (European Horizon 2020 MSCA-RISE program).

Philipp C. Sauer is an Assistant Professor in Management Engineering at the Free University of Bozen-Bolzano. He holds a B.Sc. and M.Sc. in Management Engineering and a Ph.D. in Supply Chain Management with highest honors (summa cum laude) under the supervision of Prof. Dr. Stefan Seuring. His research focuses on multi-tier supply chain management, sustainability management, circular economy as well as certification. He has received the Harry Boer Best Paper Award 2020 and was awarded Highly Commended winner of the 2020 Emerald/EFMD Outstanding Doctoral Research Award. His work has been published in internationally leading journals such as the International Journal of Production Economics, Journal of Business Ethics or Supply Chain Management: An International Journal.

Zuzana Šoltysová completed bachelor's and master's study at Technical University of Kosice, Faculty of Manufacturing Technologies with a seat in Presov, Department of Manufacturing Management in the field of Manufacturing Management. She completed her Ph.D. Study at Technical University of Kosice, Faculty of Manufacturing Technologies with a seat in Presov, Department of Manufacturing Management in the field of Industrial Technology. Her Ph.D. thesis was focused on the research of product and production complexity in terms of mass customization. Currently, she is an Assistant Professor at Faculty of Manufacturing Technologies with a seat in Presov at Department of Manufacturing Management. Moreover, her research activities include complexity, throughput, axiomatic design and production line balancing rate.

Associate Professor Dr. Apichat Sopadang was born in Chiang Mai, Thailand. He graduated from Chiang Mai University, Thailand in 1987 with a degree in industrial engineer. For several years, he worked as a maintenance planning engineer in Electricity Generator Authority of

Thailand (EGAT). He completed his Ph.D. from Clemson University, USA. in 2001. Following the completion of his Ph.D., he is working for Chiang Mai University as an Associate Professor and Head of Excellence Center in Logistics and Supply Chain Management (E-LSCM). He is a frequent speaker at industry and academic meetings. His current research areas are on Industry 4.0, Sustainability Supply Chain, Aviation Logistics, Lean Manufacturing System and Performance Measurement. Dr. Sopadang also served as a consultant in many private organizations in Thailand and international organizations such as the Asian Development Bank (ADB) and The Japan External Trade Organization (JETRO). He is author and co-author for more than 100 academic papers that include book chapters and articles.

Jakub Sopko works as an Assistant Professor at Faculty of Economics, Technical University of Košice, Slovakia. In 2019, he successfully graduated with Ph.D. in Finance from Faculty of Economics, Technical University of Košice. As part of his research activities, he focuses on banking, health finance, efficiency measurement in the healthcare sector and the banking sector using DEA models. Besides previous experience, he cooperates with Department of Industrial Engineering, Chiang Mai University, Thailand, in the evaluation of the importance of value creation and value constellations for developing the sustainable organizational and business models and readiness degree toward Industry 4.0 in industrial companies, and he is participating in the international project SME4.0 which focuses on Industry 4.0 and is funded by European Union's Horizon 2020 MSCA-RISE program.

Mr. Krisana Tamvimol was born in Ubon Ratchathani, Thailand. He graduated from Chiang Mai University, Thailand in 1996 with a degree in computer engineering. For several years, he worked as CEO in computer companies in Chiang Mai. In 2015, he received funding for research on a Plant Factory (Indoor Vertical Farm) project from the Ministry of Science. Following the completion of his research, he has established a start-up Wangree Fresh Plant Factory in Bangkok. He built the first and largest Plant Factory in Thailand. He is a frequent speaker at agriculture and academic meetings. His current research areas are on Edible vaccines and Cultured meat. Krisana Tamvimol also served as a consultant in many private organizations in Thailand such as the Ministry of Higher Education, Science, Research and Innovation and the PTT Public Company Limited. He is currently engaged in a social enterprise to help orphans and the abandoned elderly using Plant Factory Solution.

Trasapong Thaiupathump received his B.Eng. (Computer Engineering) degree from King Mongkut's Institute of Technology, Ladkrabang, Thailand in 1993, M.S. (Computer Engineering) degree from University of Southern California, USA in 1996, and Ph.D. degree from University of Pennsylvania, USA in 2002. He is currently an Associate Professor in computer engineering at Chiang Mai University, Thailand. His research interests are data communications, signal processing, and computer applications in manufacturing.

Korrakot Tippayawong graduated with B.Eng., M.Eng. and Ph.D. in Industrial Engineering from Chiang Mai University, Thailand, Swinburne University of Technology, Australia, and Tokyo Institute of Technology, Japan, respectively. She has over 20 years' experience in teaching, research and industrial consultation. Dr Tippayawong has worked with more than 300 SMEs as well as a number of large public and private enterprises. She is currently an Assistant Professor at Department of Industrial Engineering, Chiang Mai University. Her research focuses on logistics & supply chain, industrial engineering & management. She has received many major grants, including those from Thai Ministry of Industry, Ministry of Science and Technology and European Horizon 2020 MSCA-RISE program.

Monika Trojanová works as a Researcher at the Department of Industrial Engineering and Informatics, Faculty of Manufacturing Technologies with the Seat in Prešov, Technical University of Košice (Slovakia). She is currently research in the field of intelligent mechatronic systems—focusing on systems driven with fluid muscle and soft actuators. She is the co-author of 3 utility models, 1 patent model, and several publications (10) registered in the Web of Science and Scopus databases. She actively collaborates on reviews of publications, especially for impact journals such as the International Scientific-Technical Conference Manufacturing, the International Journal of Advanced Robotic Systems or Mechanical Systems and Signal Processing. During her work, she participates in several projects. She was the leader of a research project funded by the Faculty of Manufacturing Technologies with the Seat in Prešov. She is currently the head of the obtained research grant for young researchers for the year 2021, which is under the auspice of the Rector of the Technical University in Košice.

Renato Vidoni received his M.Sc. in Electronic Engineering—focus: industrial automation—from the University of Udine, Italy, in 2005. In 2009, he obtained his Ph.D. degree in Industrial and Information Engineering from the University of Udine, Italy. Currently, he is an Associate Professor in Applied Mechanics at the Free University of Bozen-Bolzano (Italy) where he is responsible of the activities in robotics and mechatronics inner the Smart Mini-Factory laboratory for Industry 4.0 and he is the Head of the Field Robotics laboratory. He is Course Director of the M.Sc. in Industrial Mechanical Engineering and Rector's delegate at the CRUI (Conference of the Rectors of Italian Universities) Foundation's for the University-Business Observatory. His research activity is documented by more than 100 scientific contributions that deal with topics of the Applied Mechanics sector both in "classical" fields as well in new and emerging domains (e.g., industry and Agri 4.0). The recent research activity can be grouped in three different research areas that fall into the "Industrial Engineering and Automation" macro-area of the Faculty of Science and Technology: High-performance automatic machines and robots, Mechatronic applications for Energy Efficiency, Mechatronics and Robotics for field activities.

Manuel Woschank is a Senior Researcher, Senior Lecturer, and the Deputy Head of the Chair of Industrial Logistics at the Montanuniversitaet Leoben and an Adjunct Associate Professor at the Faculty of Business, Management and Economics at the University of Latvia. He received a diploma degree in industrial management and a master's degree in international supply management from the University of Applied Sciences, FH JOANNEUM, Graz, Austria, and a Ph.D. degree in management sciences with summa cum laude from the University of

Latvia, Riga, Latvia. He was a visiting scholar at the Technical University of Kosice (Slovakia), and at the Chiang Mai University (Thailand). His research interests include the areas of logistics system engineering, production planning and control systems, logistics 4.0 concepts and technologies, behavioral decision making, and industrial logistics engineering education.

Kamil Židek is focused on the research in the area of image processing for manufacturing applications and knowledge extraction by algorithms of artificial intelligence. He completed his habitation thesis named "Identification and classification surface errors of mechanical engineering products by vision systems" at Faculty of Manufacturing Technologies of Technical University of Kosice. Currently, he is an Associate Professor at this faculty. He published his research titled "Embedded vision equipment of industrial robot for in-line detection of product errors by clustering-classification algorithms" in the mention area in Current Contents (IF 2016 = 0.987). He is the lead author of one Current Contents article and co-author of 3 other CC articles. He is inventor and co-inventor of 6 published patents 4 and 8 utility models. He is also co-author of 1 monograph, 2 university textbooks and 3 scripts.

Helmut Zsifkovits holds the Chair of Industrial Logistics at the Department of Economics and Business Management at Montanuniversitaet Leoben, Austria. He graduated from the University of Graz, Austria and has professional experience in automotive industry, logistics consultancy and IT. His research interests include logistics systems engineering, supply chain strategy and operations management. He is President of the European Certification Board for Logistics (ECBL), Vice-President of Bundesvereinigung Logistik Austria (BVL), and President of Logistics Club Leoben. In 2018, he was appointed as an Adjunct Professor at the University of the Sunshine Coast, Australia. He has teaching assignments at various universities in Austria, Latvia, Colombia, and Germany and is the author of numerous scientific publications and several books.

List of Figures

List of Tables

Part I

Implementing Industry 4.0 for Smart Manufacturing in SMEs

1

Status of the Implementation of Industry 4.0 in SMEs and Framework for Smart Manufacturing

Erwin Rauch and Dominik T. Matt

1.1 Introduction

Since the term Industry 4.0 was first mentioned at the Hannover Messe in 2011 (BMBF 2012), 10 years have passed. The accompanying Fourth Industrial Revolution has almost turned both research and industrial practice upside down and led to a multitude of technological innovations and the digitalization of production (Kagermann et al. 2013). A lot has happened in these 10 years. After the discussions around

E. Rauch · D. T. Matt (✉)
Industrial Engineering and Automation (IEA),
Faculty of Science and Technology, Free University of Bozen-Bolzano,
Piazza Università, Bolzano, Italy
e-mail: dominik.matt@fraunhofer.it; dominik.matt@unibz.it

E. Rauch
e-mail: erwin.rauch@unibz.it

D. T. Matt
Innovation Engineering Center (IEC), Fraunhofer Research Italia s.c.a.r.l.,
Bolzano, Italy

© The Author(s) 2021
D. T. Matt et al. (eds.), *Implementing Industry 4.0 in SMEs*,
https://doi.org/10.1007/978-3-030-70516-9_1

Industry 4.0 were initially still very much limited to Germany and later Europe, a look at the publications on this topic shows that the term was discussed worldwide, especially from 2016 onwards. The first years were largely dominated by discussing what the collective term Industry 4.0 means, how it can be defined and which core technologies support the fourth industrial revolution. From 2015 onwards, the first initiatives and national plans for the implementation and introduction of Industry 4.0 in industrial practice emerged, particularly in Europe. Such plans were mostly linked to financial support or tax relief in order to also prepare a financial incentive for companies to invest in new technologies (Matt and Rauch 2020). At the same time as the national plans and initiatives, competence centers, research laboratories or demo-labs for the transfer of advanced technologies from research to practice have also been established in many cases.

"Smart Manufacturing" or "Smart Factory" are often used as synonyms for Industry 4.0. In particular, Industry 4.0 in combination with digitalization has ultimately also contributed to the formation of the term Digital Transformation (Deloitte 2015). While Industry 4.0 refers primarily to the manufacturing industry, the concepts of digital transformation also apply to non-industrial sectors and contribute to the introduction of digital technologies and digital business models there. In recent years, the basic concepts of Industry 4.0 and digital transformation have also been transferred to many other areas. In 2016, for example, the term Society 5.0 was created with the overarching goal of increasing the well-being of society in the long term by means of new and advanced technologies and thus transferring many of the technologies into everyday life (Fukuda 2020). Furthermore, many cities and regions have been working on digitalization strategies for several years in order to increase efficiency in the public sector and sustainability (Safiullin et al. 2019). With the pressure to introduce Industry 4.0 technologies in companies, there is also a shortage of skilled workers with appropriate knowledge in these technologies. This has led most education systems to adapt their curricula and content to these new developments. Such developments have often been referred to as Education 4.0 (Hussin 2018) or, for the engineering sector, Engineering Education 4.0 (Morandell et al. 2019). The latest trend that has been emerging for a few years is the introduction of artificial intelligence in companies. In the field of manufacturing, too, opportunities are currently being sought to introduce established AI methods for the optimization of production processes and to explore the range of possible applications (Woschank et al. 2020; WMF 2020).

In this now ten-year development around Industry 4.0, research on the introduction of Industry 4.0 in SMEs has also increased significantly. Especially from 2017 onwards, partly due to the launch of an important EU H2020 MSCA RISE project entitled "SME 4.0 - Industry 4.0 for SMEs" (Rauch et al. 2018), the research work regarding SMEs has significantly increased year by year. In these years, a lot of research has been done on which technologies are particularly suitable for SMEs (Prause 2019), which prerequisites and limitations SMEs have when introducing Industry 4.0 (Masood and Sonntag 2020), and which tools facilitate a successful introduction (Matt et al. 2018; Rauch et al. 2020a). One objective of this chapter is to determine the status of Industry 4.0 implementation in SMEs. This will provide an overview of how far the implementation of Industry 4.0 has progressed and which empirical or case study-based studies exist to date that report on the impact and effects of Industry 4.0 on the performance of SMEs. Based on this and previous research by the authors, a modular framework model is created to facilitate SME manufacturers to introduce smart manufacturing in their companies. The presented framework is accompanied by a stage model that supports the step-by-step implementation in SMEs. Finally, future developments will be addressed to prepare SMEs for the challenges of the future. For this purpose, the term "Industry 4.0+" is introduced, which describes the next level of Industry 4.0 for the next five to ten years.

Based on the objectives described above, the following research questions are defined for this chapter:

RQ1: *What is the status of the application and implementation of Industry 4.0 in SMEs?*

RQ2: *What frameworks and guidelines can SMEs use to successfully implement smart manufacturing?*

RQ3: What are the medium and long-term challenges SMEs will have to face in the future?

The chapter is structured as follows. After the introduction to the topic in Sect. 1.1, Section 1.2 describes the status of the introduction of Industry 4.0 in SMEs based on literature research. Section 1.3 then shows the modular framework model for SME manufacturers. Finally, Sect. 1.4 gives an outlook on the future challenges in the context of Industry 4.0+.

1.2   Status of Industry 4.0 Implementation in SMEs

1.2.1   Review of Literature on Industry 4.0 Implementation in SMEs

In order to investigate the current state of adoption of Industry 4.0 technologies in SMEs, a literature review was conducted based on scientific papers. For the search, the database Scopus was used and the following keywords in title, abstract and keywords were searched: ("SME" OR "small and medium-sized" OR "small and medium-sized") AND (implementation OR adoption OR introduction) AND "industry 4.0". Afterwards the identified works were reduced to papers from the last two years 2019–2020 and screened for relevant papers for this research.

Table 1.1 summarizes the results of various studies on the application of Industry 4.0 technologies in SMEs. The search in literature has been conducted at the end of December 2020 (20 December 2020) therefore the search includes mostly all published works in these two years. The results of the evaluation of the individual technologies in the studies have different evaluation scales and were therefore classified by the authors for this comparison as follows: low (low application), medium (medium application), and high (high application).

Ghobakhloo and Ching (2019) conducted a study in Iranian and Malaysian SMEs to investigate the adoption of Industry 4.0 technologies. According to them, the most widely adopted Industry 4.0 technologies in SMEs are the following. First, the use of cloud data and storage by

Table 1.1 Adoption of Industry 4.0 technologies in SMEs based on surveys in different countries/regions

Technology	Ghobakhloo and Ching (2019) Iran-Malaysia	Cimini et al. (2020) Italy	Rauch et al. (2020a) Italy-Austria	Ko et al. (2020) Korea	Kilimis et al. (2019) Germany	Yu and Schweisfurth (2020) Denmark-Germany	Pech and Vrchota (2020) Czech Republic	Gergin et al. (2019) Turkey	Türkes et al. (2019) Romania	Ingaldi and Ulewicz (2020) Poland	Overall rating
Advanced Automation and Robotics	Medium	Medium	Medium	High	Medium	Medium	Medium	High	High	Medium	**Medium**
Additive Manufacturing (AM)	Low	/	Low	/	/	Low	Low	Low	Low	Medium	**Low**
Simulation	Medium	Medium	Medium	/	Medium	Medium	/	Medium	Medium	/	**Medium**
Virtual/Augmented Reality (VR/AR)	Low	Low	Low	/	/	low	Low	Low	Low	/	**Low**
Horizontal/Vertical Data Integration	High	/	High	High	High	High	High	/	High	High	**High**
Industrial Internet of Things (IIoT)	High	High	Low	High	Medium	Low	Medium	Medium	Medium	High	**Medium to high**
Cloud Computing	High	/	High	/	/	High	Medium	/	Low	/	**High**
Cybersecurity	Medium	Low	Medium	/	/	High	/	Medium	Medium	High	**Medium**
Big Data Analytics	Medium	High	Medium	Medium	/	Low	High	Low	High	/	**Medium to high**
Artificial Intelligence (AI)	Low	/	Low	Low	/	/	/	/	Low	Low	**Low**

changing inhouse software as well as data storage in cloud-based services. Secondly, many SMEs invested already in horizontal/vertical data integration by introducing basically Enterprise Resource Planning (ERP) systems for the management of data on the business process level. Many of them have already started also to introduce Manufacturing Execution Systems (MES) for production planning and control. The third technology mostly adopted by SMEs is Industrial Internet of Things (IIoT) equipping legacy machines with actuators and sensors for data collection and introducing machine and process control systems like programmable logic controllers (PLC) or supervisory control and data acquisition (SCADA). This allows SMEs to monitor the status of manufacturing systems in real-time. According to their study many SMEs are already on a good way to adopt advanced automation and robotics like mobile and collaborative robots. The same is for cybersecurity, data analytics and simulation. SMEs are already using the basic technologies available on the market (e.g. firewall, antivirus, virtual private network, etc.) to protect their businesses from cyberattacks. Further, they are moving toward data analytics although in most of the cases based on simple and commercially available data monitoring and analysis tools, while the use of machine learning or more complex artificial intelligence technologies is still in its infancy in SMEs. For simulation, two kinds of simulation need to be differentiated. While SMEs are already using advanced computer-aided design and engineering software (CAD and CAE) for doing simulations with digital models of parts and products they still do not yet use simulation software for manufacturing or logistics purposes (e.g., discrete event simulation) as they are cost-expensive. According to the authors, in addition to artificial intelligence there are other two technologies not yet exploited well by SMEs: additive manufacturing and virtual and augmented reality (VR/AR). The reason therefore might be that additive manufacturing of metal materials is still very cost-expensive while additive manufacturing of plastic materials (not rapid prototyping) is not of such great importance for many producing SMEs. The low rate of adoption of VR/AR is surprising as VR/AR headsets are available for an affordable price. The reason might be related to missing qualifications of the staff in SMEs in order to use or develop VR and AR models/environments. Further, the authors observed that implementers of AI, VR/AR, autonomous robots believe these applications provide them with organizational improvement and productivity. Results, however, showed that perceived costs have a significant negative influence on SMIDT adoption. It was observed that higher perceived cost has resulted in non-adoption of complex Industry 4.0 technologies including VR/AR, AI, additive manufacturing, ERP, industrial sensors, and machine and process controllers. This finding is in line with the majority of prior literature introducing the implementation costs as a major barrier to Industry 4.0 adoption by SMEs.

Cimini et al. (2020) conducted a study in several Italian SMEs. In their study, they include only a limited number of Industry 4.0 technologies. Basically, it is confirmed that Industrial IoT is already implemented by most of the SMEs. Different to the previous reference, the use of big data analytics is rated as high. It is further confirmed that the level of the use of advanced automation and robotics solutions and simulation can be rated as medium. It is also confirmed that VR and AR are only implemented at a very low level. In contrast to the aforementioned reference, cybersecurity is classified as low.

The authors itself conducted an assessment of Industry 4.0 in 13 SMEs from the Italy-Austrian border region (Rauch et al. 2020a). Data integration (ERP and MES) as well as cloud computing are reported as highly adopted Industry 4.0 technologies. Advanced automation and robotics, simulation, big data analytics and cybersecurity have been assessed as medium-level adopted technologies. VR/AR, additive manufacturing and artificial intelligence are (similar to other studies) technologies with a low adoption in SMEs. Further also IIoT is assessed as not widely adopted technology different to most of the other studies summarized in Table 1.1.

Ko et al. (2020) conducted a survey with responses from 113 Korean SMEs. According to their results, advanced automation and robotics like autonomous robots are highly adopted. The same is for horizontal/vertical data integration through ERP and MES systems for production planning and control and for IIoT based on tracking and tracing of products. Data analytics in sense of predictive maintenance is adopted at a medium level while artificial intelligence is at a very low level in SMEs.

Kilimis et al. (2019) describe in their work the results of a study in German SMEs. The most adopted technology is related to horizontal/vertical data integration in sense of the introduction of ERP or other advanced IT systems for production planning and control. Advanced automation and robotics, manufacturing simulation as well as digitalization technologies (IIoT) are adopted at a medium level.

Yu and Schweisfurth (2020) investigated the adoption of Industry 4.0 technologies in the Danish-German border region. The interests in additive manufacturing, VR/AR, big data analytics and IIoT are relatively low. A moderate number of SMEs use simulation and advanced robots. Those technologies, which have reached a high degree of implementation in the sample are data integration, cloud computing and cybersecurity.

Pech and Vrchota (2020) conducted a comprehensive questionnaire-based study on Industry 4.0 adoption in 186 SMEs. According to their results the introduction of ERP and MES software systems as well as data analysis are adopted in most of the SMEs. Advanced automation and robotics are implemented at a medium level. The same is for the introduction of IIoT and cloud computing. Additive manufacturing and VR/AR were implemented only on a low level in the participating companies.

Gergin et al. (2019) implemented a survey in 588 SMEs in Turkey. Advanced automation and robotics technologies in sense of automation are implemented by many SMEs (high level). Simulation, IIoT and cybersecurity are implemented at a medium level. Additive manufacturing, big data analytics and VR/AR are implemented only by a few companies and are therefore at a low level.

Türkeş et al. (2019) conducted a survey with 176 participating SMEs from Romania. Advanced automation and robotics, horizontal/vertical data integration and big data analytics are the most adopted technologies. Many companies (medium level) adopt simulation, IIoT and cybersecurity. The lowest level of adoption is reported for additive manufacturing, VR/AR, cloud computing and artificial intelligence. While the others are in line with what has been reported in other studies, there is a different opinion regarding cloud computing technologies.

Ingaldi and Ulewicz (2020) describe the results of a survey conducted in 187 SMEs from Poland. Data integration, IIoT and cybersecurity are identified as widely adopted technologies in SMEs, while advanced automation and additive manufacturing are at a medium level of adoption. As reported in many other studies, artificial intelligence is implemented only in a few SMEs.

1.2.2 Summary on the Adoption of Industry 4.0 Technologies in SMEs

Based on the results of the literature-based comparison of ten different studies on the adoption of Industry 4.0 technologies, we can summarize the following results:

- **High:** horizontal/vertical data integration, cloud computing,
- **Medium to High:** industrial internet of things, big data analytics,
- **Medium:** advanced automation and robotics, simulation, cybersecurity,
- **Low:** additive manufacturing, virtual and augmented reality, artificial intelligence.

However, with respect to this analysis, the following limitations must be considered. The presented studies in Table 1.1. were mainly conducted in European countries, while no clear data were available on other major economic powers such as China, United States, Canada, Brazil or Southeast Asia. In addition, the categorization of SMEs and their differentiation from large companies differ in some cases in the individual studies.

Nevertheless, the comparison carried out gives a good picture of the current situation with regard to the introduction of Industry 4.0 technologies in SMEs. Based on this overview, it becomes clear where SMEs have already successfully embarked on a good path and which technologies currently still need time for successful implementation. This includes advanced IT systems for data integration such as ERP systems, supply chain management systems, MES software and other technologies like RFID or new sensors that enable a seamless data flow. Cloud computing solutions are also already established and consolidated technologies that have been used at all levels of the enterprise for several years due to their ease of implementation and cost-saving potentials.

There could be identified two technologies with a great potential for SMEs. These are IIoT and big data analytics. With the use of already available data in the company, these technologies can help to achieve real-time monitoring of the status of machines and entire manufacturing systems and also uncover optimization potential using data analytics methods and thus drive data-driven innovation. Due to the current dynamics and constantly new providers of software and services in this area, it is also becoming easier for SMEs to introduce these technologies.

Furthermore, there is the group with a medium level of implementation such as advanced automation and robotics, simulation and cybersecurity. In the field of advanced automation and robotics, mobile and collaborative robots will increase in the near future, as they offer ideal conditions for SME-typical flexible production. In the area of simulation, a distinction must be made between simulation systems for production and those for product development. While there is currently already a wide range of inexpensive or free software for CAE, this range is still much smaller in the area of factory planning and production system planning software. In the future, it is expected that open or less expensive software will also enable the creation of digital twins for SMEs. Cybersecurity seems to be implemented at a very good level. It can be assumed that increasing digitization will lead to even more risks and that SMEs, together with their external IT partners, will have to constantly rethink and adapt their cybersecurity situation.

With regard to those technologies that currently have a low level of application, different arguments can be made. Based on the articles read and our own experience from research with SMEs, it is assumed that additive manufacturing will only play a greater role for SMEs once the investment costs for manufacturing equipment for additive manufacturing of metals drops significantly. Virtual and augmented reality seem to be of little interest to SMEs despite favorable hardware prices. Artificial intelligence, on the other hand, promises great potential for SMEs as well, but still requires several years of research to make the leap into broad industrial application.

Further, Masood and Sonntag (2020) conducted a survey-based study to identify benefits and complexity-based challenges related to Industry 4.0 technologies introduced in SMEs. According to their study, the following technologies provide a high benefit although linked to a high effort and complexity in introducing them: (1) data integration (like ERP/MES), (2) big data and analytics, (3) artificial intelligence (machine learning and deep learning) and (4) advanced robotics. The following technologies also provide a high benefit while complexity is low ("low hanging fruits"): (5) cyber-physical and embedded systems, (6) simulation, (7) predictive maintenance, (8) additive manufacturing and (9) sensors. Further, the following listed technologies provide a limited benefit and are easy to implement: (10) IIoT and (11) cloud computing.

Based on these findings and previous research of the authors, the following Sect. 1.3 provides guidelines and a modular framework for introducing highly adaptable and smart manufacturing systems in SMEs.

1.3   Framework and Guidelines for Smart Manufacturing in SMEs

1.3.1   Axiomatic Design Guidelines for Implementing Industry 4.0 in SMEs

Based on previous research reported in Rauch et al. (2020b), the authors developed and adapted a set of coarse design guidelines for implementing smart manufacturing in SMEs by using an Axiomatic Design based top-down approach (Brown 2020). In the following, this set of guidelines are broken down indicating also the most promising Industry 4.0 technologies. On the highest level (Level 0), the following functional requirement (FR) and design parameter (DP) have been defined:

FR_0 Create a smart and highly adaptable manufacturing system for SMEs

DP_0 Design guidelines for a smart and highly adaptable manufacturing system for SMEs

The abovementioned highest-level FR-DP pair can be further decomposed into the following top-level FR-DP pairs (Level 1).

FR$_1$ Adapt the manufacturing system very quickly in a flexible way
DP$_1$ Changeable and responsive manufacturing system
FR$_2$ Make the manufacturing system smarter
DP$_2$ Industry 4.0 technologies and concepts

The top-level FR-DP pairs, describing the main goals in sense of a highly adaptive and a more intelligent manufacturing system, can again be further decomposed into a set of FR-DP pairs on Level 2.

For FR$_1$/DP$_1$ (adaptability of the manufacturing system), the decomposition is as follows.

FR$_{1.1}$ Change and reconfigure the system with low effort
DP$_{1.1}$ Flexible and changeable SME manufacturing system (**advanced and autonomous robotics, additive manufacturing**)
FR$_{2.1}$ React immediately to changes
DP$_{2.1}$ Responsive SME manufacturing system (**computer vision**)

For FR$_2$/DP$_2$ (smartness of the manufacturing system), the decomposition is as follows

FR$_{2.1}$ Create data in manufacturing systems
DP$_{2.1}$ Multi-sensor data fusion (**sensor technologies**)
FR$_{2.2}$ Store and manage data in manufacturing systems
DP$_{2.2}$ Innovative storage systems with capabilities for big data (**cloud computing**)
FR$_{2.3}$ Connect all elements in the system to get real-time data
DP$_{2.3}$ Connectivity and interoperability to exchange data (**IIoT, horizontal/vertical data integration**)
FR$_{2.4}$ Take advantage of data in the manufacturing system
DP$_{2.4}$ Smart data analysis (**big data analytics, artificial intelligence**)
FR$_{2.5}$ Create digital models to test, monitor and control manufacturing systems

DP$_{2.5}$ Digital twin of products and manufacturing systems (**simulation**)

FR$_{2.6}$ Interact as human with the cyber-physical world

DP$_{2.6}$ Multimodal Human-machine interaction (**virtual and augmented reality**)

FR$_{2.7}$ Provide appropriate protection against cyberattacks

DP$_{2.7}$ Cybersecurity solutions for SMEs (**cybersecurity**)

1.3.2 Framework for Highly Adaptable and Smart Manufacturing in SMEs

Based on the design guidelines from the previous section, a framework is presented to help SMEs make their manufacturing systems highly adaptable and smart (see Fig. 1.1). The framework is divided into a part that represents the physical world. This means the physical manufacturing system with its machines, which must be designed to be as flexible and changeable as possible in order to meet the need for increasing individualization and thus the trend toward mass customization. On the other hand, it is divided into a cyber-world that is driven by the digital transformation and aims to achieve a smart manufacturing system. In between

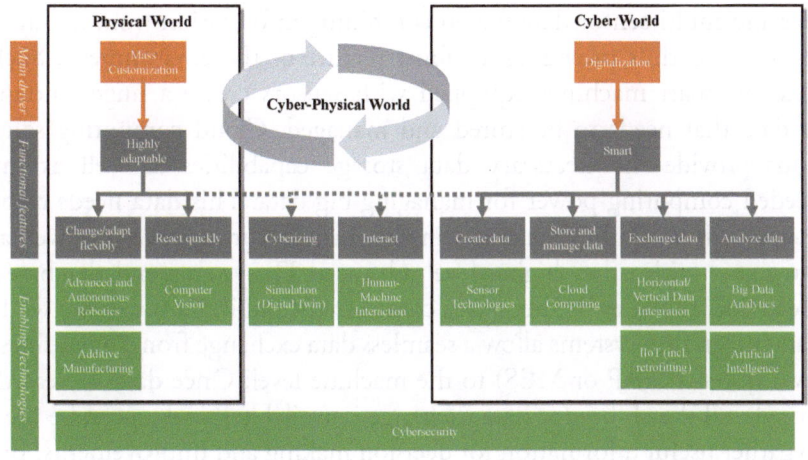

Fig. 1.1 Framework for highly adaptable and smart manufacturing systems

lies what we also call cyber-physical production systems. For all three, the framework identifies functional features of future SME manufacturing systems. They should be flexible and responsive, able to cyberize the physical allowing human interaction as well as creating, managing, exchanging and analyzing data to improve manufacturing processes.

For each functional feature, the framework indicates the most promising and enabling technologies. In the physical world, autonomous mobile and collaborative robots fulfill the requirement to adapt manufacturing processes easily to a changing environment or changing products or variants (advanced and autonomous robotics). In addition, additive manufacturing provides a possibility to produce customized and complex geometries on demand. Computer vision technologies enable context awareness as well as scene understanding capabilities in order to identify changes in the production environment immediately and therefore react with high response.

In the cyber-physical world, simulation and advanced CAE technologies provide the basis for cyberizing the physical and thus create a digital twin of products, manufacturing systems and entire factories. The human can interact with such digital models by using multimodal human-machine interaction technologies like virtual and augmented reality.

In the cyber-world, data can be created by using sensor technologies and multi-sensor data fusion combining information from different sensors and thus providing a robust picture of the environment. Such kind of smart machines equipped with sensors create a huge amount of data that needs to be stored and managed. Cloud computing solutions provide the necessary data storage capabilities as well as the needed computing power for managing big data. This data needs to be exchanged with other machines in the manufacturing system as well as integrated with other higher-level advanced IT systems and databases. Here IIoT provides the necessary connectivity and interoperability while data integration systems allow a seamless data exchange from the business level systems (ERP or MES) to the machine level. Once data is created, stored and the ability for data exchange is available this data can be used to gather useful information for decision making and improvements. Big data analysis techniques can be applied to analyze the data and artificial intelligence can be used to perform predictive analysis.

Due to connectivity and digitization, the entire smart manufacturing system must be adequately protected against cyberattacks. For this reason, cybersecurity is one of the fundamental enabling technologies for implementing Industry 4.0 in SMEs.

Many references repeatedly point out the limitation of SMEs due to financial resources (Orzes et al. 2018; Masood and Sonntag 2020). This means that most SMEs cannot completely renew their machinery in the short or medium term. In order to still be able to use Industry 4.0 technologies and in particular the advantage of a connected factory, SMEs are dependent on making legacy machines "Industry 4.0 ready". This is also referred to as "retrofitting" in the relevant literature. Digital retrofit is a process responsible for updating existing equipment by adding new software, hardware, or protocol-related components, allowing the system to extend its capabilities and meet new requirements. From an economic aspect, it avoids the incurring of large investments associated with the redesign or purchase of new equipment. In addition, the retrofit technique can assess the complete (or partial) migration to new technologies without compromising the integrity of the traditional methods employed by the company (García et al. 2020). Therefore, the need for retrofitting is also considered in the proposed framework as part of IIoT.

1.3.3   Three-Stage Model for Implementing Industry 4.0 in SMEs

To implement the presented framework, it is recommended to use a three-stage model as shown in Fig. 1.2. This model is divided into 3 stages: (i) design level, (ii) implementation level and (iii) operational level.

The design level comprises the previously presented design guidelines and the framework and supports the system designer in this design phase.

The implementation level is divided into initial pilot projects and the subsequent roll-out of the concept and technologies. In previously defined pilot projects, the technologies are initially tested and evaluated

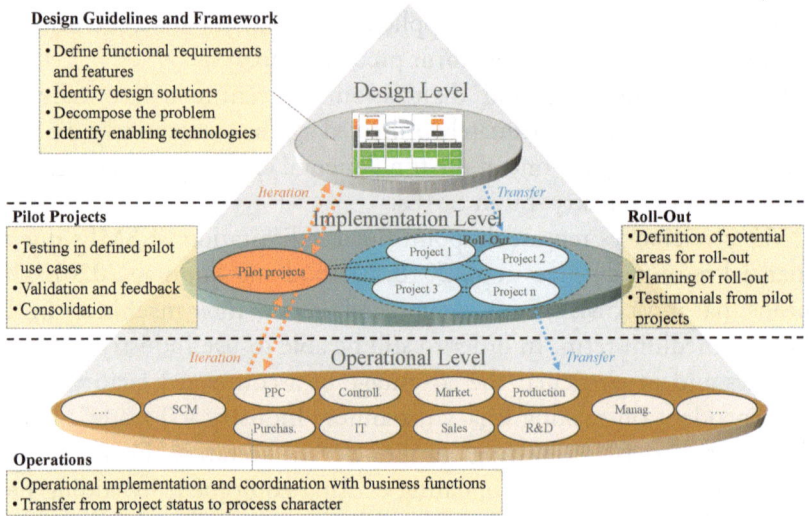

Fig. 1.2 Three-stage model for implementing Industry 4.0 in SMEs

for their suitability for operation and application. Based on the feedback, further work is then carried out to consolidate the implementation of the technologies. This is followed by the roll-out, in which the areas with the greatest potential are first planned and the timeline is prioritized. The roll-out is then completed with the support of testimonials from the pilot projects.

The operational level deals with the integration and coordination with the functional departments in the company. For both the pilot phase and the roll-out, it is necessary to involve the individual departments and thus to initiate the transfer from the project to the business process. With regard to knowledge transfer, on the one hand, new knowledge and new technologies are transferred to the operational areas in the company. On the other hand, the guidelines and framework defined at the design level are also put to the test by feedback from the operational level in an iterative process and adapted if necessary.

1.4   Industry 4.0+: An Outlook on Future Challenges for SMEs

After ten years of a continuing hype of Industry 4.0 and the effort to introduce Industry 4.0 also in SMEs we should take a look in the future to identify what kind of challenges SMEs will face in the forthcoming next ten years. Several researchers already discuss if there would be the beginning of a new industrial revolution or "Industry 5.0" (Özdemir and Hekim 2018) with a strong focus on the introduction of artificial intelligence in manufacturing. Taking a closer look to the original goals of Industry 4.0 from the beginning (Kagermann et al. 2013), we can understand that they have not yet been achieved completely. While initial objectives like cyber-physical systems, smart and connected factories, IoT and data integration are already well-researched topics ready for implementation, other objectives like intelligent manufacturing systems or self-optimizing factories are in its infancy and need still more research. Therefore, we do not see the beginning of a completely new industrial revolution but the completion of something that stared roughly ten years ago.

In this regard, we propose a concept called "Industry 4.0+" describing a next level of Industry 4.0 with new challenges for the next decade. Figure 1.3 shows the classic picture of the four industrial revolutions with the extension of Industry 4.0 by the next level of Industry 4.0+. The first level aims at achieving a smart and connected factory, while the next and future level of Industry 4.0 aims to achieve an intelligent and self-optimizing factory. The first level of Industry 4.0 is characterized by technology-driven innovation, which is also the prerequisite for the next level. In this last decade, we developed technologies making it possible to create data, collect and manage data and process large amount of data. The second level of Industry 4.0 is characterized by data and intelligence-driven innovation. This next decade of Industry 4.0 will be dominated by making sense of data and utilizing such data for not only optimizing our factories, but to bring them to a level of self-regulation and self-optimization. Going this way toward intelligent factories, we can exploit the capabilities from artificial intelligence as well as bio-inspired intelligence (Rauch 2020).

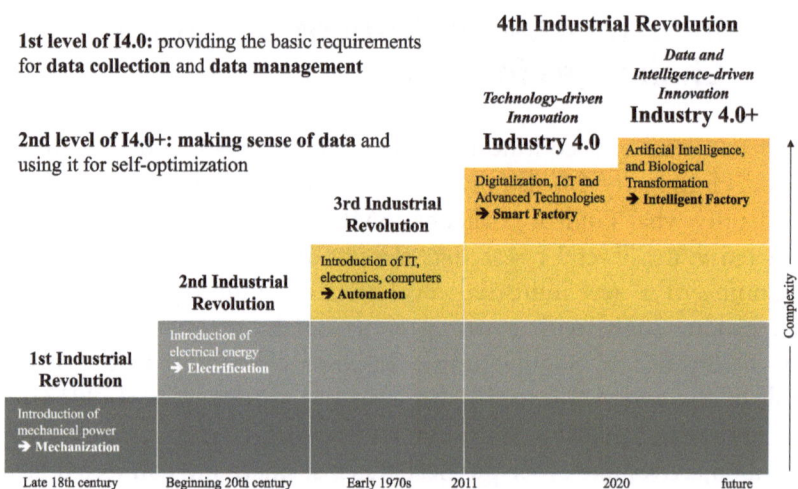

Fig. 1.3 Industry 4.0+ as the next level of Industry 4.0 (Rauch 2020)

First, we should clearly define what we mean with Industry 4.0 by differentiating the two terms "smart factory" and "intelligent factory". While a smart factory can be understood as a manufacturing system, which is capable to apply previously acquired knowledge an intelligent factory may be seen as a factory, which can autonomously acquire new knowledge and apply it for self-optimization purposes. To achieve this goal, the results from the first era of Industry 4.0 play an important role, as digitalization, connectivity and advanced manufacturing technologies are a prerequisite for this next level of Industry 4.0. In the coming years, the goal will be to fully realize the Industry 4.0 vision by equipping our manufacturing systems with intelligence using nature as an inspiration and profit from the latest advances in artificial intelligence. (Rauch 2020).

An intelligent and self-optimizing manufacturing system can be realized by using artificial intelligence (including **machine learning** and with increasing amount and complexity of data especially **deep learning**). Possibilities for the application of artificial intelligence in manufacturing are expected in automated or assisted engineering design, manufacturing system reconfiguration, production planning, predictive maintenance, quality inspection as well as in supply chain management

(Rauch 2020). The introduction of artificial intelligence in manufacturing enables manufacturing systems to become self-aware, self-comparing, self-predicting, self-optimizing and thus more resilient as traditional manufacturing systems (Lee et al. 2018).

Resilience is also one of the central characteristics of many biological systems. From a biological point of view, resilience is a property that enables a system to maintain its functions against internal and external disturbances (Van Brussel and Valckenaers 2017). Increasing technical capabilities in information processing and computer capacities have enabled a growing understanding of biological processes in our environment in recent years. It is to be expected that biology and information technology will grow closer together in the future. Therefore, biological transformation is also seen as a parallel process to digital transformation (Dieckhoff et al. 2018). According to (Miehe et al. 2018), biological transformation can be transferred in three levels to industrial production:

- **Bio-inspired manufacturing:** involves the imitation or transfer of phenomena from nature to complex technical problems.
- **Bio-integrated manufacturing:** means the integration of technological and biological processes into industrial value-added processes.
- **Bio-intelligent manufacturing:** as the combination of technical, informatics and biological systems creating robust and self-sufficient value creation systems.

This results in completely new potentials in the use of nature as a source of inspiration by not only imitating biological effects but by intelligently transferring principles from nature to various fields of application, such as manufacturing. It can be seen as a process that interacts symbiotically with digital transformation. While the first two levels of biological transformation mentioned above have already been applied in the past and present, the third level represents a groundbreaking innovation that will be able to fully unfold its full potential shortly based on the latest Industry 4.0 technologies and enhanced by the progress in artificial intelligence (Rauch 2020).

These new challenges are currently a long way off for SMEs in particular, as artificial intelligence and especially bio intelligence are topics that small companies are currently not or hardly concerned with. In the medium to long term, however, SMEs must be prepared for this future development, the necessary qualification programs must be provided and it must be identified how SMEs can best profit from this new development. This means driving forward the implementation of Industry 4.0 (see proposed framework) as the proposed enabling technologies are also the basis for the next level of Industry 4.0+. In the future, however, it will also be necessary to adapt Industry 4.0+ approaches, applications and tools to the needs of smaller companies.

Acknowledgements This project has received funding from the European Union's Horizon 2020 research and innovation program under the Marie Skłodowska-Curie grant agreement No 734713.

References

BMBF. 2012. *Zukunftsbild Industrie 4.0.* https://www.plattform-i40.de/PI40/Redaktion/DE/Downloads/Publikation/zukunftsbild-industrie-4-0.pdf?__blob=publicationFile&v=4. Accessed on March 23, 2019.

Brown, C.A. 2020. Axiomatic design for products, processes, and systems. In *Industry 4.0 for SMEs challenges, opportunities and requirements*, ed. D.T. Matt, V. Modrak, and H. Zsifkovits, 3–36. Basingstoke: Palgrave Macmillan. https://doi.org/10.1007/978-3-030-25425-4_13.

Cimini, C., A. Boffelli, A. Lagorio, M. Kalchschmidt, and R. Pinto. 2020. How do Industry 4.0 technologies influence organisational change? An empirical analysis of Italian SMEs. *Journal of Manufacturing Technology Management* (in press). https://doi.org/10.1108/JMTM-04-2019-0135.

Deloitte. 2015. Industry 4.0—Challenges and solutions for the digital transformation and use of exponential technologies. Study of Deloitte

Consulting.   https://www2.deloitte.com/content/dam/Deloitte/ch/Doc
uments/manufacturing/ch-en-manufacturing-industry-4-0-24102014.pdf.
Accessed on January 7, 2016.

Dieckhoff, P., R. Möhlmann, and J. van Ackeren. 2018. Biologische Transformation und Bioökonomie (White Paper). Munich: Fraunhofer. https://
www.fraunhofer.de/content/dam/zv/de/forschung/artikel/2018/Biologische-
Transformation/Whitepaper-Biologische-Transformation-und-Bio-Oekono
mie.pdf. Accessed on December 21, 2020.

Fukuda, K. 2020. Science, technology and innovation ecosystem transformation toward society 5.0. *International Journal of Production Economics* 220:
107460. https://doi.org/10.1016/j.ijpe.2019.07.033.

García, J.I., R.E. Cano, and J.D. Contreras. 2020. Digital retrofit: A first step
toward the adoption of Industry 4.0 to the manufacturing systems of small
and medium-sized enterprises. *Proceedings of the Institution of Mechanical
Engineers, Part B: Journal of Engineering Manufacture* 234 (8): 1156–1169.
https://doi.org/10.1177/0954405420904852.

Gergin, Z., D.A. İlhan, F. Üney-Yüksektepe, M.G. Gençyılmaz, U. Dündar,
and A.İ. Çavdarlı. 2019. Comparative analysis of the most industrialized
cities in Turkey from the perspective of Industry 4.0. Proceedings of the
International Symposium for Production Research 2019, 263–277. Vienna,
Austria. https://doi.org/10.1007/978-3-030-31343-2_23.

Ghobakhloo, M., and N.T. Ching. 2019. Adoption of digital technologies of
smart manufacturing in SMEs. *Journal of Industrial Information Integration*
16: 100107. https://doi.org/10.1016/j.jii.2019.100107.

Hussin, A.A. 2018. Education 4.0 made simple: Ideas for teaching. *International Journal of Education and Literacy Studies* 6 (3): 92–98. https://doi.
org/10.7575/aiac.ijels.v.6n.3p.92.

Ingaldi, M., and R. Ulewicz. 2020. Problems with the implementation of
Industry 4.0 in enterprises from the SME sector. *Sustainability* 12 (1): 217.
https://doi.org/10.3390/su12010217.

Kagermann, H., J. Helbig, A. Hellinger, and W. Wahlster. 2013. Recommendations for implementing the strategic initiative INDUSTRIE 4.0: Securing
the future of German manufacturing industry. Final report of the Industrie
4.0 Working Group. Forschungsunion.

Kilimis, P., W. Zou, M. Lehmann, and U. Berger. 2019. A Survey on digitalization for SMEs in Brandenburg, Germany. *IFAC-PapersOnLine* 52 (13):
2140–2145. https://doi.org/10.1016/j.ifacol.2019.11.522.

Ko, M., C. Kim, S. Lee, and Y. Cho. 2020. An assessment of smart factories in Korea: An exploratory empirical investigation. *Applied Sciences* 10 (21): 7486. https://doi.org/10.3390/app10217486.

Lee, J., H. Davari, J. Singh, and V. Pandhare. 2018. Industrial artificial intelligence for industry 4.0-based manufacturing systems. *Manufacturing Letters* 18: 20–23. https://doi.org/10.1016/j.mfglet.2018.09.002.

Masood, T., and P. Sonntag. 2020. Industry 4.0: Adoption challenges and benefits for SMEs. *Computers in Industry* 121: 103261. https://doi.org/10.1016/j.compind.2020.103261.

Matt, D.T., and E. Rauch. 2020. "SME 4.0: The role of small-and medium-sized enterprises in the digital transformation. In *Industry 4.0 for SMEs challenges, opportunities and requirements*, ed. D.T. Matt, V. Modrak, and H. Zsifkovits, 3–36. Basingstoke: Palgrave Macmillan. https://doi.org/10.1007/978-3-030-25425-4_1.

Matt, D.T., E. Rauch, and M. Riedl. 2018. Knowledge transfer of Industry 4.0 principles to SMEs: A five-step methodology to introduce Industry 4.0. In *Analyzing the impacts of Industry 4.0 in modern business environments*, ed. R.-B. Thornton and F. Martínez, 256–282. Hershey: IGI Global. https://doi.org/10.4018/978-1-5225-3468-6.ch013.

Miehe, R., T. Bauernhansl, O. Schwarz, A. Traube, A. Lorenzoni, L. Waltersmann, J. Full, J. Horbelt, and A. Sauer. 2018. The biological transformation of the manufacturing industry–envisioning biointelligent value adding. *Procedia CIRP* 72: 739–743. https://doi.org/10.1016/j.procir.2018.04.085.

Morandell, F., B.G. Mark, E. Rauch, and D.T. Matt. 2019. Engineering Education 4.0: Herausforderungen und Empfehlungen für eine zukunftsorientierte Gestaltung der Ausbildung von Fachkräften und Ingenieuren. In *Digitale Transformation – Gutes Arbeiten und Qualifizierung aktiv gestalten*, ed. D. Spath and B. Spanner-Ulmer, 273–298. Berlin: GITO.

Orzes, G., E. Rauch, S. Bednar, and R. Poklemba. 2018. Industry 4.0 implementation barriers in small and medium sized enterprises: A focus group study. Proceedings of 2018 IEEE International Conference on Industrial Engineering and Engineering Management (IEEM), 1348–1352. Macao, China. https://doi.org/10.1109/IEEM.2018.8607477.

Özdemir, V., and N. Hekim. 2018. Birth of Industry 5.0: Making sense of big data with artificial intelligence, "the internet of things" and next-generation technology policy. *Omics: A Journal of Integrative Biology* 22 (1): 65–76. https://doi.org/10.1089/omi.2017.0194.

Pech, M., and J. Vrchota. 2020. Classification of small-and medium-sized enterprises based on the level of Industry 4.0 implementation. *Applied Sciences* 10 (15): 5150. https://doi.org/10.3390/app10155150.

Prause, M. 2019. Challenges of Industry 4.0 technology adoption for SMEs: The case of Japan. *Sustainability* 11 (20): 5807. https://doi.org/10.3390/su11205807.

Rauch, E. 2020. Industry 4.0+: The next level of intelligent and self-optimizing factories. In *Advances in design, simulation and manufacturing III. DSMIE 2020. Lecture notes in mechanical engineering*, ed. V. Ivanov, J. Trojanowska, I. Pavlenko, J. Zajac, and D. Peraković, 176–186. Cham: Springer. https://doi.org/10.1007/978-3-030-50794-7_18.

Rauch, E., D.T. Matt, C.A. Brown, W. Towner, A. Vickery, and S. Santiteerakul. 2018. Transfer of industry 4.0 to small and medium sized enterprises. *Transdisciplinary Engineering Methods for Social Innovation of Industry* 4: 63–71. https://doi.org/10.3233/978-1-61499-898-3-63.

Rauch, E., M. Unterhofer, R.A. Rojas, L. Gualtieri, M. Woschank, and D.T. Matt 2020a. A maturity level-based assessment tool to enhance the implementation of Industry 4.0 in small and medium-sized enterprises. *Sustainability* 12 (9): 3559. https://doi.org/10.3390/su12093559.

Rauch E., A.R. Vickery, C.A. Brown, and D.T. Matt. 2020b. SME requirements and guidelines for the design of smart and highly adaptable manufacturing systems. In *Industry 4.0 for SMEs challenges, opportunities and requirements*, ed. D.T. Matt, V. Modrak, and H. Zsifkovits, 3–36. Basingstoke: Palgrave Macmillan. https://doi.org/10.1007/978-3-030-25425-4_2.

Safiullin, A., L. Krasnyuk, and Z. Kapelyuk. 2019. Integration of Industry 4.0 technologies for "smart cities" development. *IOP Conference Series: Materials Science and Engineering* 497 (1): 012089. https://doi.org/10.1088/1757-899X/497/1/012089.

Türkeş, M.C., I. Oncioiu, H.D. Aslam, A. Marin-Pantelescu, D.I. Topor, and S. Căpuşneanu. 2019. Drivers and barriers in using Industry 4.0: A perspective of SMEs in Romania. *Processes* 7 (3): 153. https://doi.org/10.3390/pr7030153.

Van Brussel, H., and P. Valckenaers. 2017. Design of holonic manufacturing systems. *Journal of Machine Engineering* 17 (3): 5–23.

Woschank, M., E. Rauch, and H. Zsifkovits. 2020. A review of further directions for artificial intelligence, machine learning, and deep learning in smart logistics. *Sustainability* 12 (9): 3760. https://doi.org/10.3390/su12093760.

WMF. 2020. World Manufacturing Report 2020. https://worldmanufacturing. org/activities/report-2020/. Accessed on November 30, 2020.

Yu, F., and T. Schweisfurth. 2020. Industry 4.0 technology implementation in SMEs—A survey in the Danish-German border region. *International Journal of Innovation Studies* 4 (3): 76–84. https://doi.org/10.1016/j.ijis. 2020.05.001.

2

Computational Intelligence in the Context of Industry 4.0

Alexander Hošovský, Ján Piteľ, Monika Trojanová, and Kamil Židek

2.1 Introduction

Artificial Intelligence (AI) has been in the forefront of research interest for relatively long time. Since its very beginning, the expectations about possible achievements had been quite high. Despite the fact that even today we are a long way from achieving human capabilities in solving

A. Hošovský · J. Piteľ · M. Trojanová (✉) · K. Židek
Department of Industrial Engineering and Informatics,
Faculty of Manufacturing Technologies With a Seat in Prešov,
Technical University of Košice, Prešov, Slovakia
e-mail: monika.trojanova@tuke.sk

A. Hošovský
e-mail: alexander.hosovsky@tuke.sk

J. Piteľ
e-mail: jan.pitel@tuke.sk

K. Židek
e-mail: kamil.zidek@tuke.sk

© The Author(s) 2021
D. T. Matt et al. (eds.), *Implementing Industry 4.0 in SMEs*,
https://doi.org/10.1007/978-3-030-70516-9_2

27

any kind of problems with similar success, in many particular areas these capabilities have already been surpassed. As pointed out clearly in Fulcher (2008), there is not a single, universally approved definition of what (artificial) intelligence actually is. In one of the broadest sense, it can be seen as "the study of making computers or programs to mimic thought processes, like reasoning and learning" (Munakata 2008). To help with this general definition, one can list some of the commonly accepted characteristics and/or capabilities that an intelligent system should have (Karray and Silva 2004):

- ability to deal with unfamiliar situations,
- learning and knowledge acquisition,
- ability to infer from incomplete or approximate information,
- sensory perception,
- pattern recognition,
- inductive reasoning,
- common sense or emotions.

As a result of intense research in this field, we have been able to achieve significant advancements in all of these points—possibly the last two are lagging behind the others, but from an industrial viewpoint, they are the easiest to overlook. At least two fundamentally different approaches have formed in the area: **symbolic AI** and **subsymbolic AI** (Munakata 2008). Symbolic AI works on a higher level of abstraction and is sometimes considered to be a traditional AI (Fulcher 2008). It tries to mimic our way of thinking using logic, reasoning, symbols, and models. The subsymbolic AI works differently and it is this part that started to take the dominant position in the field since the 1980s. It is instead more inspired by Nature itself or lower level of functioning in a human body in providing solutions to various kinds of problems— some of the best examples are nervous or immune systems or evolution and/or genetics. What is even more important from technical point of view is that subsymbolic AI is primarily data-based, making it very suitable for current implementations of intelligent systems where enormous amounts of data are generated, collected, and analyzed. Usually, a term **Computational Intelligence (CI)** is applied to the collection of methods in subsymbolic AI. Another term sometimes considered equivalent and coined by Lotfi Zadeh (father of fuzzy logic) is **Soft Computing (SC)**. Without any attempts to contribute to academic debates regarding what exactly constitutes CI and SC and if it is indeed equivalent, suffice it to say that the three basic paradigms are considered major building blocks of CI or SC: *neural networks, fuzzy logic and evolutionary computation* (to be described later in the text).

An interesting attempt at CI classification can be found in Sumathi et al. (2018), where the four main areas of computational intelligence are distinguished: *machine learning and connectionist systems, global search and optimization algorithms, approximate reasoning,* and *conditioning approximate reasoning* (Fig. 2.1 right). Using this classification, the major three pillars of CI (neural networks, evolutionary computation, and fuzzy logic) could be put into the first three areas respectively. The category of GSOA (Growing Self-Organizing Array) can be tricky to classify since quite a number of various nature-inspired algorithms has been already introduced see, e.g., Xing and Gao (2014), but most of them are not well-established and the approach of searching for biological inspiration in developing new algorithms may be a bit counterproductive if the underlying mathematical foundations are disregarded (Lones 2014). Conditioning approximate reasoning category includes methods like hidden Markov models, Bayesian belief networks, or graphic models (Sumathi et al. 2018).

An important aspect and advantage over traditional AI can be observed from Fig. 2.1. It is the possibility to trade the quality of solution for lower computational load expressed through the availability

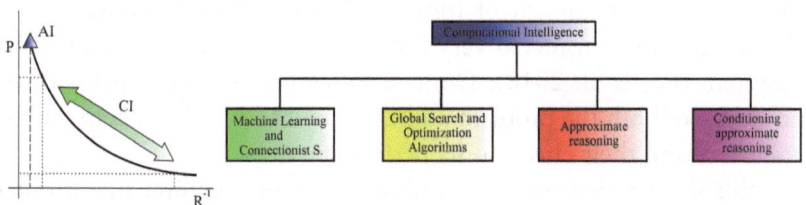

Fig. 2.1 Performance versus resources in AI and CI (P—performance, R— resources) and major CI classification

of computational resources (Fig. 2.1) (Fulcher 2008). In contrast to traditional AI, where high-quality solutions are sought at the expense of heavy computational burden, CI allows us to obtain solutions of possibly lower quality but with the reduced requirements for resources. This is significant since in many cases (e.g. NP-hard problems—Nondeterministic Polynomial hard problems) even good and not necessarily optimal solution may be acceptable.

It is now important to emphasize how the CI relates to Industry 4.0. The term Industry 4.0 is now ubiquitous as far as the area of manufacturing is concerned. It is quite understandable as its central aspect is "Smart Manufacturing for the Future" (Demir et al. 2019). As stated in the paper, its main objective is to increase productivity and achieve mass production using innovative technology. Industry 4.0 relies on several key concepts like Internet of Things, big data, cyber-physical systems, and others (Dilberoglu et al. 2017). As a result, massive amounts of data are exchanged between multitudes of devices and this very fact makes the use of a data-driven approach like CI obvious. Equipped with a plethora of powerful paradigms, CI allows any of the key concepts mentioned above to be endowed with many of the characteristics of intelligent systems.

In an attempt to emphasize the strong link between AI in general and the concept of smart manufacturing under Industry 4.0, a new term—Industrial AI (IAI)—was coined (Lee et al. 2018). Even though the authors stress its infancy in terms of clearly defined structure and methodologies, they at least set the key elements in IAI denoted with "ABCDE": Analytics technology, big data technology, cloud or cyber-technology, domain knowhow, and evidence. The first three are well-known in the context of Industry 4.0, but domain knowhow and evidence are also considered very important for the development of IAI Ecosystem (Lee et al. 2018). Domain know-how is concerned with the knowledge of both the problem addressed by IAI and the system with its parameters and their effect on its performance.

In this short review, we concentrate mainly on the three major pillars of computational intelligence, i.e., neural network—NN, fuzzy logic—FL, and evolutionary computing. In addition to that, swarm intelligence and artificial immune systems approaches are also included since these are also well-established methods in the field and their potential for Industry 4.0-based systems and cybersecurity is promising. Separate sections than contain some of the key concepts of Industry 4.0 (big data, cyber-physical systems), where the latest research in the field of CI use in these concepts is highlighted. The sections start with a short introduction to a given topic followed by the literature survey of the most recent research in CI-based approaches for Industry 4.0.

Figure 2.2 shows the basic structure of the chapter together with all the links between its sections. We used a classification similar to Sumathi et al. (2018) but we limited the range to three major classes—machine learning and connectionist systems, global search and optimization algorithms, and approximate reasoning (classification level). Within these three major classes, we used the division to several basic CI paradigms, with two paradigms for machine learning and connectionist systems, three paradigms for global search and optimization algorithms, and one for approximate reasoning. Therefore, the paradigm level contains six parts (deep learning, neural networks, evolutionary computation, swarm intelligence, artificial immune systems, and fuzzy logic and fuzzy systems), each corresponding to a single section within the text. In addition to the basic CI paradigms at Paradigm Level, three more sections were added to extend the focus of the chapter. The concepts of big data and cyber-physical systems are of crucial importance in the framework of Industry 4.0 and are also likely candidates to benefit from the use of computational intelligence. These can be found in separate sections

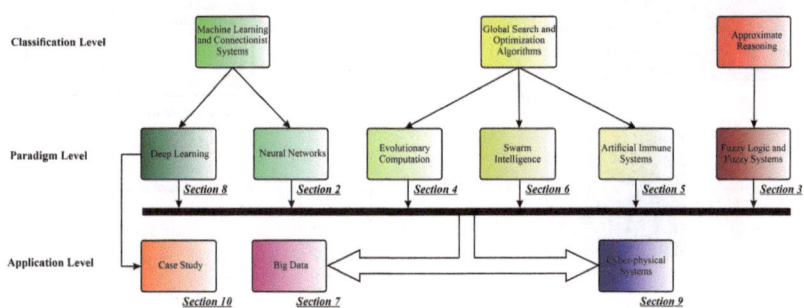

Fig. 2.2 The basic structure of the chapter

(Sects. 2.7 and 2.9) at the application level and since almost any CI paradigm can be used within their context, they are linked to all sections at the paradigm level. In order to further increase the application value of the chapter, a case study of using convolutional neural network (CNN) in object recognition during the assembly process was included as the last chapter.

2.2   Neural Networks

2.2.1   Fundamentals of Neural Networks

Some two to three decades ago the tasks of pattern recognition or obstacle avoidance were typical examples in which humans definitely excelled over the computers. Their way of processing, e.g., visual information, the capability of learning as well as performing their tasks in an unknown environment has been a source of inspiration for a longer time. Our brain relies on the parallel activity of a huge number of nerve cells called *neurons* (Fig. 2.3). This basic architecture—the interconnection of a large number of computational elements—became the idea behind artificial neural networks (simply known as neural networks—NNs). Needless to say, this inspiration is very loose and extremely simplified

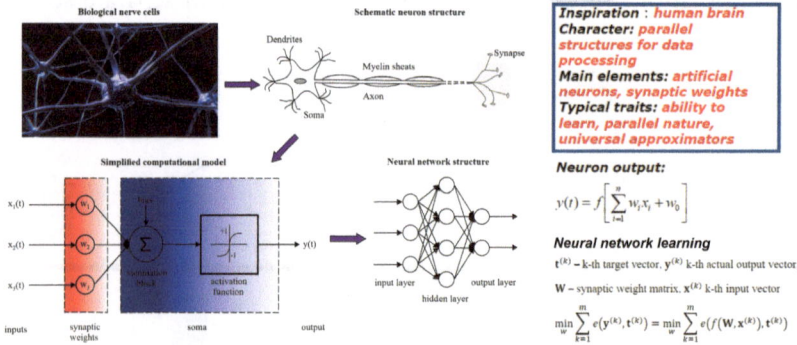

Fig. 2.3 Neural network inspiration and basic operation

compared to its biological counterpart, but still proved remarkably effective for solving various kinds of problems. The main property that is of interest is **learning**—i.e., acquiring knowledge and using this knowledge to infer the right decisions for unknown situations (Karray and Silva 2004). In the case of the NNs, this learning is known as *numerical learning*—the capability of adjusting its parameters (*synaptic weights*) in response to the training signals. Basically, three types of learning are distinguished: **supervised learning** (correct answers are known), **unsupervised learning** (finding patterns in the data), and **reinforcement learning** (it is known if the answer is correct or not).

The neurons are organized in layers and their particular organization within a given network determines their *topology*. If connections are allowed only in one direction, this is known as *feedforward topology*, if feedback connections are present, this is known as *recurrent topology*. The power of nonlinear capabilities of NNs lies in their *activation function*, typically of sigmoid or hyperbolic tangent type in classical (shallow) architectures. Some typical neural network models include (Karray and Silva 2004):

- **Multi-Layer Perceptrons (MLP),**
- **Radial Basis Function Networks (RBFN),**
- **Kohonen's Self-Organizing Networks (KSON),**
- **Hopfield Networks (HN).**

MLPs are one of the most widespread classical neural network models with feedforward architecture and typically three layers (input, hidden, and output). This class of NNs could be used either for regression or classification tasks (Haykin 2009). RBFNs are a special class of feedforward NNs inspired by the biological receptive fields of the cerebral cortex (Karray and Silva 2004) mainly developed for nonlinear function approximation tasks. In contrast to MLPs and RBFNs, KHONs are typical unsupervised neural networks, where the parameters are updated without the knowledge of correct answers (Haykin 2009). They produce a low-dimension representation of the input space by retaining the original ordering (Karray and Silva 2004). The Hopfield NNs are a special class of networks with a recurrent topology that is primarily intended as content-addressable memories with a number of locally stable states (Haykin 2009).

The performance of any of these networks is (besides other factors) highly dependent also on used training algorithm. The most famous of all is the backpropagation training algorithm typically used for MLP-like networks (Kim 2017). This is a gradient-based technique where the errors in a network are propagated backward. In the case of radial basis function neural networks (RBFNNs), it is typical to use least-squares to determine the weights once the locations of node centers as well as the widths of their RBFs are known (Liu 2013). Other methods like competitive "winner takes all" strategy or Hebbian learning rule are also possible for KSONs and Hopfield networks (Karray and Silva 2004).

2.2.2 Use of Neural Networks in the Context of Industry 4.0

The networks like MLP, RBFNN, KSON, or HNs are now considered classical types of networks that experienced a boom mostly around 2000. Since then, also thanks to the significant advancements in computer hardware, deep architectures (Section 2.8 Deep Learning) started to dominate the field due to their powerful capabilities, mainly in object and voice recognition areas but also others. With its almost implicit reliance on huge amounts of data, this fact is even more pronounced in Industry 4.0 concept. However, they still hold potential for specific applications in particular fields, especially when hybridized with other CI paradigms or when, for some reason, limited data is available.

In one of the more recent works, Yang et al. (2019) used online learning RBFNN to compensate for the unmodeled effects of the system. Together with an accurate inverse kinematic model, this network was used for a disturbance observer design. The authors used this approach for 3-PRR (Prismatic-Revolute-Revolute) compliant parallel manipulator with variable thickness flexure pivots. The use of compliant mechanisms is in line with the current trends in robotics to be used in smart factories, where the human-machine interaction is of crucial importance. A very interesting application of RBFNNs in the food industry can be found in Shi et al. (2019), where the researchers developed RBFNN for estimating freshness of fish fillets under non-isothermal conditions. To achieve this, they selected nine optimal wavelengths from hyperspectral imaging based on successive projections algorithms to monitor important freshness parameters.

Automated Guidance Vehicles (AGV) are considered an important part of a smart factory concept to provide higher flexibility in manufacturing and they are used for transporting goods or materials to various parts of a factory (Mehami et al. 2018). Wong and Yu (2019) used an optimization algorithm to minimize path following error based on Lyapunov direct method controller with RBF neural network estimator. This solution was proposed to address the problems of vision-based simultaneous localization and mapping when disturbances are occurring.

Optimal operation of power systems is also a significant factor in modern factories and achieving this optimality is becoming more difficult in view of the stringent requirements assumed by Industry 4.0 concept. This particular problem was addressed in Veerasamy et al. (2020), where authors used a new approach for solving non-linear transcendental power flow equations using Runge–Kutta-based modified Hopfield neural network. This was compared to the conventionally used Newton–Raphson method and showed a lower computational load with highly accurate results. Similarly, Djedidi and Djeziri in (2020) developed a new type of power estimator for ARM-based (Advanced RISC Machine) embedded systems with granularity at the level of components. This estimator was based on nonlinear autoregressive with eXogenous input (NARX) neural network and authors were able to achieve mean absolute percentage error (MAPE) of 2.2%. The results are important for IoT area, where the power consumption of embedded systems is of crucial importance.

2.3   Fuzzy Systems

2.3.1 Fundamentals of Fuzzy Systems

It is a well-known fact that the usefulness of binary logic that is fundamental in our computers is severely compromised when applied to the possible explanation of human thinking. Our ways of communication are in stark contrast with crisp and rigorous expressions needed for the proper functioning of computers. However, we are capable of solving complex problems as well as processing incomplete, uncertain, and contradicting information. This served as a powerful source of inspiration for the father of fuzzy logic Lotfi Zadeh, who introduced the concept with his seminal paper in 1965.

While the values of binary logic are restricted to 0 and 1, the fuzzy logic is multivalued, and given input may belong to a given set with any membership function value between 0 and 1. In addition to that, the particular input may also belong to more than one set with different values of the membership function. Using this concept it is possible to express the degree of truth (Antão 2017), which is defined for certain types of membership functions assigned with linguistic labels. These are known as fuzzy sets and may be assigned labels like "*low*", "*high*", "*very low*", "*very high*" and similar. These fuzzy sets have specific shapes, with triangular, trapezoid, or Gaussian being the most common. Besides the process of fuzzification of crisp values, it is also the use of *If–Then* rules in fuzzy systems that make it possible to mimic the way of human reasoning when, e.g., controlling systems or processes. Using the definition of such rules, it is possible to incorporate the expert's knowledge in certain areas into a fuzzy system (Pedrycz and Gomide 2007). *If* parts of the rules are known as *antecedents* and *Then* parts as consequents and based on the form of consequents two most common types of fuzzy systems are distinguished:

1. **Mamdani**
2. **Takagi–Sugeno**

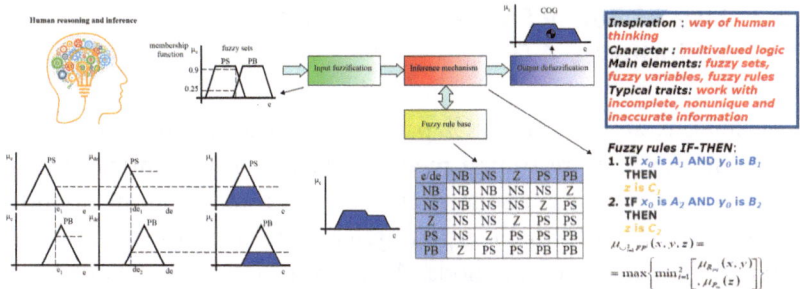

Fig. 2.4 Fuzzy logic inspiration and basic operation

The first type uses fuzzy sets in consequent parts of the rules while the second one uses a linear function of its inputs. In general, every fuzzy system contains four basic parts: *fuzzifier, rule base, inference system,* and *defuzzifier* (Antão 2017). The purpose of a fuzzifier is to convert the crisp value of a given variable into a fuzzy domain. The rule base stores the knowledge in the form of *If–Then* rules (Fig. 2.4) which can be extracted from an expert or numerical data. Using the inference engine, it is possible to make basic algebraic manipulations with fuzzy sets, while defuzzifier calculates the crisp value at the output of a fuzzy system based on the aggregated results from all active rules.

It was recognized very soon that even though fuzzy systems were designed for handling the uncertainty in data, once all parameters are determined it becomes completely certain (Antão 2017). To address this, Type 2 fuzzy systems were introduced where the value of membership function is uncertain and specified also using a fuzzy set with a value from [0,1] interval. All possible values of membership functions are then limited from above with *upper membership function* and from below with *lower membership function* and the area between them is known as a *Footprint of Uncertainty (FOU).*

2.3.2 Use of Fuzzy Systems in the Context of Industry 4.0

Fuzzy logic is an important CI paradigm, which belongs to the category of approximate reasoning methods (Fig. 2.1). In contrast to other purely data-driven CI methods, fuzzy logic (FL) stands on the boundary of traditional AI, of which expert systems are a prime example, and subsymbolic, data-based methods. As a result of this, they have certain advantages in being able to incorporate the expert knowledge in their structure, while also taking care of uncertainty and imprecision of this knowledge.

With advances in Industry 4.0 concept implementation in the manufacturing process, the reality of smart manufacturing becomes imminent. In that case, the interconnection of many elements that share enormous amounts of data is one of the central points to consider when designing the control part of this network. Researchers in Huo et al. (2020) proposed to use a fuzzy control system to provide real-time analysis of information on an assembly line. To improve the performance, two types of a fuzzy controller were used: one of them of Type1 and the second one of Type 2. The former handled the situations where the need for re-balancing the assembly line for satisfying demands was decided. The latter one's purpose was to adjust the production rate in order to eliminate blockages and increase the utilization of machines. More specifically, authors in Lu and Liu (2018) tried to address the issue of keeping the quality of a manufactured product within acceptable bounds based on Taguchi methods. For this, they developed a fuzzy nonlinear programming model based on a fuzzy signal-to-noise ratio. By using this approach, they were able to obtain optimal solutions of lower and upper bound fuzzy S/N (Signal/Noise) ratio.

As one of the principal technologies under Industry 4.0, Internet of Things may certainly benefit from the application of fuzzy logic on many levels. One of them is the use of Wireless Sensor Networks (WSNs) used for sensing the environment and collection/sending of data to the base station for analysis (Thangaramya et al. 2019). In this area, both the intelligent routing and energy optimization are aspects that need to be addressed to keep quality of service in the network at an acceptable level. The authors proposed the use of neuro-fuzzy rule-based cluster formation and the routing protocol to handle these issues and also employed a convolutional neural network for rule formation on discovering energy-efficient routing. Fuzzy logic-based methods can also help with the processing of vast amounts of data generated by a large number of interconnected devices. In Bu (2018), authors propose a high-order tensor fuzzy c-mean algorithm, which was said to achieve much higher clustering efficiency compared to a traditional algorithm.

With the vast amounts of data becoming ubiquitous, the concept of big data and methods for its efficient handling is central to Industry 4.0. The principles of fuzzy logic hold great potential for applications in big data analytics. In Shukla et al. (2020), researchers proposed the use of interval type-2 fuzzy sets for handling the veracity issue in big data to prevent the unusability of data. The problem of handling big data was addressed also in Zhang et al. (2020b), where a quantitative model and method based on fuzzy DEcision MAking Trial and Evaluation Laboratory (DEMATEL) were proposed. As reported in the paper, it could be used as a theoretical basis for handling big data by industry or government. Likewise Chen et al. (2020) used DEMATEL for determining the criteria weights in a smart supply chain. However, the authors identified problems simultaneous manipulation of internal and external uncertainties. For this, they proposed a hybrid rough-fuzzy DEMATEL-TOPSIS (Technique for Order of Preference by Similarity to Ideal Solution) approach for sustainable supplier selection in a smart supply chain.

2.4 Evolutionary Computation

2.4.1 Fundamentals of Evolutionary Computation

Evolutionary computation is a paradigm that is widely regarded as one of the main pillars in the field of computational intelligence. Taking its inspiration from neo-Darwinism, this collection of computational methods makes use of basic principles of evolutionary biology, natural selection process, and genetic variations (Castro 2006). The genetic variations happen at the level of chromosomes with their basic functional units named *genes*. These form the genetic makeup of an individual termed *genotype*. This makeup affects the observable characteristics or *traits* of an organism. It is through these traits that an individual can show its better adaptation to the environment and thus increase the probability of survival and reproduction—this is known as the *fitness* of given individual (Fulcher 2008).

It was natural to adapt these principles to a computational form, where it is possible to search for the solutions to optimization problems. These methods were collectively named as *evolutionary computation* and include three different approaches (Castro 2006):

- **genetic algorithms,**
- **evolution strategies,**
- **evolutionary programming.**

In general, evolutionary algorithms maintain a population of individuals which themselves represent solutions to the problem with various forms of encoding. In analogy with the main inspiration, the individuals within the population are evaluated according to their fitness—i.e., how well they are adapted to their environment or, in terms of optimization problem solving what is the value of an objective function. From this population, a certain number of individuals is selected to mating pool, where the *crossover* (recombination of their genetic information), as well as mutation (alteration of existing genetic information), take place, with all these processes corresponding to one *generation* (Fig. 2.5).

The genetic Algorithm (GA) is one of the most widespread evolutionary algorithms. In its well-known form, it uses solutions encoded as binary numbers (bitstrings), which are usually known as *chromosomes* (Sumathi et al. 2008). Each place in this chromosome is known as *locus* and its possible value at this position is called an *allele*. The whole chromosome then represents one solution to the problem at hand and its quality is evaluated using the fitness function. The parents for mating can be selected using *roulette wheel selection* or *tournament*. The form of crossover operation is in the simplest case one- or two-point, which means the number of positions at which the chromosomes are cut and recombined. In binary genetic algorithm (GA), the mutation can be

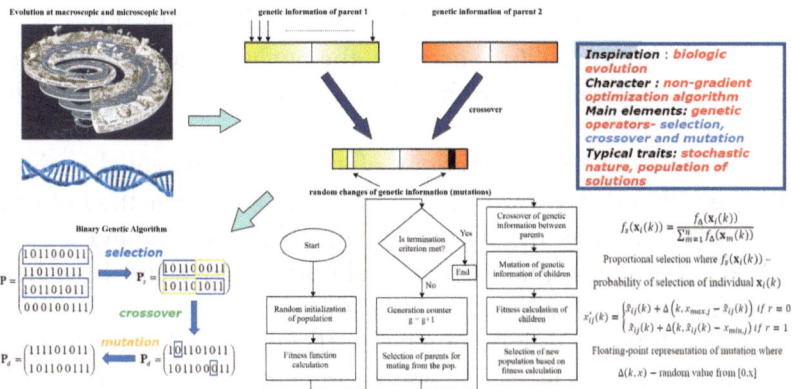

Fig. 2.5 Evolutionary computation inspiration and basic operation

carried out as simple flipping of the original value (from 0 to 1 or vice versa). This has to be done with small probability so that the valuable genetic information contained in the population is not destroyed by excessive random modifications.

Since the introduction of GA, a very high number of variations have been developed to address various aspects of the original implementation. This includes messy GA (Goldberg et al. 1995), island GA (Cantú-Paz 1998), niching GA capable of locating multiple solutions (dynamic niche sharing [Miller and Shaw 1996], nondominated sort GA [Srinivas and Deb 1991]), coevolutionary shared niching (Goldberg and Wang 1997) and many others.

Differential Evolution (DE) is an evolutionary algorithm that basically differs from the genetic algorithm in that instead of crossover, the mutation is applied first to generate the so-called *trial vector*. Only after this step, the crossover operator is applied to produce one offspring. In addition to that mutation step sizes are not sampled from a prior known probability distribution function (Engelbrecht 2007). The working principle of differential evolution is based on the concept of *difference vectors*, which correspond to the magnitudes of distances between individuals in the population. If those distances are large (individuals are far away from each other), the search space should be explored (taking large steps). However, if the opposite is true, it is reasonable to exploit the search space and look for the solutions only in the close vicinity of the current position (Feoktistov 2006). Therefore, the mutation steps are calculated as weighted differences between individuals that are selected in random (Engelbrecht 2007).

In addition to classic variations of the basic differential evolution (DE) algorithm denoted with DE/x/y/z, where x is the method of target vector selection, y is the number of difference vectors and z is the method of crossover, many other modifications have also been introduced. These modifications include gradient-based hybrid DE (Chiou and Wang 1998), DE-hybridized GA (Hrstka and Kučerová 2004), DE-hybridized particle swarm optimization (PSO) (Hendtlass 2001), dynamic self-adaptive DE (Chang and Xu 2000), angle modulated differential evolution (Pampara et al. 2006) and others.

Coevolution is a special type of evolution, where the complementary interaction between species is considered (Engelbrecht 2007). This can happen, e.g., in *predator–prey* interaction, where one of the species evolves to be better in escaping a predator while the other one evolves to be better in catching this prey. That is, this interaction is complementary because the failure of one of the species naturally means the success of the other one. The main difference compared to the standard evolution-based algorithm is that in coevolution type algorithms one does not use the absolute fitness function to evaluate the optimality but attempts to achieve optimality through defeating opponent Engelbrecht (2007). There are basically two types of coevolution:

- **competitive,**
- **cooperative.**

The predator–prey model of coevolution can be considered a competitive type since, as mentioned above, the success of one of the species leads to the failure of the other one. In contrast, a cooperative type of coevolution involves the possible improvement of both or one of the species.

2.4.2 Use of Evolutionary Computation in the Context of Industry 4.0

Evolutionary computation is a well-established CI paradigm, which, either in its original or hybridized form, has been used successfully in many applications. Due to its population character, it lends itself to a parallel implementation to make it more effective. Its possible use in Industry 4.0 concept is manifold—if the problem at hand can be cast as an optimization problem, evolution-based algorithms can be used for a search of the solution. These problems can range from the controller design and/or neural networks training to job-shop scheduling and supply chain optimization.

The need to apply advanced computational methods in the area of logistics and supply chain management as a part of smart manufacturing under the concept of Industry 4.0 is evident. In this scenario, the problem of resource-constrained job scheduling is an important one and bio-inspired computational methods are often applied to address it. The researchers in Nguyen et al. (2019) used a hybrid optimization method based on differential evolution, iterated greedy search, mixed integer programming as well as parallel computing to solve the problem of resource-constrained job scheduling for large-scale instances. The problem of supply chains was tackled also in Saif-Eddine et al. (2019), where specifically the total supply chain cost was optimized. Since this belongs to the group of NP-hard problems, an improved genetic algorithm was designed and used to address the problem. It was shown that this modification outperformed classical GA for two instances (10 and 30 customers).

Cognitive Radio Networks (CRN) are an important type of networks with applications ranging from wireless sensor networks to Medical Body Area Networks, and are thus an important part of communication framework within Industry 4.0. In CRN the energy efficiency issue is of utmost importance, which is addressed in Tang and Xin (2016) through the use of new energy efficiency metric. The optimization problem itself is solved using a chaotic particle swarm algorithm and coevolution methodology, which helps to decompose the original problem into several smaller ones.

The concept of distributed manufacturing is of great interest in meeting current demands on quick responses to the market changes and the sharing of resources. On the other hand, using this concept requires to address the problem of job assignment to different shops as well as its sequencing. The researchers in Zheng et al. (2020) used a cooperative coevolution algorithm for multi-objective fuzzy distributed hybrid flow shop. The coevolution part of the algorithm is proposed to achieve a proper balance between the exploration and exploitation capabilities based on the information entropy and elite solutions diversity.

Additive manufacturing is considered to be a crucial part of Industry 4.0-based manufacturing, which assumes the integration of intelligent production systems and advanced information technologies (Dilberoglu et al. 2017). Following this, Mele and Campana (2020) used evolutionary computing for addressing the problem of parts build orientation, based on the life-cycle impact assessment indicators used for modeling the Pareto front of environmentally non-dominated solutions. Likewise, Ewald and colleagues (Ewald et al. 2018) adapted evolutionary algorithm for varying the size, orientation, and position of wrought material in hybrid manufacturing strategy that combined laser metal deposition and milling or turning.

2.5   Swarm Intelligence

2.5.1   Fundamentals of Swarm Intelligence

Swarm intelligence has been a subject of interest among researchers in technical fields almost since the introduction of this term. Even before that, the behavior of the collection of certain animals was found intriguing and could serve as a remarkable source of inspiration Fig. 2.6. As a matter of fact, many systems in the industry can be viewed as a collection of simple agents cooperating among themselves in the same environment. In this regard, swarm intelligence can be defined as "the emergent collective intelligence of groups of simple agents" (Bonabeau et al. 1999; Tan et al. 2010; Nayyar et al. 2018). By observing the behavior of those simple agents in nature (be it birds, ants, bees, or

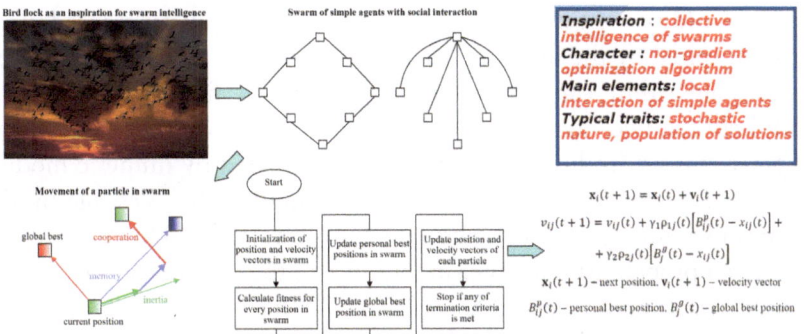

Fig. 2.6 Swarm intelligence inspiration and basic operation

others), we see the emergence of properties which are not inherent in any of those individuals. To describe two fundamental properties of swarm intelligence, one can refer to two different animal species—ants and birds. In both species, the self-organization as one of those properties can be observed. According to Blum and Merkle (2008), self-organization is "a process in which patterns at the global level of a system emerge solely from numerous interactions among the lower-level components of the system". In addition to that, the behavior of ants includes another fundamental property, which is a division of labor viewed as parallel execution of different tasks by agents in a swarm (Nayyar et al. 2018).

While many algorithms inspired by the behavior of swarms have been developed, two of them form the backbone of so-called swarm intelligence algorithms, i.e., particle swarm optimization and ant colony optimization (Engelbrecht 2007).

Particle Swarm Optimization (PSO) is a population-based search algorithm inspired by the behavior of birds when flying in a flock (Engelbrecht 2007; Yang et al. 2013). In the form of an algorithm, the number of particles (members of the population) flies through hyperdimensional search space in an attempt to find an extremum of a given function, possibly under certain constraints. The position of each of the particles is changed based on its own experience as well as the experience of its neighbors through the cognitive and the social components (Engelbrecht 2007).

Ant Colony Optimization (ACO) is also a population-based algorithm inspired by the foraging behavior of real ants. The problem solved by the algorithm can be cast as a problem of finding the shortest path between two nodes, which is achieved through so-called *stigmergy*. Stigmergy is an "indirect communication mediated by numeric modifications of environmental states which are only locally accessible by the communicating agents" (Dorigo and Di Caro 1999). The ants choose paths in a probabilistic manner in response to the amount of pheromone concentration on a given path.

From the inception of both types of algorithms many modifications have been proposed including social-based PSO (Messerschmidt and Engelbrecht 2004), GA-PSO hybrid algorithm (Angeline 1998), NichePSO (Agrafiotis and Cedeño 2002), craziness PSO (Kennedy and Eberhart 1995), quantum-behaved PSO (Fang et al. 2010), ant colony system algorithm (Ippolito et al. 2004), max–min AS (Stützle and Hoos 2000), Ant-Q algorithm (Mariano and Morales 1999), Antabu (Fonlupt et al. 2006) and many more. All these modifications addressed various problems of the original algorithms either in general or in specific applications, where they contributed to the improved performance.

2.5.2 Use of Swarm Intelligence in the Context of Industry 4.0

Similar to evolutionary computing, swarm intelligence-based algorithms have become quite popular as non-gradient optimization techniques applied to hard optimization problems. The basic PSO is quite simple but further modifications increased its complexity and improved its performance. The newer implementations like quantum-behaved PSO are quite powerful, even for high-dimensional problems where other optimization techniques might fail. The cooperation of simple agents as seen in swarm intelligence has not been used solely in optimization algorithms but served as an inspiration for many other approaches proposed for the smart manufacturing area and/or Industry 4.0 in general.

In Sun et al. (2018), researchers used swarm intelligence for community detection, which is a task of critical value in the analysis of complex networks. This is especially important for dynamic networks, where the properties of decentralized, self-organized, and self-evolving systems are of importance. The use of swarm intelligence also addresses the problem of overlapping community detection since it can handle the joining of the vertex into multiple communities and also the addition or deletion of a vertex dynamically. In particular, a particle swarm optimization algorithm was used in Gill et al. (2018) for the problem of cloud resource scheduling, which requires the mapping of cloud resources to cloud workloads. By using PSO, the parameters of Quality of Service (execution cost, time, and energy consumption in particular but also others) could be significantly reduced. Taking into account the specific properties of the ACO algorithm and its variants, it is natural to consider it for solving the problems of routing as a part of Internet of Things implementations. This was used in Thapar and Batra (2018), where the network of sensor nodes was seen as a colony of ants. RPL protocol (Routing Protocol for Low-power and lossy networks) was used to build a destination-oriented directed acyclic graph using the objective function, which was responsible for fixation of the rank of node and selection of best directed acyclic graph using ACO. This algorithm was also adapted to resource distribution optimization in Hong et al. (2019) in the form of resource indexing optimization. The velocity and position of cluster resource indexing were updated based on the ant colony trajectory with constraint condition on the minimum variance of the fitness function.

With an increasing number of various embedded devices in IoT, the scheduling tasks can be considered NP-hard problems for which no polynomial-time algorithms might be available. Therefore, the use of metaheuristics like PSO can be suitable to obtain good (or even close to optimal) solutions. Authors in Xie et al. (2019) used PSO for the problem of workflow scheduling in a cloud-edge environment. They introduced a Directional and Non-local-Convergent PSO (DNCPSO), which employed non-linear inertia weight where the selection and the mutation operations were performed using the directional search process.

The problem of path planning for mobile robots is relevant also for the concept of smart manufacturing, where the extensive use of AGVs is expected. Depending on the conditions, this is usually a computationally demanding task, where the bio-inspired computational methods can be of great benefit. Dewang et al. (2018) used adaptive particle swarm optimization (APSO) for the path planning of a mobile robot. This was showed to be faster than using a conventional PSO algorithm.

2.6 Artificial Immune Systems

2.6.1 Fundamentals of Artificial Immune Systems

Similar to the human brain and nervous system in general, the immune system is well-known for its remarkable properties in maintaining the balance of internal state of humans, especially in response to the invasion of external harmful agents (e.g., viruses and bacteria). This system is extremely complex and sophisticated, being in constant interplay with other systems within the human body to achieve homeostasis (dynamic state of equilibrium). Even there are several layers of body protection against invaders, the most important division of the immune system in terms of its function and properties is to the *innate immune system* and *adaptive immune system.*

The innate immune system is known to be able to mount a response against harmful agents by recognizing their generic molecular patterns not present in the cells of a host but only in invading pathogens (Castro 2006). When those agents damage the cells of a host, the innate immune system provides co-stimulatory signals, needed, e.g., for the action of the adaptive immune system. Moreover, it is the innate immune system that provides a faster response to the invasion, while the adaptive system starts to act. On the other hand, it is the adaptive immune system that is capable of fighting even against the invaders never seen before. Furthermore, when these pathogens are present again, the adaptive immune system can mount a faster response through the "immune memory" (Engelbrecht 2007; Castro 2006).

It is an adaptive immune system in particular that became the main source of inspiration for developing algorithms loosely inspired by its function. According to Engelbrecht (2007), some of the capabilities of the natural immune system usable in computational tools are the following:

- The immune system can distinguish between self and foreign/non-self cells (and knows their structure).
- Foreign cells can be *dangerous* or *non-dangerous.*
- Lymphocytes (a certain type of white blood cells) are subject to cloning and mutation to adapt to the structure of foreign cells, which leads to the formation of *memory.*
- Lymphocytes have coordination and co-stimulation among them, forming *immune networks* as a result.

An artificial immune system algorithm is a population-based algorithm that can be used for clustering and/or optimization problems. In its basic form, this algorithm uses artificial lymphocytes that form the population of the solution to a given problem. After selecting a subset of this population, the affinity (the measurement of similarity or dissimilarity) between this subset and antigen is calculated. The calculation of affinity can be applied also to ALCs (Artificial LymphoCytes) themselves in analogy to immune networks. Then, based on these results, some of the ALC can be selected (through negative or positive selection) to be cloned and mutated to find ALCs with an even better affinity with antigen. Some of them can be selected to become memory cells for the secondary response of artificial immune system (AIS) when similar antigens are encountered.

Artificial immune networks are a special type of artificial immune system model, where the main difference compared to clonal selection-based models is their characterization as dynamic systems capable of functioning also without the antigen stimulation (Castro 2006). From a mathematical point of view, it is natural to describe such systems using ordinary differential equations allowing easy incorporation of real immune system (IS) properties like learning, memory, self-tolerance, and network interactions (Castro and Timmis 2002). In this kind of network, it possible to achieve stimulation of cells by another cell or a foreign antigen, while its suppression occurs due to recognition of self only (Castro 2006).

In addition to the basic AIS algorithm, many other modifications appeared some of which were based on clonal selection theory models (like CLONALG—CLOnal selection ALGorithm (Castro and Zuben 2000), AIS with dynamic clonal selection (Kim and Bentley 2002) or multi-layered AIS (Knight and Timmis 2002)), immune network theory models (AINE—Artificial Immune NEtwork (Timmis and Neal 2001)), EAINE—Enhanced Artificial Immune NEtwork (Nasraoui et al. 2002), aiNet (Castro and Zuben 2002) or danger theory models (Aickelin and Cayzer 2002).

2.6.2 Use of Artificial Immune Systems in the Context of Industry 4.0

Artificial immune systems as a paradigm loosely inspired by the functioning of the natural immune system offer interesting capabilities like adaptability, self-learning, and robustness that can be used for various tasks in data processing, system modeling, and control, fault detection or cybersecurity. All these aspects make it a suitable paradigm for addressing the problems in the context of Industry 4.0.

An interesting approach is used in Wang et al. (2018b) for optimizing the manufacturing process through energy monitoring as well as re-scheduling of manufacturing. The researchers did not use a standard approach where the conditions of the manufacturing process are known in advance but recorded and analyzed the data of energy consumption using neural networks and statistical tools. AIS algorithm was then used to tackle the situations with highly variable conditions in the manufacturing process. Researchers in Semwal and Nair (2020) realized that the centralization approach for the implementation of networked environments in cyber-physical systems is expensive concerning the increasing number of devices and the flow of information. While decentralization and distribution of the architecture address this issue, it remains a challenge to find the best solutions for problems distributed across devices. Inspiration is therefore taken from the function of the immune system (immune networks, danger theory, and clonal selection), which works as a decentralized system with capabilities of adaptivity, self-learning, and self-organization. The concept of AIS in the context of Industry 4.0 can be used also for addressing the issue of cybersecurity threats. This was applied in Zhang et al. (2011), where a distributed intrusion detection system in smart grids was used. The communication was provided through several wireless mesh networks with 802.15.4, 802.11, and World Interoperability for Microwave Access (WiMAX) standards, which presented cybersecurity threats. To tackle these problems, researchers proposed a distributed intrusion detection system for smart grids where several analyzing modules based on support vector machines and artificial immune system paradigms were used. These modules helped to detect and classify malicious data as well as possible cyberattacks.

The latest study of Aldhaheri et al. (2020) provides a review of the literature and recommendations for further research in the field of AIS application to secure IoT. This work tries to fill the gaps in the coherent and systematic presentation of AIS capabilities for Internet of Things, especially in terms of cybersecurity. By offering an exhaustive survey of the past as well as recent research in the area, this paper corroborates the fact that this paradigm holds the potential for current and future applications within Industry 4.0 concept.

Autonomous mobile robots are an important part of the smart factory concept, where one of the tasks may be the transport of materials across the factory floor while avoiding obstacles and identifying pickups or material dropoff in real-time. This presents a problem of navigation in an uncertain and/or unstructured environment, where the properties of adaptivity and self-organization of natural immune systems can help. Authors in Akram and Raza (2018) proposed the concept of the Robot Immune System to maintain the robot's internal health-equilibrium (analogy to homeostasis). For this, a robot uses health indicators (e.g., energy and temperature) to detect any abnormalities in its function. This eventually leads to the state of inflammation, which activates the first innate and subsequently adaptive immune system.

The monitoring of the proper operation of systems may be critical under the concept of Industry 4.0. Should any fault occur in a system, it is not only important to detect it but also to try to locate it. These approaches are known as *Fault Detection and Isolation (FDI)* and can be typically model-based or signal processing-based. In Costa Silva et al. (2017) authors presented a review of three different AIS approaches to FDI—the toll-like receptor algorithm, the dendritic cell algorithm, and the danger theory-based algorithm.

2.7 Big Data

The importance of data in general within Industry 4.0 cannot be stressed enough—it is simply of one the pillars upon which this concept is built. From the smallest and least complex systems to smart factories, the existence of massive amounts of data need to be taken into account and processed accordingly. While sometimes it may not be utterly clear what exactly should be considered big data and what should be not, there is some consensus regarding the main aspects according to which we could evaluate data in question. In Iqbal et al. (2020a), the five Vs of big data are presented:

1. **Volume**
2. **Velocity**
3. **Variety**
4. **Veracity**
5. **Value**

Without doubts, the *volume* should be the single most obvious aspect of big data. Enormous amounts of data are generated and stored in an instant and these volumes can reach even zettabytes for the whole Internet (Iqbal et al. 2020a). Even though amounts generated, e.g., in a smart factory as a whole will be much smaller, they still have large enough volume to necessitate a special approach for its analysis. *Velocity* refers to the speed at which this data is generated but also streamed or stored (Iqbal et al. 2020b). Since the source of data can be as diverse as social media or data from sensors, it is obvious that the structure of data can differ significantly—this falls under the category of *variety* of data. Due to the very large volumes of data generated at high velocity, a lot of noise is contained in the data and its trustworthiness may be impaired and this aspect is known as *veracity* of data. The last V of the list named *value* is related to the meaningful insight of data associated with its usefulness in identifying important patterns (Iqbal et al. 2020a). It should be noted that in some sources only three Vs (volume, velocity, and variety) as the main characteristics of big data are given (Khan et al. 2019).

After data is collected, it rarely can be used in its raw format and has to undergo some form of pre-processing and analysis (Fig. 2.7). As variety aspect of the data implies, it can be from different sources be it log files, industrial sensors, webpages, etc. Moreover, in its raw form, it includes a lot of unimportant information as well as some redundancy which has to be tackled in some way. This helps to maintain the data consistency as well as optimize its storage requirements, which can be quite important when vast amounts of data are considered. When considered as a whole, the methods in this step allow uncertain and incomplete data to be modified and/or removed while also making the dataset free of repetitive or superfluous information (Khan et al. 2019).

It is mainly the last step (titled Big Data Analytics in Fig. 2.7), where the use of machine learning (or specifically computational intelligence) techniques may be beneficial. Machine learning as an approach, where mathematical models derived from sampled data for making predictions are used, is suitable for finding patterns in the datasets. These methods are usually statistical in their nature and the existence of models derived using them is important for the process of decision making. On the other hand, computational intelligence represents a collection of typically

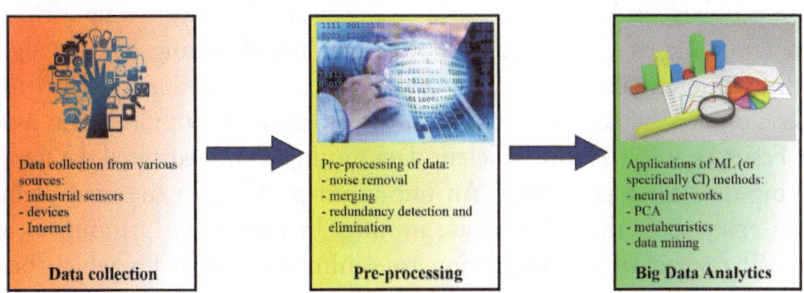

Fig. 2.7 Three basic steps in big data analysis process

nature-inspired approaches to problem-solving, often offering good solutions but usually without guarantees of optimality. However, by using these approaches previously intractable problems can be addressed efficiently with acceptable results. The problem of big data analytics is nowadays encountered in many fields and quite often the results can be generalized to some extent from one field to another.

Neural networks in general are one of those CI paradigms that are used quite often in Big Data Analytics. Hernandez et al. in (2020) used two new hybrid neural architectures in which morphological neurons and perceptrons were combined. Both types were used for feature extraction and trained by a stochastic gradient optimization technique. It was shown in the paper that multi-layer neural network (MLNN) required a lower number of learning parameters than other architectures. The problem of big data analysis was addressed also in Anbarasan et al. (2020), where authors used a combination of IoT and CNN in big data scenarios with the flood detection system, giving better results than other competitive methods. In addition to feature extraction, feature selection is also one of the key points in machine learning to eliminate the redundancy of data or avoiding the curse of dimensionality. Nature-inspired metaheuristics appear particularly suitable for this kind of problems, due to their capabilities of finding good solutions for NP-hard problems. Researchers in Abdi and Feizi-Derakhshi (2020) extended Search Manager for multi-objective problems and used it for EEG signal (ElectroEncephaloGraphy) analysis with reportedly good results. Likewise, this problem (of feature selection) was also studied in Nguyen et al. (2020), where the authors reviewed various approaches to feature selection problems in big data scenarios using swarm intelligence algorithms. An ensemble of three methods (non-dominated sorting genetic algorithm, differential evolution, and multi-objective evolutionary algorithm based on dominance and decomposition) combined with CNN was used in Essiet et al. (2019) for efficient data mining from dedicated databases of big data for a gas sensor. An important problem in the field of big data analytics is that of database mining. In particular, Djenouri et al. (2019) researched the association rule mining problem for which bees swarm optimization was considered effective but too computationally demanding. The authors developed Graphics Processing Unit (GPU)-based bees swarm optimization miner, where the GPU was used as a co-processor and found the method to be 800× faster than CPU-based (Central Processing Unit) method.

2.8 Deep Learning

2.8.1 Fundamentals of Deep Learning

As was mentioned in Section 2.2 neural networks, the neural networks consist of several simple computing units called neurons organized in layers and interconnected by synaptic weights. It is through the modification of these weights that neural networks are capable of learning (whether in a supervised or unsupervised manner). The term "deep" is related to the depth of neural network architecture, which is used whenever the network contains at least two hidden layers (Fig. 2.8). Thus, in principle, the deep architectures are closely related to their shallow counterparts from a structural viewpoint. However, the increase in the number of hidden layers offers significant improvements in network

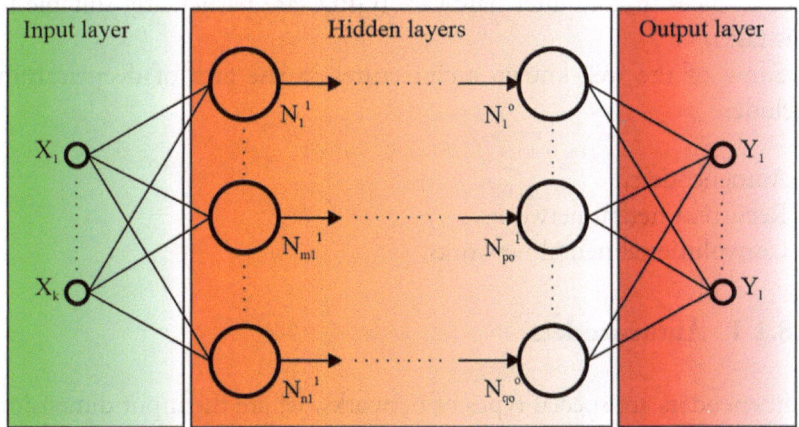

Fig. 2.8 Structure of deep feedforward neural network with *k* inputs, *l* output, *o* hidden layers and generally different number of neurons in each hidden layer

performance that cannot be matched by increasing the number of neurons in only a single hidden layer (Aggarwal 2018).

It is clear that the idea of including at least one additional hidden layer to a network to improve its performance is not a new one. Yet, the successes achieved by using deep neural networks are more recent. The reasons for that can be found in three main problems that were effectively solved in the last fifteen years (Kim 2017): *vanishing gradient problem, overfitting,* and *computational load.*

The problem of vanishing gradient refers to the drop in gradient values, which can happen when multiple layers are present. Actually, the updates in weights for earlier layers become almost negligibly small which equals the stop of the training process. The use Rectifield Linear Unit (ReLU) function the derivative of which is a constant address this problem efficiently (Kim 2017). Also, by using a large number of layers with many parameters the risk of overfitting becomes much more imminent—therefore, new effective methods for prevention of overfitting were needed. A simple but very powerful method is called *dropout*, where some neurons are set to zero during the training and this encourages learning sparse representations. Moreover, the larger the network is the longer time it takes to have it trained (given the same hardware). That is why the effective training of deep neural networks was conditioned by the advances in computer hardware (GPUs are particularly suitable for this task).

Some of the well-known architectures in the area of deep learning include:

- Autoencoders,
- Recurrent neural networks,
- Convolutional neural networks.

2.8.1.1 Autoencoders

Autoencoders are special types of networks, where the input dimensionality is the same as the output dimensionality. The main point of their function is that it assumed that the number of neurons in the hidden layer is lower than the number of inputs/outputs, thus allowing for a

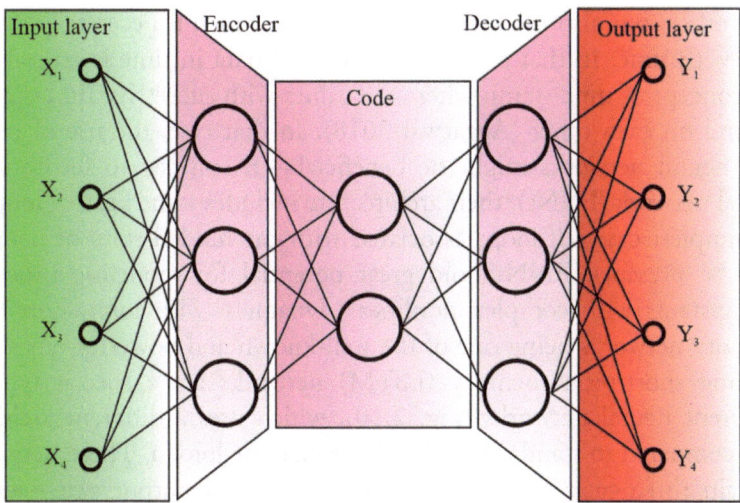

Fig. 2.9 Structure of deep autoencoder with four inputs/outputs and 3-2-3 neurons in three hidden layers

more compact representation of input data (Aggarwal 2018; Sengupta et al. 2020). Since data is passed through a structurally constricted part of a network, the result is information loss which can be expressed through a common error criterion (e.g., SSE—Sum of Squared Errors) (Aggarwal 2018). Autoencoders can be also trained using standard backpropagation training algorithms (Sengupta et al. 2020). The process of getting a compact representation of input using a constricted structure of an autoencoder is called *encoding* (and this part *encoder*) while the process of reconstruction of original data from the encoder is called *decoding* (and this part *decoder*). When multiple hidden layers are used in an autoencoder, it can be called *deep autoencoder*. While not necessarily so, the hidden layers of deep autoencoder are typically symmetrically structured (Fig. 2.9) (Aggarwal 2018).

2.8.1.2 Recurrent Neural Networks (RNN)

Sometimes it is important to take into account not only the relationship between the input and output data (without their explicit dependence on each other) but also the sequential character, which is necessarily associated with time. In that case, the ordering of data in time is crucial and the concept of time-stamp where the values with successive time-stamps depend on each other (Aggarwal 2018). In that case, the use of recurrent neural networks might be beneficial—in contrast to feedforward neural network (FNNs), their architecture includes some kind of looping (in simplest case self-loops associated with the hidden state of neurons may be present). RNNs hold great potential for modeling processes and systems with complex nonlinear dynamics, with long short-term memory networks being one of the well-known and powerful types.

Long short-term memory (LSTM) networks are a special type of recurrent neural networks (Fig. 2.10), which uses a different architecture compared to standard RNNs like Elman or Jordan. These networks contain three gates—forget gate, input gate, and output gate—which provide fine-grain control over data written into long-term memory (Sengupta et al. 2020). The training of RNNs is known to be difficult mainly due to the issues with vanishing and exploding gradients as well as highly varying sensitivities of the error surface to different temporal layers (Aggarwal 2018). Since weights are shared and these networks can become very deep after unfolding in time, successive multiplication with weights smaller than 1 in gradient calculation tends to zero (vanishing gradient) while for weights larger than 1 it tends to very large values (exploding gradient). This issue is addressed by using the above-mentioned three gates, where to forget gate controls the amount of information to be removed from previous cell state c_{t-1} whereas input

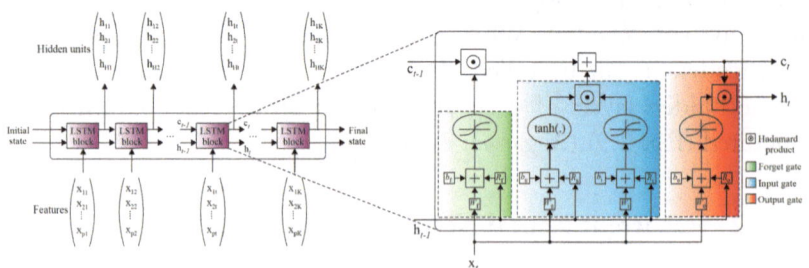

Fig. 2.10 Basic structure of a long short-term memory network

gate decides with what amount of information contained in cell state candidate $\tilde{c}_t$ should a new cell state c_t be updated (Bianchi et al. 2017). The output gate then selects which part of the cell state would be returned as output. Four nonlinearities in total are used in LSTM structure, two of them are placed in the input gate (hyperbolic tangent function and sigmoid function) and one is placed in both forget gate and output gate.

2.8.1.3 Convolutional Neural Networks (CNN)

Convolutional neural networks belong to a special type of neural network particularly suited for the tasks of image recognition (Fig. 2.11). Their architecture is quite different compared to previously mentioned neural networks and is inspired by how images are processed in the visual cortex of the brain (Goodfellow et al. 2016; Kim 2017). It is only during the last years that their potential started to be explored and used in the area of machine vision with triumphant dominance over other techniques. Previous approaches were based on methods demanding extreme expenditure in terms of cost and time for development and offering inconsistent performance (Kim 2017; Goodfellow et al. 2016). The reason for that lies in the need for feature extractor design, which could be specific for a given application and therefore lacking the properties of the general-purpose image recognition tool. This issue is specifically addressed in CNNs, where the design of feature extractor is a part of the training process and can thus be used generally (Kim 2017; Sengupta

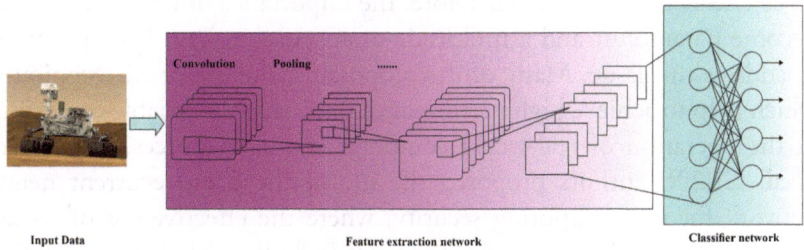

Fig. 2.11 Basic structure of a convolutional neural network

et al. 2020). The *feature extraction network* consists basically of two types of layers: **convolutional layer** and **pooling layer** (Goodfellow et al. 2016).

The function of convolutional layers is based on a mathematical operation called (quite expectedly) *convolution.* This is performed in 2D and acts as a set of digital filters (Kim 2017) to produce so-called feature maps, where some of the features of the original image are enhanced. The pooling layers then serve as dimension reduction elements by merging neighboring pixels into one based on either max or averaging operation. The last part of CNN structure is a *classification network*, which is typically represented by the fully connected network with a number of outputs corresponding to the number of classification classes (Sengupta et al. 2020).

2.8.2   Use of Deep Learning in the Context of Industry 4.0

Deep learning is one of the most perspective CI paradigms for the concept of Industry 4.0 as a whole. Recent advancements in this field confirmed its usefulness for a wide spectrum of problems. It is well-established that their success depends on the availability of huge amounts of data as well as high-performance hardware. Provided these requirements are met, the capabilities of DL in solving certain types of problems can even surpass that of humans.

Considering the crucial position of cloud and fog computing in Industry 4.0 framework, the risk of cyberattacks with potentially disastrous effects is very high. Therefore, the importance of cybersecurity has become paramount and a multitude of approaches have been proposed to address this issue. Many of the proposals make use of AI techniques, which help to achieve high performance under highly variable conditions of the operation of real-world computer system resources. In Almiani et al. (2020), authors proposed the use of the deep recurrent neural network for fog computing security, where the effectiveness of its use was demonstrated using various metrics including Mathew correlation and Cohen's Kappa coefficients.

Although powerful in the applications where a limited amount of data is available for characterizing the properties of various systems, more typical paradigms of computational intelligence like shallow neural networks, support vector machines, logistic regression, etc. may have limited performance under the assumption of massive amounts of data. This assumption is of crucial importance in, e.g., smart manufacturing, where the use of deep architecture models may be beneficial. This topic is researched in Wang et al. (2018a), where an extensive treatise of the methods and applications of deep learning in the field of smart manufacturing is presented. The advantageous use of deep learning methods within the concept of smart manufacturing can happen on many different levels. A good example of this is presented in Andersen et al. (2019), where deep reinforcement learning is used for industrial robots to cope with natural variations in the brine injection process during the production of a meat product. The prospect of deep learning application in the field of robotics is further emphasized in Wang et al. (2020a) where it is used in a multi-robot scenario. In this work, a multi-robot cooperative algorithm using deep reinforcement learning is designed based on the use of Duel network structure, where two streams representing the state value function and state-dependent action advantage function appear and their results are merged.

In particular, the processes in manufacturing themselves may benefit significantly from applying deep learning concepts—be it for their analysis or, quite typically for current trends, or visual inspection. Researchers in Wang et al. (2020b) prepared a tutorial for researchers on how to apply (and also understand) deep learning in manufacturing, with welding used as an example. They discussed two of the most typical techniques, namely the convolutional neural network and recurrent neural networks. Similarly, in Xia et al. (2020) defects in Keyhole Tungsten Inert Gas welding were inspected using Resnet (a type of CNN) to recognize different welding states, including burn through, undercut, incomplete penetration and others. Similarly, defect inspection based on deep learning and Hough Transform (HT) was studied in Wang et al. (2019), where researchers used the Gaussian filter for limiting the random noise in obtained images and then used HT for extracting a

Region of Interest clear of useless background in the image. The identification module used a convolutional neural network and the method was reported to be a good balance between the accuracy and computational load.

2.9   Use of Computational Intelligence in Cyber-Physical Systems

The notion of a *Cyber-Physical System (CPS)* is pervasive to every kind of Industry 4.0 concept description. As such, it was introduced some years prior to the introduction of I4.0 term itself (2006 vs. 2011). Both terms are just natural outcomes of the increased extent of digitalization within the industry (and other areas) in general. While original definitions of "systems using computation and communication deeply embedded in and interacting with physical processes to add new capabilities to physical systems" (CPS report 2008; Song et al. 2016) were appropriate, more refined definitions were deemed necessary to better distinguish between CPS and non-CPS systems. An interesting approach to this issue is offered in Song et al. (2016), where four key aspects are taken into account when characterizing CPS:

- **technical emphasis,**
- **cross-cutting aspects,**
- **level of automation,**
- **life-cycle integration.**

Those aspects are not size-related and CPS systems may include miniature systems as well as large-scale and complex ones.

The first of the aforementioned aspects is also one of the most obvious since the term itself implies the interaction of the physical and cybernetic world. It has to be noted that the interaction between physical (in this case mechanical) part of a system and cybernetic (in this case computational) part of a system to enhance its capabilities has been known for a long time in *mechatronic systems*. However, a new dimension was added to this by implicit inclusion of *connectivity* of those systems to allow for their mutual communication. The sheer extent of connectivity in CPS makes it necessary to consider many aspects from different (even nontechnical) fields, including *security* and *legislation* which are a part of cross-cutting aspects. In addition to that, it is obvious that CPS is designed with a significant degree of automation in its functionality but the human input at a certain level is always expected. This is incorporated into the level of automation aspect of CPS. Since CPS encompasses a very large spectrum of various systems with connectivity capabilities as one of their main properties, they can be characterized also by different levels of integration into the management of products, services, and data (Song et al. 2016; Napoleone et al. 2020).

As pointed out in Panetto et al. (2019) cyber-physical systems in manufacturing (but with some generalization also in other fields) face many challenges under the concept of Industry 4.0 like highly customized supply network control, creation of resilient enterprise to better cope with possible risks, scheduling, and control of digital manufacturing networks or collaborative control. To meet such requirements, it is necessary to apply techniques that allow systems to adapt or learn together with the possibility of self-organization, fault-tolerance as well as handling uncertainty at various levels. With the assumption of vast amounts of data generated by CPS and the necessity to meet previously mentioned requirements, the benefits of the use of computational intelligence for cyber-physical systems are obvious.

This is well summarized in Delicato et al. (2020), where the paradigm of smart cyber-physical systems covering intelligent, self-aware, self-managing, and self-configuring pervasive systems is analyzed. As a part of the cross-cutting aspects of CPS, the security of those systems in view of their connectivity is of crucial importance. This issue is often addressed using a computational intelligence-based approach— researchers in Ding et al. (2018) provides a review of recent advances in security control and attack detection of industrial CPS. In addition to statistics-based machine learning methods, authors present also other methods belonging to the area of computational intelligence (reinforcement learning, neural networks, fuzzy systems).

The problem of scheduling is associated with various aspects of CPS and often needs to be handled with advanced computational techniques to achieve high performance. With application in wireless sensors, this issue was addressed in Leong et al. (2020), where scheduling of sensor transmissions to estimate the states of multiple remote processes was studied. This was formulated as a Markov decision process and Deep Q-Network was used as a solution. The scheduling problem was also researched in Yi et al. (2020) for tasks in multi-processor distributed systems, but this time authors used an ant colony optimization algorithm to enhance the local search ability and improve the quality of the solution.

More application-oriented research related to the use of computational intelligence in cyber-physical systems was presented in Hou et al. (2020), where the CPS framework is introduced to track truckloads in a highway corridor and to trigger the structural health system for bridges. The linking of bridge response to truck weights is carried out using convolutional neural networks and very good performance is reported. Likewise, the cyber-physical framework is used in Zhang et al. (2019) for structural optimization of complex structures in Real-Time using Hybrid Simulations (RTHS). RTHS is used for evaluation of candidate designs and particle swarm optimization algorithm is used for solving an optimization problem. As noted in Zhang et al. (2020a) current islanded microgrids are turning into CPS, which brings with it various kinds of problems like upload interruption problem. In the work, this is addressed with the use of a secondary control strategy based on improved growing and pruning-radial basis function neural network, leading to improved voltage and frequency stability. In Wang et al. (2018c), researchers used a hybrid fuzzy-PID controller which adapts parameters based on environmental and process variables for controlling the secondary loop of a Lead–Bismuth Eutectic eXperimental Accelerator Driven System (LBE-XADS). This system is viewed as a CPS where physical process variables are monitored and processed intelligently to keep the values of safety parameters in the safety range.

2.10 Case Study: Industrial Parts Recognition by Convolutional Neural Networks for Assisted Assembly

The Industry 4.0 concept defines its supporting technologies for example: digital twin, Radio Frequency IDentification (RFID) technology, virtual and augmented reality, cooperative robotics, big data, deep learning and advanced vision systems. The main idea is the implementation of these technologies for full digitalization in the design of production lines and the necessity of changing and deploying asynchronous assembly lines instead of synchronous. Applications of automatized lines can be found in several areas of the industry: consumer electronics, furniture, clothing, and automotive production. Because of the variation in production, it is almost unnecessary for human stuff to interact with the machines during the assembly process. For the Industry 4.0 concept, cooperative robots with advanced vision systems for knowledge extraction were defined as the main element suitable for cooperation with workers as described, for example, in Liu and Wang (2017). Nowadays, the trend is to have highly variable subassemblies, which must be manually assembled due to unmanageable automation and its implementation. In the case of manual assembly of highly variable parts, it is appropriate to use a Virtual (Augmented) Reality (VR/AR) in combination with image processing to simplify and check the assembly process. For example, an anchoring support system using with AR toolkit is described in Takaseki et al. (2015). There is also possibility to use Computer-Aided Design (CAD) 3D models in the approach from CAD assemblies toward knowledge-based assemblies using an intrinsic knowledge-based assembly model (Vilmart et al. 2018). VR/AR is a direct or indirect view of the physical environment with monitored parts. The field of view for workers can be extended with some additional digital data, mostly as text or image. This additional graphical information must be relevant to the object we are looking at. The visible information can be combined from the vision system or other sources for example integrated industrial sensors, RFID systems, or MEMS units (MicroElectroMechanical Systems).

This case study describes a new approach to parts recognition that are not fixed position (different 2D placement and field of view, large scale range with 3D rotation) by convolutional neural networks. Standard industrial vision systems usually cooperate with conveyor systems and recognized parts are placed on the conveyor belt with a fixed distance from the camera lens and they are digitalized only from one side (usually top). These vision systems can cover some invariance, but in a very limited range (2D rotation with placement and very limited scale). Assisted assembly process based on virtual or augmented reality devices has advanced requirements to recognition robustness. It is necessary to reliable recognize and identify parts from every side with different distances and angles from the camera lens. Convolutional neural networks can help solve this complex task without extra demanding on programming as it was presented in Židek et al. (2019a). There are also two novel neural networks, fire-FRD-CNN (Feature Reuse Detection-Convolutional Neural Network) and mobile-FRD-CNN described in Li et al. (2019). A nice review on recent advances in small object detection based on deep learning can be found in Tong et al. (2020). The most problematic part of the usability of convolutional neural networks is the preparation of the input training image set. This monotonous task can be simplified by the automatized generation of the image set from 3D virtual models which was solved in Židek et al. (2019b). This problem is also described in Socher et al. (2012), Su et al. (2015), Sarkar et al. (2017), and Tian et al. (2018). CNN model trained with general samples can be used after transfer learning also for other recognition tasks. So it is possible to use these pre-trained models for recognition of the industrial part with a significant decrease in training time. For example, an interesting applications for recognition of bearing errors using artificial neural networks are described in Pavlenko et al. (2019a, b). Other applications are in the field of quality prediction of manufacturing processes (Hrehova 2016), validation of serviceability of manufacturing systems (Lazár and Husár 2012), intelligent systems in the railway freight management (Balog et al. 2019) and so on.

The main idea of this case study is a combination of standard machine vision algorithms (thresholding with the Region of Interest) and CNN algorithms for reliable small part recognition in images with higher resolution (HD or 4K). The pretrained models of CNN networks can be used for industrial parts recognition, as for example:

- Inception V2, 3, 4 with SSD extension,
- MobileNet V2, 3 with SSD extension,
- ResNet-50,
- Xception,
- Inception-ResNet-V2.

The methodology of industrial parts recognition for assisted assembly and its implementation divided into three mains steps is explained in the block diagram in Fig. 2.12:

- I. step: generation of training samples from virtual 3D models and implementation of standard machine vision algorithms for identification Region Of Interest (ROI),
- II. step: training (evaluation) of CNN models by virtual samples and testing in embedded systems with Accelerated Processing Units (APU),
- III. step: transfer of trained convolutional neural network models to virtual or augmented devices for assisted assembly tasks.

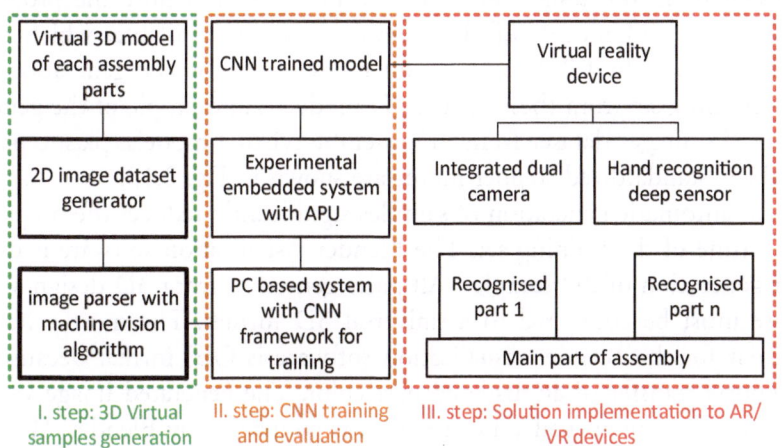

Fig. 2.12 Main steps of CNN implementation to the assisted assembly process

The main novelty in the field of convolutional networks is the methodology of recognition for industrial parts, preparation of samples from virtual 3D models, and increasing reliability for small parts recognition by identification Regions of Interest in images. This methodology after implementation to embedded devices can be transferred to assisted assembly systems based on VR/AR devices for these three main tasks:

- to train employer for new assembly tasks,
- to help operator marking parts for next assembly step,
- to real-time check of a manual assembly task.

2.10.1 Input Samples Generation from 3D Virtual Models

The assembly's parts can be divided into two basic groups: nonstandard parts (machined parts) and standardized parts (nuts, bolts, washers, etc.). The base part of the assembly is the stepper motor. Next there are two plastic parts with different colors produced by rapid prototyping technology connected to the main part by standardized parts (bolts, nuts, washer, and spring). The standardized parts have small dimensions and are assembled to nonstandard parts. All these parts are usually created or generated in 3D design software and are available before the production starts. 3D models of all assembly parts are available, which can help to train the CNN recognition model faster by the generation of the training set from these virtual 3D models. An example of the generated 2D images dataset from 3D assembly virtual models: plastic parts and the standardized stepper motor are shown in Fig. 2.13.

An automatic generation of samples significantly reduces the preparation time of the training set. The Blender visualization software is used for generation of 2D samples. All parts generated from 3D design software must be converted to a universal 3D format. The most suitable format for the Blender visualization software is OBJ format because it supports transfer of an assigned part color. The generated image variation of movement, scale, and rotation is controlled by Blender API via Python script.

Fig. 2.13 Generated 2D image dataset from the virtual 3D virtual model of the assembly

2.10.2 Identification of a Region of Interest for Recognition of Small Parts

The main limitation for the recognition of small parts by CNN models is low input resolution ($224 \times 224 \times 3$ or $299 \times 299 \times 3$). For larger objects, CNN models work reliably. Small objects lose details during the recognition process because high-resolution images (4K or 8K) are automatically downsampled to the default input layer with low resolution. This is one reason why the recognition of small industrial parts in assembly objects is difficult. The inspiration for solving this problem can be taken from the human brain, which solves the same problem with the recognition of small objects by changing the distance from a recognized object. The next very interesting feature of the human brain is its ability to ignore the areas with plain color in a recognition process and focus mainly on places with some pattern. This problem is solved very simply by changing the position and distance from the recognized object. But this approach is not suitable for industrial tasks.

Two much more effective methods for industry are useful:

1. The vision system with an automated optical zoom (suitable for recognition of parts for long distances).
2. The vision system with a high-resolution camera and integrated identification of the Region of Interest.

The first approach is not suitable for the assisted assembly process because the field of view is usually very near to assembly and the optical zoom procedure is a very time-consuming task. This approach can be used in automatized security camera systems because the detected object can be very far. The second approach based on high-resolution camera, for example, 4K or more, combined with parsing image to set of small regions of interest is much more effective for industrial part recognition. This method has a prerequisite of reliable object detection with minimal delays. Standardized parts with minimal dimensions (e.g., screws, nuts, washers, holes, threads, etc.) used in assembly or before assembly process can be recognized. It is also more easily implemented to virtual devices for assisted assembly tasks.

The process of extraction Region of Interest to identify where industrial parts are located can be realized by these standard machine vision algorithms:

- the Gradient algorithm to isolate clusters of pixels,
- the Contours algorithm to define borders of objects,
- the Closing Square algorithm to increase objects size,
- the Thresholding to reduce noise pixels from image,
- the Region of Interest to localize places of clusters.

An example of the testing input image, its processed image and the final image with thresholding and regions is shown in Fig. 2.14. There are detected six regions where some small parts can be located. This operation reduces image resolution for the CNN model input to 30% and increases input resolution for every feature during the detection process.

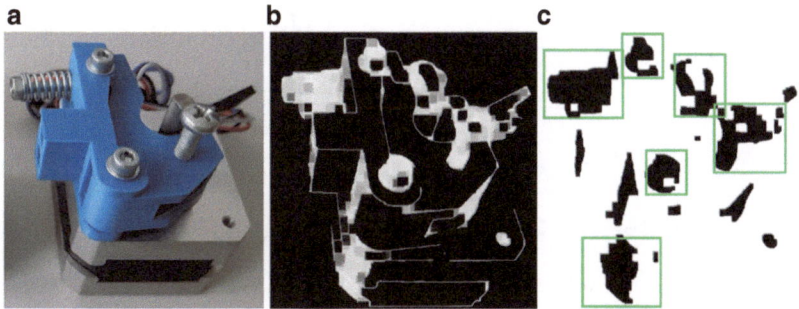

Fig. 2.14 Real image of assembly (**a**), the processed image (**b**), the final image after thresholding with identification of ROIs (**c**)

2.10.3 Convolutional Network Transform Learning

The Inception CNN model was selected for the experimental testing. Two separate CNN models (Faster RCNN Inception V2 SSD trained by Common Objects in COntext (COCO) dataset) were tested. The first CNN model was used for training on non-standardized parts and the second one on DIN-standardized parts. The timelines of the training process for classification and position losses for both CNN models are shown in Fig. 2.15.

An example of recognition results from testing with virtual photorealistic images and real part images is shown in Fig. 2.16.

The results of the recognition process after CNN transfer learning are shown in Table 2.1.

Training times has been significantly reduced under 2 hours because transfer learning techniques were used. The minimal recognition classification precision decreases after testing with real part images about 30%, which is still acceptable for assisted assembly tasks.

2.10.4 Implementation into Devices for Assisted Assembly

A standard CPU doesn't have enough power to process tasks as image capture, basic filtering, and CNN model execution. So the first step is

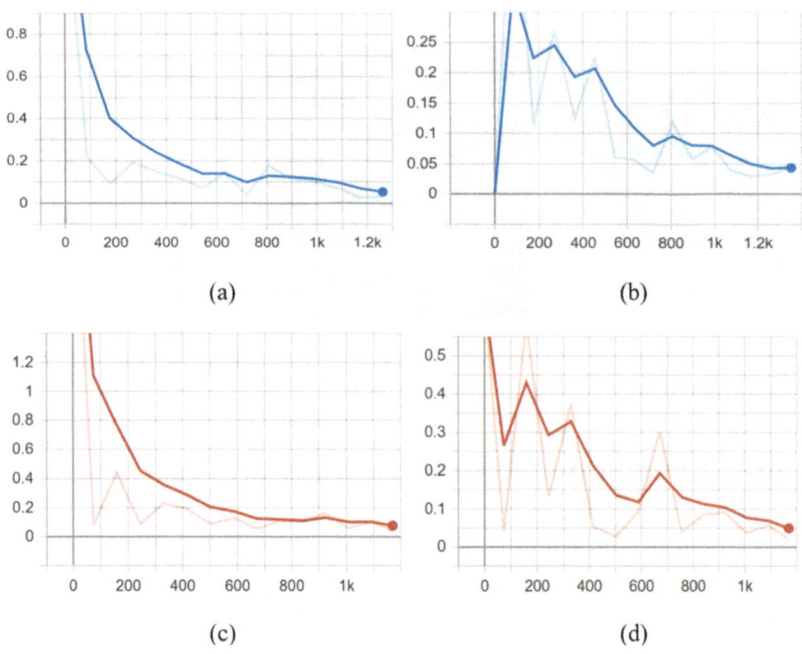

Fig. 2.15 Training process for: classification (**a**)/position (**b**) loss of unstandard-ized assembly parts, classification (**c**)/position (**d**) loss of standardized assembly parts

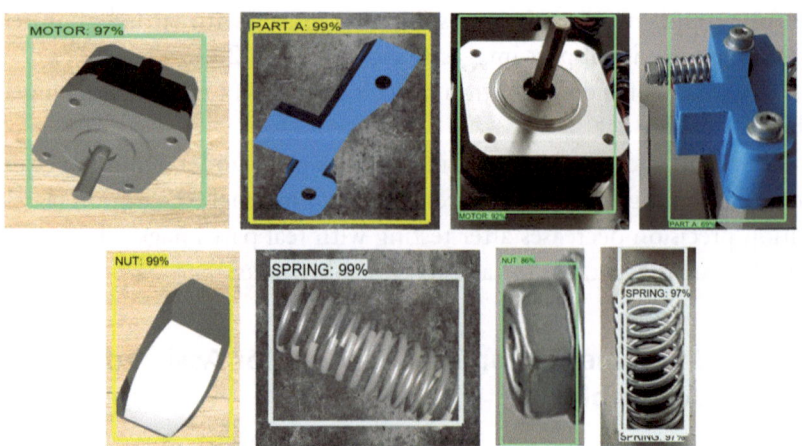

Fig. 2.16 Results of experiments with recognition reliability of trained CNN models

Table 2.1 Table with recognition results from all tested CNN models

CNN model used for transfer learning	Training time (TL) hours [h]	Minimal recognition precision virtual parts [%]	Minimal recognition precision real parts [%]
Inception v2 SSD for nonstandard parts	1.3	97	69
Inception v2 SSD for standardized parts	1.5	96	73

testing trained CNN models in embedded devices with support of acceleration neural network execution unit. The next step is visualization in virtual or augmented reality devices.

2.10.4.1   Implementation into Embedded Devices

The first testing platform was an embedded board with integrated APU (GPU with Tensors) Nvidia Xavier development kit with Ubuntu OS Linux distribution as is shown in Fig. 2.17a. The 4K images are acquired by E-Cons dual-camera system with 13Mpix resolution as Continual Serice Improvement (CSI) module, mounted in the experimental stand, and rapid prototyped holders. The second testing platform is embedded board Raspberry PI4, which doesn't include any APU unit. Additional

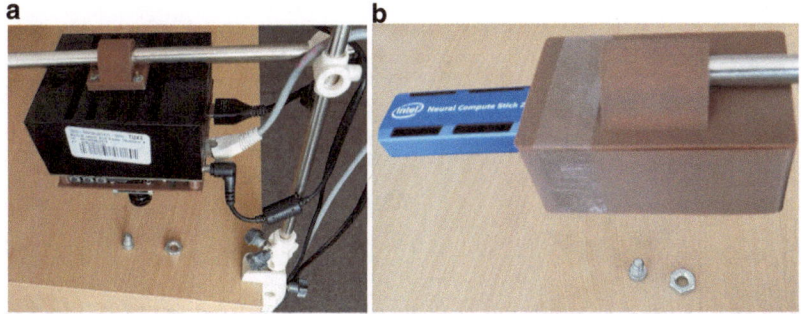

Fig. 2.17 Embedded devices with the implementation of the convolutional neural network (**a**) Nvidia AGX with E-Cons dual-camera (**b**) Raspberry PI 4 with CSI camera

computing power for CNN acceleration is acquired by the USB Neural Compute Stick Movidius 2 special module from Intel, which is shown in Fig. 2.17b.

The TensorFlow Framework from Google version 1.15 was used for training all CNN models. Nvidia provides SDK manager with Tensor RT library for trained CNN model to accelerate NVIDIA Xavier embedded device during execution of the CNN model. Intel offers Open VINO toolkit for CNN model acceleration by Intel Movidius USB compute stick combined with Raspberry PI 4. The Open CV library version 4.1 is a universal framework and is used on both platforms for the Region of Interest detection.

2.10.4.2 Implementation to VR/AR Devices

The validated CNN model can be implemented into virtual device HTC Vive Pro for assisted assembly tasks, which provides higher performance for CNN model execution because it uses standard PC with the dedicated graphics card. Standalone augmented devices can be used for simpler assisted assembly tasks, as for example Epson Moverio BT350 with integrated Android board. Both solutions are shown in Fig. 2.18.

The visualization data from the recognition process is realized by the Unity 3D engine, which doesn't have direct support for the CNN model, but it can communicate with the OpenCV framework by the

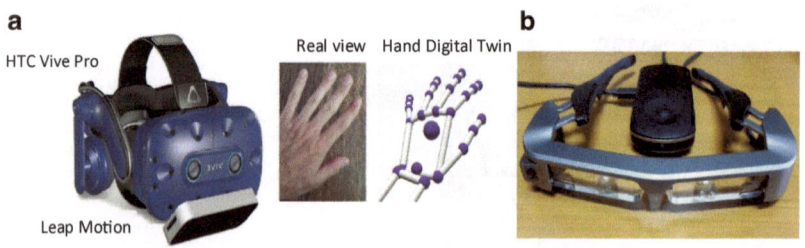

Fig. 2.18 Assisted assembly **(a)** Virtual Reality device HTC Vive Pro and Leap Motion, **(b)** Augmented Reality device Epson Moverio BT350

deep neural network (DNN) library. The Unity 3D creates a PC application for virtual reality device HTC Vive Pro and Android application for augmented reality device Epson Moverio BT350.

To summarize, two CNN models have been designed and tested: the one for nonstandard parts and the second for small standardized parts with single-shot detection algorithm for localization in the plane. The main reason for the preparation of two different CNN models is the reusability CNN model for the standardized part, which can be used for other assemblies. The first convolutional neural network model acquires precision with real parts classification minimum of about 69%. The second CNN model had better accuracy in classification after extraction of the Region of Interest with a minimum of 73%. The future works will be implementation of the Segmentation algorithm included in the TensorFlow version 2, which replaces simple Single-Shot Detection (SSD) algorithms to help detect the exact shape of the object for precise orientation detection in the workspace.

2.11 Discussion

In this chapter, we focused on the use of specific CI paradigms in the context of Industry 4.0. Since both of these areas can be considered very large, we limited to only the most important concepts. It is important to note that the definite consensus of what exactly constitutes each of these fields is lacking. To establish a basic framework for the chapter, we used the major classification presented in Sumathi et al. (2018) and on the lower level we identified six well-established paradigms—neural networks, fuzzy logic, evolutionary computation, swarm intelligence, artificial immune systems, and deep learning—each of which holds significant potential for the design of intelligent systems (Table 2.2). The inclusion of a high number of various novel nature-inspired metaheuristics was avoided since in many cases the benefit of using them compared to the better-established techniques may be questionable. Similarly, the concept of Industry 4.0 encompasses several major technologies and a number of components, where the use of advanced computational techniques is naturally assumed to meet the stringent requirements for high

Table 2.2 Summary of application areas within Industry 4.0 using computational intelligence

CI paradigm	Application area within I4.0	References	CI paradigm	Application area within I4.0	References
Neural Networks	Smart Manufacturing	Mehami et al. (2018) Shi et al. (2019) Wong and Yu (2019)	Swarm Intelligence	CPS	Yi et al. (2020) Zhang et al. (2019)
	CPS	Zhang et al. (2020a)		IoT	Xie et al. (2019) Sun et al. (2018) Gill et al. (2018) Thapar and Batra (2018) Hong et al. (2019)
	Robotics	Yang et al. (2019)			
	IoT	Djedidi and Djeziri (2020) Veerasamy et al. (2020)		Smart Manufacturing	Dewang et al. (2018)
	Big Data	Hernández et al. (2020)		Big Data	Djenouri et al. (2019) Abdi and Feizi-Derakhshi (2020) Nguyen et al. (2020)
Fuzzy logic	IoT	Thangaramya et al. (2019) Bu (2018)	Artificial Immune Systems	Smart Manufacturing	Wang et al. (2018b) Akram and Raza (2018)

CI paradigm	Application area within I4.0	References	CI paradigm	Application area within I4.0	References
Evolutionary Computation	Big Data	Shukla et al. (2020) Zhang et al. (2020b)	Deep Learning	IoT	Semwal and Nair (2020)
	Smart Manufacturing	Huo et al. (2020) Lu and Liu (2018) Chen et al. (2020)		Cybersecurity	Zhang et al. (2011) Aldhaheri et al. (2020)
	CPS	Wang et al. (2018c)		Fault Detection and Isolation	Costa Silva et al. (2017)
	IoT	Tang and Xin (2016)		Smart Manufacturing	Wang et al. (2018a) Wang et al. (2020b) Xia et al. (2020) Wang et al. (2019)
	Smart Manufacturing	Dilberoglu et al. (2017) Mele and Campana (2020) Ewald et al. (2018) Zheng et al. (2020) Nguyen et al. (2019)		Cybersecurity	Almiani et al. (2020)
	Big Data	Saif-Eddine et al. (2019)		CPS	Andersen et al. (2019) Wang et al. (2020a) Leong et al. (2020) Hou et al. (2020)
				Big Data	Wang et al. (2020b) Anbarasan et al. (2020) Essiet et al. (2019)

performance. As shown in Table 2.2. again a smaller number of such application areas was identified so that the use of CI in those works can be easier to generalize. On the other hand, a conceptual similarity of those areas was not taken into account—so some of them may be more general than the others. The most important application areas within Industry 4.0 in terms of their connection with computational intelligence techniques are smart manufacturing, Internet-of-Things, CPS, and Big Data Analytics.

In the case of neural networks, shallow and deep architectures were evaluated separately with the DNNs included in "Deep Learning" Sect. 2.8. Even though the DNNs are currently one of the most promising CI paradigms for many types of problems, "classical" (i.e., shallow) neural networks are still used for various applications. The works summarized in Table 2.2. confirm that the use of network types like RBF or Hopfield still offers attractive properties, e.g., for the regression problems in the context of Industry 4.0. The development of Type-2 fuzzy systems allowed for better handling of uncertainty, for which fuzzy logic is typically applied. Their use, whether in the form of Type-1 or Type-2, ranges from the nonlinear control to machine learning techniques like clustering. These methods are especially suitable for big data analytics, where advanced data mining techniques for finding patterns in vast amounts of data are of crucial importance. As a metaheuristic, evolutionary computation is a fine candidate for the optimization problems with no special requirements for their knowledge. The reviewed works show that this can be used in a wide range of problems, from the supply chain management through the optimization of energy efficiency in wireless networks to additive manufacturing. Swarm intelligence techniques like particle swarm optimization can be actually used in a similar way to EC methods like genetic algorithms or differential evolution. On the other hand, their source of inspiration (bird flocks or ant colonies, etc.) is a tempting solution also for decentralized bio-inspired control of many simple agents (like in networks). Particular tasks within areas like IoT or smart manufacturing in general include workflow scheduling, analysis of complex networks, or even path planning of AGVs in smart factories. Another type of bio-inspired computational paradigm is artificial immune systems, which, in addition to the previously mentioned optimization or data mining problems, is also used for the area of cyber-security. This use is quite natural and conforms to the idea of natural immune systems providing a defense against harmful pathogens, and may be of benefit in complex networks (e.g., smart grids and similar).

As mentioned several times in the text, the concept of deep learning, in general, is at this time considered one of the most perspective CI techniques for applications where large amounts of data are present. Therefore, it is closely tied to the very idea of Industry 4.0 and can be expected to become even more powerful with further progress in the hardware. In addition to the use of CNNs, which are the networks typically used in computer vision applications, deep reinforcement learning is particularly interesting, e.g., for robot control in an uncertain environment or handling large sensor networks.

To illustrate the benefit of using deep neural networks in the product manufacturing scenario, a case study of CNNs used for parts recognition in assisted assembly task was introduced. The main advantage of this approach was the possibility of creating training datasets using virtual models. After proper training of the CNN, the solution was implemented in AR/VR devices. The results confirm the viability of the proposed method for the tasks of assisted assembly.

2.12   Conclusion and Future Prospects

This short review attempts at summarizing the use of certain computational intelligence paradigms in the concept of Industry 4.0. However, due to the limited space, only some fundamental paradigms were addressed since the spectrum of bio-inspired computation methods that are applicable within I4.0 is very large. What we tried to address were some of the well-known approaches in CI, which proved to be effective in many different fields and hold significant potential for the use in smart manufacturing. We need to be aware of the fact that the area of computational intelligence is subject to very intense research, making it difficult to capture all its capabilities in a given instant. What remains firmly set and important for the concept of I4.0 is a data-driven aspect of CI methods, which makes it naturally suited for key aspects of I4.0

like cyber-physical systems and big data. Huge amounts of data associated with the use of countless interconnected devices make the methods and models capable of processing it and extracting meaningful information for either finding solutions to the problems or making decisions almost indispensable. In this regard, deep learning is currently one of the most perspective paradigms for many applications in I4.0. Rapid advancements in this particular paradigm have been caused mainly by the availability of very powerful hardware (like GPGPUs—General-Purpose computing on Graphics Processing Units) as well as the aforementioned huge amounts of data. Although there is nothing fundamentally new about the deep neural networks, the lack of powerful enough hardware together with the absence of effective training methods for very large networks made it difficult to obtain good results with them. Fuzzy logic and fuzzy systems have also solidified their position in future applications through the recent developments in Type-2 fuzzy logic, which helps to better tackle uncertainty in data. As such, they can offer a very important advantage over the purely black-box approaches, i.e. the interpretability of the results, which can be of great importance in many fields.

What has not been emphasized in the chapter but is also extremely important with regard to computational intelligence techniques, is their performance boost through the hybridization. Starting from the neuro-fuzzy approaches with which we can obtain interpretable models with NN-like learning and possibly ending with the search of (quasi)optimal parameters or hyperparameters of the models like neural networks, fuzzy systems, support vector machines, and others with bio-inspired non-gradient optimization methods. The methods like particle swarm optimization, genetic algorithm, differential evolution, artificial immune system offer a way to attack many various problems with minimal knowledge. Even though it might be sometimes difficult to explain why they actually work, many NP-hard problems are intractable using conventional computational techniques, making the prospect of having at least some (acceptable) solution attractive. It is of note that many of those methods may serve as an inspiration also on another level—swarm intelligence-based methods are of interest due to the cooperation of simple agents that gives rise to very tempting features like self-organization and self-learning. Such features are certainly more than desirable in the context of a multitude of embedded devices communicating with each other.

To show a possible application of some of the recent CI techniques, we presented a case study of deep learning paradigm in computer vision. This is one of the most striking examples of successful use of deep neural networks in the area of manufacturing, where the tasks of product inspection for possible defects are of extreme importance. The development of hardware specially designed for handling the tasks of DNN training in these applications allows us to achieve high performance, required for effective use in the industrial area.

A number of the most recent works in I4.0-related research included in this short review attest to the great interest of researchers in CI paradigms. This fact is fully in accordance with the importance of AI within the concept of I4.0—actually AI is so deeply rooted in the basic idea of I4.0 that we can safely say that it is one of its pillars. With this in mind, it is obvious that the actual implementation of I4.0 in current and future factories is also dependent on the success of the implementation of some of the CI paradigms in given applications.

A number of the most recent works in I4.0-related research included in this short review attest to the great interest of researchers in CI paradigms. This fact is fully in accordance with the importance of artificial intelligence within the concept of I4.0—actually AI is so deeply rooted in the basic idea of I4.0 that we can safely say that it is one of its pillars. With this in mind, it is obvious that the actual implementation of I4.0 in current and future factories is also dependent on the success of the implementation of some of the CI paradigms in given applications.

Even though we are still a relatively long way from having the concept of I4.0 implemented in majority of enterprises, even more advanced concepts keep springing up in academic sources. In one of the visions (Demir et al. 2019) for I5.0, very close interaction between humans and robots is assumed. While this is becoming reality also today through a gradual use of collaborative robots, we still cannot talk about the Human-Robot Interaction (HRI) as a natural aspect of the manufacturing process. It is obvious that any advances in the field of HRI are closely bound with the advances in artificial intelligence since in this interaction we are certainly looking for machines that are safe and close to us in our abilities to adapt and learn.

Each of the commonly accepted three basic pillars of CI (neural networks, fuzzy logic and evolutionary computation) has been subject to intense research in the last decades. Nevertheless, it seems that deep neural networks and DL in general are currently the paradigms that are seen as holding the greatest potential for future applications of intelligent systems. Despite the fact that the best results probably have been achieved in the fields of vision systems as well as voice recognition, possible benefits of DL application can be found in many other fields. On the other hand, possible hybridization of various CI techniques makes it possible to further enhance the performance of intelligent systems in which they are applied. This is especially true for any kind of NP-hard problems found in many applications within I4.0 (or more advanced) concepts, where metaheuristics can be effectively used. In addition to that, DL techniques can be potentially hybridized with fuzzy logic to form so-called deep fuzzy neural networks that fuse the capabilities of neural networks with our way of reasoning. Together with the availability of huge amounts of data, such powerful fusions allow us to take the capabilities of future intelligent machines much closer to humans.

References

Abdi, Y., and M.-R. Feizi-Derakhshi. 2020. Hybrid multi-objective evolutionary algorithm based on Search Manager framework for big data optimization problems. *Applied Soft Computing* 87 (February 1): 105991.

Aggarwal, C.C. 2018. Neural networks and deep learning: A textbook [Internet]. NY, USA: Springer International Publishing; 2018 [cited 2020 June 18]. Available from: https://www.springer.com/gp/book/978331994 4623.

Agrafiotis, D., and W. Cedeño. 2002. Feature selection for structure—Activity correlation using binary particle swarms. *Journal of Medicinal Chemistry* 45 (March 1): 1098–1107.

Aickelin, U., and S. Cayzer. 2002. The danger theory and its application to artificial immune systems. Proceedings of the First International Conference on Artificial Immune Systems, 141–148.

Akram, M., and A. Raza. 2018. Data for: Towards the development of robot immune system: A combined approach involving innate immune cells and T-lymphocytes. BioSystems [Internet]. Mendeley, August 30 [cited 2020 June 17]; 1. Available from: https://data.mendeley.com/datasets/568 4z3wztx.

Aldhaheri, S., D. Alghazzawi, L. Cheng, A. Barnawi, and B. Alzahrani. 2020. Artificial immune systems approaches to secure the internet of things: A systematic review of the literature and recommendations for future research. *Journal of Network and Computer Applications* 157 (May 1): 102537.

Almiani, M., A. AbuGhazleh, A. Al-Rahayfeh, S. Atiewi, and A. Razaque. 2020. Deep recurrent neural network for IoT intrusion detection system. *Simulation Modelling Practice and Theory* 101 (May 1): 102031.

Anbarasan, M., B. Muthu, C.B. Sivaparthipan, R. Sundarasekar, S. Kadry, S. Krishnamoorthy, et al. 2020. Detection of flood disaster system based on IoT, big data and convolutional deep neural network. *Computer Communications* 150 (January 15): 150–157.

Andersen, R.E., S. Madsen, A.B.K. Barlo, S.B. Johansen, M. Nør, R.S. Andersen, et al. 2019. Self-learning processes in smart factories: Deep reinforcement learning for process control of robot brine injection. *Procedia Manufacturing* 38 (January 1): 171–177.

Angeline, P.J. 1998. Using selection to improve particle swarm optimization. 1998 IEEE International Conference on Evolutionary Computation Proceedings. IEEE World Congress on Computational Intelligence (Cat. No.98TH8360). Anchorake, USA. pp. 84–89.

Antão, R. 2017. Type-2 Fuzzy logic: Uncertain systems' modeling and control [Internet]. Springer Singapore [cited 2020 June 18]. Available from: https://www.springer.com/gp/book/9789811046322.

Balog, M., H. Sokhatska, and A. Iakovets. 2019. Intelligent systems in the railway freight management. *Lecture Notes in Mechanical Engineering*, 390–405.

Bianchi, F.M., E. Maiorino, M.C. Kampffmeyer, A. Rizzi, and R. Jenssen. 2017. *Recurrent neural networks for short-term load forecasting: An overview and comparative analysis* [Internet]. Springer International Publishing [cited 2020 June 18]. Available from: https://www.springer.com/gp/book/978331 9703374.

Blum, C., and D. Merkle (eds.). 2008. *Swarm intelligence: Introduction and applications* [Internet]. Berlin Heidelberg: Springer-Verlag [cited 2020 June 15]. Available from: https://www.springer.com/gp/book/9783540740889.

Bonabeau, E., G. Theraulaz, and M. Dorigo. 1999. *Swarm intelligence: From natural to artificial systems*, 1st ed. New York, NY: Oxford University Press.

Bu, F. 2018. An efficient fuzzy c-means approach based on canonical polyadic decomposition for clustering big data in IoT. *Future Generation Computer Systems* 88 (November 1): 675–682.

Cantú-Paz, E. 1998. A survey of parallel genetic algorithms. *Calculateurs Paralleles, Reseaux et Systems Repartis* 10 (2): 141–171.

Castro, L.N., and J. Timmis. 2002. An artificial immune network for multi-modal function optimization. Proceedings of the 2002 Congress on Evolutionary Computation. CEC'02 (Cat. No.02TH8600), pp. 699–704, Vol. 1.

Castro, L.N., and F.J.V. Zuben. 2000. The clonal selection algorithm with engineering applications. Workshop Proceedings of GECCO, 7.

Chang, C.S., and D.Y. Xu. 2000. Differential evolution based tuning of fuzzy automatic train operation for mass rapid transit system. *IEE Proceedings—Electric Power Applications* 147 (3, May): 206–212.

Chen, Z., X. Ming, T. Zhou, and Y. Chang. 2020. Sustainable supplier selection for smart supply chain considering internal and external uncertainty: An integrated rough-fuzzy approach. *Applied Soft Computing* 87 (February 1): 106004.

Chiou, J.-P., and F.-S. Wang. 1998. A hybrid method of differential evolution with application to optimal control problems of a bioprocess system. 1998 IEEE International Conference on Evolutionary Computation Proceedings. IEEE World Congress on Computational Intelligence (Cat. No.98TH8360), pp. 627–632.

Costa Silva, G., W.M. Caminhas, and R.M. Palhares. 2017. Artificial immune systems applied to fault detection and isolation: A brief review of immune response-based approaches and a case study. *Applied Soft Computing* 57 (August 1): 118–131.

CPS report. 2008. CPS-Summit. Holistic approaches to cyber-physical integration [Internet]. CPSWeek report [cited 2020 June 18]. Available from: https://iccps2012.cse.wustl.edu/_doc/CPS_Summit_Report.pdf.

de Castro, L.N. 2006. *Fundamentals of natural computing: Basic concepts, algorithms, and applications*, 1st ed. Boca Raton: Chapman and Hall/CRC.

de Castro, L.N., and F.J.V. Zuben. 2002. *aiNet: An artificial immune network for data analysis* [Internet]. Data mining: A heuristic approach. IGI Global [cited 2020 June 15]. pp. 231–260. Available from: www.igi-global.com/cha pter/data-mining-heuristic-approach/7592.

Delicato, F.C., A. Al-Anbuky, and K.I.-K. Wang. 2020. Editorial: Smart cyber–physical systems: Toward pervasive intelligence systems. *Future Generation Computer Systems* 107 (June 1): 1134–1139.

Demir, K.A., G. Döven, and B. Sezen. 2019. Industry 5.0 and human-robot co-working. *Procedia Computer Science* 158 (January 1): 688–695.

Dewang, H.S., P.K. Mohanty, and S. Kundu. 2018. A robust path planning for mobile robot using smart particle swarm optimization. *Procedia Computer Science* 133 (January 1): 290–297.

Dilberoglu, U.M., B. Gharehpapagh, U. Yaman, and M. Dolen. The role of additive manufacturing in the era of Industry 4.0. *Procedia Manufacturing* 11 (January 1): 545–554.

Ding, D., Q.-L. Han, Y. Xiang, X. Ge, and X.-M. Zhang. 2018. A survey on security control and attack detection for industrial cyber-physical systems. *Neurocomputing* 275 (January 31): 1674–1683.

Djedidi, O., and M.A. Djeziri. 2020. Power profiling and monitoring in embedded systems: A comparative study and a novel methodology based on NARX neural networks. *Journal of Systems Architecture* 111 (December 1): 101805.

Djenouri, Y., D. Djenouri, A. Belhadi, P. Fournier-Viger, J. Chun-Wei Lin, and A. Bendjoudi. 2019. Exploiting GPU parallelism in improving bees swarm optimization for mining big transactional databases. *Information Sciences* 496 (September 1): 326–342.

Dorigo, M., and G. Di Caro. 1999. Ant colony optimization: A new meta-heuristic. Proceedings of the 1999 Congress on Evolutionary Computation-CEC99 (Cat. No. 99TH8406), pp. 1470–1477, Vol. 2.

Engelbrecht, A.P. 2007. *Computational intelligence: An introduction*, 2nd ed. West Sussex: Wiley.

Essiet, I., Y. Sun, and Z. Wang. 2019. Big data analysis for gas sensor using convolutional neural network and ensemble of evolutionary algorithms. *Procedia Manufacturing* 35 (January 1): 629–634.

Ewald, A., T. Sassenberg, and J. Schlattmann. 2018. Evolutionary-based optimization strategy in a hybrid manufactured process using LMD. *Procedia CIRP* 74 (January 1): 163–167.

Fang, W., J. Sun, Y. Ding, X. Wu, and W. Xu. 2010. A review of quantum-behaved particle swarm optimization. *IETE Technical Review* 27 (4, July 1): 336–348. Taylor & Francis.

Feoktistov, V. 2006. *Differential evolution: In search of solutions* [Internet]. Springer US [cited 2020 June 15]. Available from: https://www.springer.com/gp/book/9780387368955.

Fonlupt, C., J.-K. Hao, E. Lutton, E. Ronald, and M. Schoenauer. 2006. Artificial evolution: 4th European conference. AE'99 Dunkerque, France, November 3–5, 1999 Selected Papers. Springer.

Fulcher, J. (ed.). 2008. *Computational intelligence: A compendium* [Internet]. Berlin Heidelberg: Springer-Verlag [cited 2020 June 15]. Available from: https://www.springer.com/gp/book/9783540782926.

Gill, S.S., R. Buyya, I. Chana, M. Singh, and A. Abraham. 2018. BULLET: Particle swarm optimization based scheduling technique for provisioned cloud resources. *Journal of Network and Systems Management* 26 (2, April 1): 361–400.

Goldberg, D., K. Deb, H. Kargupta, and G. Harik. 1995. Rapid, accurate optimization of difficult problems using fast messy genetic algorithms. *Morgan Kaufmann*, pp. 56–64.

Goldberg, D.E., and L. Wang. Adaptive niching via coevolutionary sharing. In *Genetic algorithms and evolution strategy in engineering and computer science*, Chapter 2, 21–38. West Sussex: Wiley.

Goodfellow, I., Y Bengio, and A. Courville. 2016. *Deep learning*. MIT Press.

Haykin, S.S. 2009. *Neural networks and learning machines*. New York: Prentice Hall.

Hendtlass, T. 2001. A combined swarm differential evolution algorithm for optimization problems. In *Engineering of intelligent systems*, ed. L. Monostori, J. Váncza, and M. Ali, 11–18. Berlin, Heidelberg: Springer.

Hernández, G., E. Zamora, H. Sossa, G. Téllez, and F. Furlán. 2020. Hybrid neural networks for big data classification. *Neurocomputing* 390 (May 21): 327–340.

Hong, Y., L. Chen, and L. Mo. 2019. Optimization of cluster resource indexing of Internet of Things based on improved ant colony algorithm. *Cluster Computing* 22 (3, May 1): 7379–7387.

Hou, R., S. Jeong, J.P. Lynch, and K.H. Law. 2020. Cyber-physical system architecture for automating the mapping of truck loads to bridge behavior using computer vision in connected highway corridors. *Transportation Research Part C: Emerging Technologies* 111 (February 1): 547–571.

Hrehova, S. 2016. Predictive model to evaluation quality of the manufacturing process using Matlab tools. *Procedia Engineering*, 149–154. Elsevier Ltd.

Hrstka, O., and A. Kučerová. 2004. Improvements of real coded genetic algorithms based on differential operators preventing premature convergence. *Advances in Engineering Software* 35 (3, March 1): 237–246.

Huo, J., F.T.S. Chan, C.K.M. Lee, J.O. Strandhagen, and B. Niu. 2020. Smart control of the assembly process with a fuzzy control system in the context of Industry 4.0. *Advanced Engineering Informatics* 43 (January 1): 101031.

Ippolito, M.G., E.R. Sanseverino, and F. Vuinovich. 2004. Multiobjective ant colony search algorithm optimal electrical distribution system planning. Proceedings of the 2004 Congress on Evolutionary Computation (IEEE Cat. No.04TH8753). Portland, USA, IEEE, pp. 1924–1931, Vol. 2.

Iqbal, R., F. Doctor, B. More, S. Mahmud, and U. Yousuf. 2020a. Big data analytics and computational intelligence for cyber-physical systems: Recent trends and state of the art applications. *Future Generation Computer Systems* 105 (April 1): 766–778.

Iqbal, R., F. Doctor, B. More, S. Mahmud, and U. Yousuf. 2020b. Big data analytics: Computational intelligence techniques and application areas. *Technological Forecasting and Social Change* 153 (April 1): 119253.

Karray, F.O., and C.W.D. Silva. 2004. *Soft computing and intelligent systems design: Theory, tools and applications*, 1st ed. Harlow, UK and New York: Addison-Wesley.

Kennedy, J., and R. Eberhart. 1995. Particle swarm optimization. Proceedings of ICNN'95—International Conference on Neural Networks. Perth, Australia: IEEE, pp. 1942–1948, Vol. 4.

Khan, M., B. Jan, and H. Farman. 2019. Deep learning: Convergence to Big Data analytics [Internet]. Springer Singapore [cited 2020 June 17]. Available from: https://www.springer.com/gp/book/9789811334580.

Kim, J., and P.J. Bentley. 2002. Towards an artificial immune system for network intrusion detection: An investigation of dynamic clonal selection. Proceedings of the 2002 Congress on Evolutionary Computation. CEC'02 (Cat. No.02TH8600), pp. 1015–1020, Vol. 2.

Kim, P. 2017. MATLAB deep learning. Seoul, Korea: Apress.

Knight, T., and J. Timmis. 2002. A multi-layered immune inspired approach to data mining. Proceedings of the 4th International Conference Recent Advances in Soft Computing pp. 266–271.

Lazár, Ivan, and J. Husár. 2012. Validation of the serviceability of the manufacturing system using simulation. *Journal on Efficiency and Responsibility in Education and Science* 5 (4): 252–261. Czech University of Life Sciences Prague.

Lee, J., H. Davari, J. Singh, and V. Pandhare. 2018. Industrial artificial intelligence for Industry 4.0-based manufacturing systems. *Manufacturing Letters* 18 (October 1): 20–23.

Leong, A.S., A. Ramaswamy, D.E. Quevedo, H. Karl, and L. Shi. 2020. Deep reinforcement learning for wireless sensor scheduling in cyber–physical systems. *Automatica* 113 (March 1): 108759.

Li, W., K. Liu, L. Yan, F. Cheng, Y.Q. Lv, and L.Z. Zhang. 2019. FRD-CNN: Object detection based on small-scale convolutional neural networks and feature reuse. *Scientific Reports* 9 (1, December 1). Nature Publishing Group.

Liu, H., and L. Wang. 2017. An AR-based worker support system for human-robot collaboration. *Procedia Manufacturing* 11: 22–30.

Liu, J. 2013. Radial Basis Function (RBF) neural network control for mechanical systems: Design, analysis and Matlab simulation [Internet]. Berlin Heidelberg: Springer-Verlag [cited 2020 June 24]. Available from: https://www.springer.com/gp/book/9783642348150.

Lones, M. 2014. *Metaheuristics in nature-inspired algorithms*, 4. Vancouver: Canada.

Lu, T., and S.-T. Liu. 2018. Fuzzy nonlinear programming approach to the evaluation of manufacturing processes. *Engineering Applications of Artificial Intelligence* 72 (June 1): 183–189.

Mariano, C.E., and E. Morales. 1999. A multiple objective Ant-Q algorithm for the design of water distribution irrigation networks. *Technical Report HC-9904*, 41.

Mehami, J., M. Nawi, and R.Y. Zhong. 2018. Smart automated guided vehicles for manufacturing in the context of Industry 4.0. *Procedia Manufacturing* 26 (January 1): 1077–1086.

Mele, M., and G. Campana. 2020. Sustainability-driven multi-objective evolutionary orienting in additive manufacturing. *Sustainable Production and Consumption* 23 (July 1): 138–147.

Messerschmidt, L., and A.P. Engelbrecht. 2004. Learning to play games using a PSO-based competitive learning approach. *IEEE Transactions on Evolutionary Computation* 8 (3, June): 280–288.

Miller, B.L., and M.J. Shaw. 1996. Genetic algorithms with dynamic niche sharing for multimodal function optimization. Proceedings of IEEE International Conference on Evolutionary Computation, pp. 786–791.

Munakata, T. 2008. *Fundamentals of the new artificial intelligence: Neural, evolutionary, fuzzy and more* [Internet], 2nd ed. London: Springer-Verlag [cited 2020 June 24]. Available from: https://www.springer.com/gp/book/978184 6288388.

Napoleone, A., M. Macchi, and A. Pozzetti. 2020. A review on the characteristics of cyber-physical systems for the future smart factories. *Journal of Manufacturing Systems* 54 (January 1): 305–335.

Nasraoui, O., D. Dasgupta, and F. González. 2002. The promise and challenges of artificial immune system based web usage mining. Proceedings of the Second SIAM International Conference on Data Mining, August 11, pp. 29–32.

Nayyar, A., D.-N. Le, and N.G. Nguyen. 2018. Advances in Swarm intelligence for optimizing problems in computer science, 1st ed. Boca Raton, FL: Chapman & Hall/CRC.

Nguyen, B.H., B. Xue, and M. Zhang. 2020. A survey on swarm intelligence approaches to feature selection in data mining. *Swarm and Evolutionary Computation* 54 (May 1): 100663.

Nguyen, S., D. Thiruvady, A.T. Ernst, and D. Alahakoon. 2019. A hybrid differential evolution algorithm with column generation for resource constrained job scheduling. *Computers & Operations Research* 109 (September 1): 273–287.

Pampara, G., A.P. Engelbrecht, and N. Franken. 2006. Binary differential evolution. 2006 IEEE International Conference on Evolutionary Computation, pp. 1873–1879.

Panetto, H., B. Iung, D. Ivanov, G. Weichhart, and X. Wang. 2019. Challenges for the cyber-physical manufacturing enterprises of the future. *Annual Reviews in Control* 47 (January 1): 200–213.

Pavlenko, I., V. Simonovskiy, V. Ivanov, J. Zajac, and J. Pitel. 2019a. Application of artificial neural network for identification of bearing stiffness characteristics in rotor dynamics analysis. Advances in Design, Simulation and Manufacturing, pp. 325–335.

Pavlenko, I., J. Trojanowska, V. Ivanov, and O. Liaposhchenko. 2019b. Scientific and methodological approach for the identification of mathematical models of mechanical systems by using artificial neural networks. Innovation, Engineering and Entrepreneurship, pp. 299–306.

Pedrycz, W., and F. Gomide. 2007. *Fuzzy systems engineering: Toward human-centric computing*, 1st ed. Hoboken, NJ: Wiley-IEEE Press.

Saif-Eddine, A.S., M.M. El-Beheiry, and A.K. El-Kharbotly. 2019. An improved genetic algorithm for optimizing total supply chain cost in inventory location routing problem. *Ain Shams Engineering Journal* 10 (1, March 1): 63–76.

Sarkar, K., K. Varanasi, and D. Stricker. 2017. Trained 3D models for CNN based object recognition. Proceedings of the 12th International

Joint Conference on Computer Vision, Imaging and Computer Graphics Theory and Applications (Visigrapp 2017), Vol. 5. German Research Center Artificial Intelligence DFKI, Kaiserslautern, Germany. German Res C, pp. 130–137.

Semwal, T., and S.B. Nair. 2020. A decentralized artificial immune system for solution selection in cyber-physical systems. *Applied Soft Computing* 86 (January 1): 105920.

Sengupta, S., S. Basak, P. Saikia, S. Paul, V. Tsalavoutis, F. Atiah, et al. 2020. A review of deep learning with special emphasis on architectures, applications and recent trends. *Knowledge-Based Systems* 194 (April 22): 105596.

Shi, C., J. Qian, W. Zhu, H. Liu, S. Han, and X. Yang. 2019. Nondestructive determination of freshness indicators for tilapia fillets stored at various temperatures by hyperspectral imaging coupled with RBF neural networks. *Food Chemistry* 275 (March 1): 497–503.

Shukla, A.K., M. Yadav, S. Kumar, and P.K. Muhuri. 2020. Veracity handling and instance reduction in big data using interval type-2 fuzzy sets. *Engineering Applications of Artificial Intelligence* 88 (February 1): 103315.

Socher, R., B. Huval, B. Bhat, C.D. Manning, and A.Y. Ng. 2012. Convolutional-recursive deep learning for 3D object classification. NIPS.

Song, H., D.B. Rawat, S. Jeschke, and C. Brecher. 2016. *Cyber-physical systems: Foundations, principles and applications.* Burlington, MA: Morgan Kaufmann.

Srinivas, N., and K. Deb. 1991. Multiobjective optimization using nondominated sorting in genetic algorithms. *Evolutionary Computation* 2 (3): 221–248.

Stützle, T., and H.H. Hoos. 2000. MAX–MIN ant system. *Future Generation Computer Systems* 16 (8, June 1): 889–914.

Su, H., C.R. Qi, Y. Li, and L.J. Guibas. 2015. Render for CNN: Viewpoint estimation in images using CNNs trained with rendered 3D model views. Proceedings of the IEEE International Conference on Computer Vision.

Sumathi, S., T. Hamsapriya, and P. Surekha. 2008. *Evolutionary intelligence: An introduction to theory and applications with Matlab*, 1st ed. Berlin: Springer.

Sumathi, S., L.A. Kumar, and P. Surekha. 2018. Computational intelligence paradigms for optimization problems using MATLAB®/SIMULINK®, 1st ed. CRC Press.

Sun, H., W. Jie, J. Loo, L. Wang, S. Ma, G. Han, et al. 2018. A parallel self-organizing overlapping community detection algorithm based on swarm intelligence for large scale complex networks. *Future Generation Computer Systems* 89 (December 1): 265–285.

Takaseki, R., R. Nagashima, H. Kashima, and T. Okazaki. 2015. Development of anchoring support system using with AR Toolkit. 2015 7th International Conference on Emerging Trends in Engineering & Technology (ICETET). IEEE, pp. 123–127.

Tan, Y., Y. Shi, and K.C. Tan. 2010. Advances in swarm intelligence: First international conference, ICSI 2010, Beijing, China, June 12–15. Proceedings, Part I [Internet]. Berlin Heidelberg: Springer-Verlag [cited 2020 June 15]. Available from: https://www.springer.com/gp/book/9783642134944.

Tang, M., and Y. Xin. 2016. Energy efficient power allocation in cognitive radio network using coevolution chaotic particle swarm optimization. *Computer Networks* 100 (May 8): 1–11.

Thangaramya, K., K. Kulothungan, R. Logambigai, M. Selvi, S. Ganapathy, and A. Kannan. 2019. Energy aware cluster and neuro-fuzzy based routing algorithm for wireless sensor networks in IoT. *Computer Networks* 151 (March 14): 211–223.

Thapar, P., and U. Batra. 2018. Implementation of ant colony optimization in routing protocol for internet of things. In Innovations in computational intelligence : Best selected papers of the third international conference on REDSET 2016 [Internet], ed. B. Panda, S. Sharma, and U. Batra, 151–164. Singapore: Springer [cited 2020 June 15]. Available from: https://doi.org/10.1007/978-981-10-4555-4_10.

Tian, Y., X. Li, K. Wang, and F.Y. Wang. 2018. Training and testing object detectors with virtual images. *IEEE/CAA Journal of Automatica Sinica* 5 (2): 539–546.

Timmis, J., and M. Neal. 2001. A resource limited artificial immune system for data analysis. In *Research and development in intelligent systems XVII*, ed. M. Bramer, A. Preece, and F. Coenen, 19–32. London: Springer.

Tong, K., Y. Wu, and F. Zhou. 2020. Recent advances in small object detection based on deep learning: A review. Image and Vision Computing, p. 103910. Elsevier Ltd.

Veerasamy, V., N.I.A. Wahab, R. Ramachandran, B. Madasamy, M. Mansoor, M.L. Othman, et al. 2020. A novel RK4-Hopfield Neural network for power flow analysis of power system. *Applied Soft Computing* 93 (August 1): 106346.

Vilmart, H., J.-C. Léon, and F. Ulliana. 2018. From CAD assemblies toward knowledge-based assemblies using an intrinsic knowledge-based assembly model. *Computer-Aided Design and Applications* 15 (3, May 4): 300–317.

Wang, J., P. Fu, and R.X. Gao. 2019. Machine vision intelligence for product defect inspection based on deep learning and Hough transform. *Journal of Manufacturing Systems* 51 (April 1): 52–60.

Wang, J., Y. Ma, L. Zhang, R.X. Gao, and D. Wu. 2018a. Deep learning for smart manufacturing: Methods and applications. *Journal of Manufacturing Systems* 48 (July 1): 144–156.

Wang, S., Y.C. Liang, W.D. Li, and X.T. Cai. 2018b. Big Data enabled Intelligent Immune System for energy efficient manufacturing management. *Journal of Cleaner Production* 195 (September 10): 507–520.

Wang, W., F. Di Maio, and E. Zio. 2018c. Hybrid fuzzy-PID control of a nuclear Cyber-Physical System working under varying environmental conditions. *Nuclear Engineering and Design* 331 (May 1): 54–67.

Wang, D., H. Deng, and Z. Pan. 2020a. MRCDRL: Multi-robot coordination with deep reinforcement learning. *Neurocomputing* 406 (September 17): 68–76.

Wang, Q., W. Jiao, P. Wang, and Y. Zhang. 2020b. A tutorial on deep learning-based data analytics in manufacturing through a welding case study. *Journal of Manufacturing Processes* [Internet]. May 27 [cited 2020 June 17]. Available from: https://www.sciencedirect.com/science/article/pii/S1526612520302668.

Wong, S.F., and Z. Yu. 2019. The mobile robot anti-disturbance vSLAM navigation algorithm based on RBF neural network. *Procedia Manufacturing* 38 (January 1): 400–407.

Xia, C., Z. Pan, Z. Fei, S. Zhang, and H. Li. 2020. Vision based defects detection for Keyhole TIG welding using deep learning with visual explanation. *Journal of Manufacturing Processes* 56 (August 1): 845–855.

Xie, Y., Y. Zhu, Y. Wang, Y. Cheng, R. Xu, A.S. Sani, et al. 2019. A novel directional and non-local-convergent particle swarm optimization based workflow scheduling in cloud–edge environment. *Future Generation Computer Systems* 97 (August 1): 361–378.

Xing, B., and W.-J. Gao. 2014. Innovative computational intelligence: A rough guide to 134 Clever algorithms [Internet]. Springer International Publishing [cited 2020 June 24]. Available from: https://www.springer.com/gp/book/9783319034034.

Yang, M., Z. Du, L. Sun, and W. Dong. 2019. Optimal design, modeling and control of a long stroke 3-PRR compliant parallel manipulator with variable thickness flexure pivots. *Robotics and Computer-Integrated Manufacturing* 60 (December 1): 23–33.

Yang, X.-S., Z. Cui, R. Xiao, A.H. Gandomi, and M. Karamanoglu (eds.). 2013. *Swarm intelligence and bio-inspired computation: Theory and applications*, 1st ed. Amsterdam and Boston: Elsevier.

Yi, N., J. Xu, L. Yan, and L. Huang. 2020. Task optimization and scheduling of distributed cyber–physical system based on improved ant colony algorithm. *Future Generation Computer Systems* 109 (August 1): 134–148.

Zhang, R., B.M. Phillips, P.L. Fernández-Cabán, and F.J. Masters. 2019. Cyber-physical structural optimization using real-time hybrid simulation. *Engineering Structures* 195 (September 15): 113–124.

Zhang, B., C. Dou, T. Zhang, and Z. Zhang. 2020a. An IGAP-RBFNN-based secondary control strategy for islanded microgrid-cyber physical system considering data uploading interruption problem. *Neurocomputing* 397 (July 15): 422–437.

Zhang, X., X. Ming, and D. Yin. 2020b. Application of industrial big data for smart manufacturing in product service system based on system engineering using fuzzy DEMATEL. *Journal of Cleaner Production* 265 (August 20): 121863.

Zhang, Y., L. Wang, W. Sun, R.C. Green II., and M. Alam. 2011. Distributed intrusion detection system in a multi-layer network architecture of smart grids. *IEEE Transactions on Smart Grid* 2 (4, December): 796–808.

Zheng, J., L. Wang, and J. Wang. 2020. A cooperative coevolution algorithm for multi-objective fuzzy distributed hybrid flow shop. *Knowledge-Based Systems* 195 (April 22): 105536.

Žídek, K., A. Hosovsky, J. Piteľ, and S. Bednár. 2019a. Recognition of assembly parts by convolutional neural networks. Lecture Notes in Mechanical Engineering.

Žídek, K., P. Lazorík, J. Piteľ, and A. Hošovský. 2019b. An automated training of deep learning networks by 3D virtual models for object recognition. *Symmetry. MDPI AG* 11 (4, April 1).

3

AI and ML for Human-Robot Cooperation in Intelligent and Flexible Manufacturing

Manuel A. Ruiz Garcia, Erwin Rauch, Renato Vidoni, and Dominik T. Matt

3.1   Introduction

The last decades were marked by the further development of modern production systems and the introduction of Industry 4.0 in manufacturing. The concepts and technologies of Industry 4.0 are mostly aimed at the networking of production and the efficient design of production systems. The logic of Industry 4.0 foresees humans and

M. A. Ruiz Garcia · E. Rauch (✉) · R. Vidoni · D. T. Matt
Industrial Engineering and Automation (IEA), Faculty of Science and Technology, Free University of Bozen-Bolzano, Piazza Università, Bolzano, Italy
e-mail: Erwin.Rauch@unibz.it

M. A. Ruiz Garcia
e-mail: ManuelAlejandro.RuizGarcia@unibz.it

R. Vidoni
e-mail: Renato.Vidoni@unibz.it

D. T. Matt
e-mail: Dominik.Matt@unibz.it

robots as indistinguishable parts of a larger heterogeneous body of distributed autonomous and cooperative entities. Under such a perspective, robots are endowed with self and environment awareness and are able to smartly interact with both humans and other machines (Ruiz Garcia et al. 2019). Consequently, and in contrast to the third industrial revolution, machines are not intended to substitute humans in industry, but to work with them in synergy.

In collaborative industrial scenarios, safety greatly depends on the reciprocal understanding between the human operator and the robotic system. In particular, the most dangerous risk specific to robots is the unexpected collisions between the robot and the environment (Siciliano and Khatib 2016). When an unexpected exertion occurs between a collaborative robot and its surrounding environment, impact forces are eased thanks to their lightweight design and compliant mechanisms and control. However, avoiding unexpected force exertions implies foreseeing dangerous situations, and thus it relies on sensing, situational awareness, planning and decision-making capabilities. Therefore, without suitable exteroceptive sensing a collaborative robot cannot be considered as a safe companion in the context of human-robot cooperation (HRC). In other words, to safely interact with a human operator and the environment, a collaborative robot must predict and prevent any risky circumstances based on its own situational awareness. That is, the robot must identify, understand and forecast operator's actions and environmental changes to promptly react and safely adapt to either expected or unexpected operative conditions. On the other hand, the operator needs to be aware of the collaborative robot's motion to guarantee him or her own safety. Therefore, in the context of HRC, beyond the sensing capabilities a collaborative robot also needs to be endowed with suitable means of interaction so to constantly inform the human operator about what are the current and future goals and actions to be reached and performed, respectively, on a finite time horizon.

D. T. Matt
Innovation Engineering Center (IEC), Fraunhofer Research Italia s.c.a.r.l., Bolzano, Italy

Another important aspect is that Industry 4.0 is also seen as an enabler for the flexibilization of production systems and, thus, it potentially represents an important milestone for multi-variant manufacturing. In this concern, mass customization can be defined as the capability to deliver products and services that best meet individual customers' needs with near mass production efficiency (Tseng et al. 1996). Such a diversification in production requires to manage not only the inner product variety, but also the induced process variety due to differences in assembly sequence and the necessary changes of the manufacturing system required to handle them. A natural way to achieve this goal is through the flexibilization of manufacturing systems, such to allow changing from one product to the other without the need to stop the production for a changeover nor including other manual adaptations of the manufacturing system. It is worth noticing that such an automated adaptation in the context of collaborative manufacturing greatly resembles the ones required by a collaborative robot to stablish a safe HRC. Indeed, the understanding and the forecasting of the operator's actions and environmental changes, in terms of the current product variant, provide all the necessary information required for the definition of such an adaptation. On the other hand, the automation of such an adaptation relies on planning and decision-making capabilities.

Therefore, in abstract terms, the definition of a safe HRC and the automated adaptation of a multi-variant collaborative manufacturing system represent two particular instances of a general problem. This chapter is devoted to the deconstruction of such a general problem in terms of three smaller perceptive and cognitive issues: *scene monitoring*, *task modelling* and *planning*.

3.2   Artificial Intelligence and Machine Learning

Since the beginning of the twenty-first century, there has been a widespread use of ML techniques, specially Deep Learning (DL) ones, in the analysis of large amounts of data so to automatically drawn conclusions from it. Since then ML and DL, together with AI, are now terms belonging to the common imagination. However, there seems to be a common believe that AI, ML and DL refer to the same—or nearly the same—concept. In some particular rhetorical circumstances, this could be the case, but in general terms such a concept overlap is totally misleading. The aim of this section is twofold. On the one hand, to briefly clarify what is the scope of each research field and to highlight the relationships between them. On the other, to identify the key general problems that such techniques can potentially solve in the context of collaborative manufacturing.

3.2.1 What's Artificial Intelligence?

As a starting point, one can state that DL is a subset of ML, and that at the same time ML *seems to be* subset of AI. Therefore, it comes natural to start with the definition of AI. However, due historical reasons that fall beyond the scope of the present chapter, it is not possible to provide a "gold-standard" definition of AI unless one assumes that some background on the field is already known. So let us start instead with a brief digression on what an artificial system should do to be considered as *intelligent*. First of all, it is worth noticing that intelligence can be conceived either in terms of reasoning (thinking) or behaviour (acting). On the other hand, one can build a comparison metric of intelligence with respect to the human performance or with respect to an ideal model of intelligence, commonly known as *rationality*. Therefore, an artificial system can be considered as intelligent if it (Russell and Norvig 2010):

1 *Acts like a human* (Turing test approach). An artificial system acting like a human should be able to fool a human interrogator, who cannot distinguish if the answers are being provided by a computer or by a human. However, this evaluation mechanism implicitly assumes that the artificial system is already equipped with all the necessary means for communicating naturally and understanding the interrogator questions. Clearly, this approach doesn't logically scale, since providing such necessary means would require solving some general AI problems beforehand.

2 *Thinks like a human* (cognitive modelling approach). Whether an artificial system is able to think or not like a human, depends on the availability of an accurate theory or model of the mind, which can only be defined by experimental evaluation and validation either with human or animals. Although closely related to AI, all of such cognitive research efforts are totally out of scope.

3 *Thinks rationally* (laws of thought approach). To understand if an artificial system thinks rationally, an irrefutable reasoning process needs to be known. In this regard, the *formal logic* was introduced to study the inference in abstract (or formal) content. Based on such theories, the classical AI approach assumes that intelligent systems can be built on top of computer programs that search without exhaustion for a solution of given a set of problems stated in logical notation. Unfortunately, one key limitation of this approach is that it is difficult to model knowledge uncertainty, thus reality. On the other hand, computational resources can be easily exhausted when performing some (general) reasoning steps.

4 *Acts rationally* (rational agent approach). An artificial systems act rationally when focused on achieving a goal given a set of beliefs. Therefore, acting rationally implies perceiving, then acting, or equivalently, it implies mapping perceptual inputs or *percepts* into *actions*. Any artificial system able to perceive and act is what is called an *agent*. Here rationality is concerned with a success expectation in terms of what has been perceived—in contrast to the laws of thought approach, where rationality implies making correct inferences. As a result, a *rational agent* performs actions that are expected to maximize a *performance measure*, given a designated goal, a sequence of percepts and whatever *built-in knowledge* it may have. We observe that *causality* is a necessary condition for rationality.

Based on the latter approach, AI can be defined as the branch of computer science concerned with the study and development of rational agents. In particular, AI deals with the different ways to represent and implement how a rational agent maps percepts into actions. Consequently, AI aims to develop algorithms that, given the properties of the environment and the agent's structure, produce rational behaviours. In the rest of the section, *agent* will always refer to a *rational agent* unless stated otherwise. The environmental properties can be summarized as follow:

- **Observability**: an environment is *fully observable* if the agent's sensors allow reconstructing the whole state of the environment at each time instant; *partially observable* if the only part of the state can be reconstructed; *unobservable* if the agent has no sensors.
- **Predictability**: an environment is said to be *deterministic* if its next state can be uniquely determined in terms of its current state and the executed action by the agent; *nondeterministic* otherwise. One particular case of nondeterministic environments is the *stochastic* one, were the possible outcomes of actions are characterized by probabilities. In most practical scenarios partially observable environments are treated as stochastic ones. Therefore, an environment is *uncertain* if it is either partially observable or non-deterministic.
- **Staticness**: an environment is said to be *dynamic* if it changes while the agent is deliberating; *static* otherwise. It is worth noticing that a dynamic the environment may change either autonomously (time-variant) or due to the actions executed by the agent. If the environment changes only due to the agent's actions, then it is said to be *semidynamic*.
- **Discreteness**: the environment's state evolution can be either *continuous* or *discrete*. In total analogy, also the agent's percepts and actions can be of either type.
- **Knowledge**: in a *known* environment, the consequences of executing an action (either the outcomes itself or the outcomes' probabilities) are well understood by the agent. That is, in a known environment the agent understands the "laws" governing the environment's evolution. When those "laws" are missing from the agent's knowledge, the environment is said to be *unknown*. The environment's knowledge property is independent from its observability. It is worth noticing that in the case of an unknown environment the agent must *learn* the way it works to able to make decisions.
- **Episodicness**: an *episodic* environment does not depend on the actions taken previously; such an environment can be split as a series of

independent one-shot actions overtime. In contrast, a *sequential* environment depends on previous actions. That is, its current state is determined by past actions.

- **Agency**: an environment can be single agent or multi-agent. In the latter case, agents can either *cooperate* to reach a common goal or *compete* to conclude their individual goals or a mix of both.

As an example, an autonomous driving agent deals with a partial observable (it is not always possible to fully observe all pedestrians, vehicles or other entities on the road), stochastic (it is not possible to fully predict how such entities are going to move next), dynamic (entities' states evolve in time), continuous (likewise the rest of our world), known (pedestrians or other vehicles are not expected to fly), sequential (as a result of its continuity) and multi-agent (entities act on their own free will) where agents follow both common (e.g. avoid collisions) and individual goals (e.g. reach home on time).

The agent structure is defined by the way percepts are mapped into actions in order to achieve a goal. In particular, one can identify:

- **Reflex agents**: this type of agents execute one single action a time, given either the current percept or the whole percepts sequence. When the reflex agent relies only on the current percept to make a decision, the agent's structure is defined by a set of *condition-action rules*. On the other hand, when the agent deliberates what to do next based on the whole (or partial) percept sequence, its structure is given by an *internal model* representation of the environment together with a set of condition-action rules. Therefore, reflex agents are not concerned with the implications of their actions, the simply act as prescribed by their built-in rules. In such a sense, the goal of a reflex agent is implicit and uniquely determined for each environmental state.
- **Goal-oriented agents**: agents of this type are provided with some extra information, specifying what's the expected final or target configuration of the environment. Therefore, goal-oriented agents cannot rely on a set of condition-action rules to make decisions. On the other hand, they necessarily need to be aware about the implications each of their actions could lead. Also, they may need to execute more than one

action to actually achieve one particular goal. In general, however, it is not possible to guarantee that a goal-based agent will succeed with all given goals, even through the execution of an infinite number of actions. First, some goals may not be *reachable* from the current environmental state (unfeasible or due uncertainty). Second, goals may be conflicting in between. Also, when multiple action sequences allow to reach the same target, the goal based agent lacks a rational way to decide which sequence to execute.

- **Utility-based agents**: in the aforementioned cases where a goal based agent fails to succeed, the agent can, instead of exactly achieving a set of goals, try to execute the set of actions that maximize a given *utility* function, which specifies the appropriate trade-offs between them. Such an utility function represents the agent's internalization of the rationality's performance measure. It is worth noticing that in the case when different sets of action sequences allow to reach the same result, the utility function can be used to discern what's the best sequence among them.

Not all agent structures are appropriate for dealing with all types of environments. On the other hand, not always an utility-based agent will perform better than another agent with a simpler internal structure. This will depend mostly on environmental properties and the agent's adaptation to the environmental changes—in practice, it is impossible to have a perfect built-in knowledge of the environment. For example: modern collaborative robots implement a reflex agent to suddenly stop the robot motion when the external force exertions are above a predefined threshold to guarantee a safe physical interaction; in this applicative context the reflex agent guarantees the smallest decisional latency, thus minimizing the risk of damage to the environment or robot. In contrast, a trajectory planner implementing a reflex agent based on artificial potential fields may fail to reach the desired goal when getting trapped on a local minima.

Agent structures and environment states can be decomposed into a finite set of fundamental units or blocks. For example, one can encapsulate all perceptive aspects of an agent into a sensing unit. Each such an unit can be seen either as black box (*atomic representation*) or as a set of variables and attributes (*factored representation*) or as a set of inter-acting objects (*structured representation*). Based on the environmental properties, agent's structure and the ways of representing them, it is now possible to identify what are the basic AI problems and the algorithms and techniques to solve them.

Planning agents seek to identify and execute a sequence of actions to reach their objectives. In terms of the environmental properties, we can identify four major categories of planning agents:

- *Problem-solving agents*: use an atomic or a domain dependent factored representation of the environment. This kind of agents rely on general searching algorithms: depending on the environmental properties, the agent can use blind search, heuristic search, local search or adversarial search; in the case of factored representations, the problem-solving agent can take advantage of constraint satisfaction search. A clear limi-tation of atomic representations is that the searching algorithm cannot exploit any knowledge contained on atomic black boxes, that is, there's no room for *inference*. Example of problems that can be solved with this type of agents is the VLSI layout design and the classical travelling salesman.

- *Logical agents*: take advantage of a domain-independent structured representation of the environment. This allows to split the agent's structure into a representation unit (*knowledge base*) and a reasoning unit (*inference engine*). The knowledge base (KB) contains all domain-specific content, but it is stored as a set of formal (abstract, logical) *sentences* or statements expressed according to the syntax of a *repre-sentation language*. Each sentence can result either true or false, depending on the *model* used to evaluate it. Models are the mathe-matical abstraction of any possible environmental state. The inference engine allows to derive new sentences from the old ones in terms of *logical entailment*, that is, new sentences logically follow form the old ones. Such a logical reasoning can be done either in terms of *model checking* or *theorem proving*.

- *Classical planning agents*: in contrast to logical agents, which rely on a structured variable-free representation, planning agents use a factored representation of the environment in terms of *state variables*. This leads to a more flexible and succinct representation for actions, goals and plans, through the introduction of specific planning languages for representing the KB. This kind of agents relies on specific searching algorithms that, depending on the environmental properties, can be state-space search, planning graphs or hierarchical search.

- *Rational planning agents*: when dealing with uncertain environments, all previously described agents keep track of what is called the *belief state*, that is, the set of all possible environmental states logically explaining the observations. In turns, solving a planning task on an uncertain environment implies considering all possible explanations, no matter how *unlikely* they might be. Clearly, finding solutions on large search spaces becomes unfeasible with such agents. Another important limitation is given by the *qualification problem*: in logical terms it is not possible to specify all preconditions required for an action to succeed. In other words, it cannot be deduced whether an unexpected exception happens or not and, when such an exception happens, the plan's outcome cannot be inferred. Therefore, a rational decision must take into account both, the relative significance between goals (*utility*) and the prospect whether they will be achieved or not (*probability*). In particular, rational decisions maximize the expected utility when averaged over all of the possible outcomes of the action. These represent the bases of the *probabilistic reasoning*. Basic algorithmic approaches for implementing such type of reasoning are Bayesian networks, sampling-based methods for approximate inference and fuzzy logic. In case of partial observability, one can take advantage of hidden Markov models, Kalman filter or dynamic Bayesian networks to reconstruct the current environmental state. Rational agents immerse in episodic environment can make use of decision networks or their dynamic extension in case of sequential environments, which are modelled as (partially observable) Markov decision processes.

Perception is the process of extracting information about the environment from the sensors data. Although there's a large variety of sensing technologies providing sensory modalities, the most of the AI research efforts have been focused on *vision* (computer vision) and *speech* (natural language processing). Agents require perception to improve their knowledge of the environment and thus to achieve their goals; perception is not an end by itself. In general terms, an agent needs to identify what aspects of the perceptual stimulus actually bear or not relevant information. In general, there are three different approaches that can lead to this identification:

- *Feature extraction*: feature extraction refers to the process where raw data measurements are converted into a low-dimensional vector of numerical values, bearing the same informative content of the original measurements. Due to the dimensionality reduction, features are intended to be not only informative but also non-redundant. Nowadays manual or hand-crafted feature extraction is no longer a common practice in applied sciences, due to the advancements of machine learning algorithms (some of them listed on Sect. 3.2.2) together with the availability of large public datasets. Classical examples of feature extraction procedures could be the identification of the principal axes of a data cluster and the computation of the intensity histogram of an image.
- *Pattern recognition*: implies the automatic identification of regularities on data that are representative of some properties of the environment. Depending on the application context and nature of the perceptual information, a pattern recognition strategy can be applied directly to the raw measurements or to the features representations. As in the case of features extraction, nowadays pattern recognition problems are solved by means of machine learning algorithms. Some common examples of pattern recognition applications include automatic tumour identification from medical images, speech recognition, spam filtering and face detection.
- *Reconstruction*: refers to the direct inference of physical properties of the environment in terms of the measured data. For example, in the case of images, a reconstruction problem could be to infer the depth of each pixel. In the case of audio signals, to localize the source given a distributed array of measurements. Also the agent's velocity estimation given a sequence of range scans is a particular instance of a reconstruction problem (*state estimation*). In general, reconstruction

problems require specific algorithms to be solved. Despite, there are many successful application of machine learning algorithms on specific reconstruction tasks.

Natural language processing (NLP) deals with structured representations of the language and aims either, to acquire knowledge from data (audio or text) given in natural language, or to naturally communicate with humans or other agents. *Information-seeking tasks* rely on a *language model* (*n-gram*) based on characters or words, to predict the probability distribution of the language expressions. *Categorization* of documents can be effectively implemented using naive Bayes n-gram models or general classification algorithms (some of them listed on Sect. 3.2.2). *Information retrieval* is the task of finding documents that are relevant to a given information query and can be effectively achieved with a bags of words modelling. *Information extraction* consists of the automatic knowledge acquisition from documents; using a primitive notion of language's syntax and semantics, successful information extraction systems have been implemented using finite-state machine, hidden Markov model and conditional random fields. *Natural communication*: require more complex *grammatical models* and reasoning algorithms that takes into account the syntax, semantics and pragmatics of the language. Machine translation and speech recognition represent the most outstanding achievements of NLP in natural communication.

Robotics represents one of the most active and successful fields of AI research. Robots are complex physical agents that perform tasks on the physical world. Robotic system can exhibit distinct levels of autonomy depending on its learning and deliberating capabilities. In particular, AI methods are widely used the highest planning levels, that is, *action planning* and *path planning*. Action planning refers to the identification of a sequence of actions aimed to satisfy a given goal; task that can be addressed with any of the previously described methods for classical and stochastic planning agents. Path planning aims to identify a sequence of collision-free configurations that allow reaching a destination pose in the environment; this task can be solved by geometric algorithms, Markov decision process, sampling-based search, artificial potential fields, rapidly-exploring random trees, among others. Other low-level aspects affecting the behaviour, like for example trajectory planning, motion planning, trajectory following and motion control can be tackled either by classical methods and techniques found on the automation and control systems literature or through the application of machine learning techniques.

Knowledge representation studies what information or facts about the world should be included on the KB and how such information should be represented. The knowledge abstraction is built in terms of a *conceptualization* of the individuals and their relations in the environment, that is, a map that assigns to each one of them a symbol or a set of symbols in a computer program (the set of symbols is commonly known as *vocabulary*). The *ontology* provides the specification of a conceptualization (Poole and Mackworth 2017). In other words, an ontology specifies the meanings of symbols in terms of the environment under study. The specification provided by the ontology includes what entities can be modelled (*categories*), their properties, relationships (*hierarchy*) and clarifications (restrictions) on the meanings of some of the symbols in the form of axioms. Considering the central role of categories in any large-scale KB, algorithms for reasoning with categories has been also developed: semantic networks and description logics.

As already mentioned, together with the perceptual stimuli, an agent also relies on its built-in or prior knowledge of the environment. **Learning** refers to the ability of an agent to update, upgrade or deprecate any prior knowledge based on its own percepts sequence. Therefore, the behaviour of a learning agent can become effectively independent of its prior knowledge after sufficient experience. As a consequence, any learning agent is inherently *autonomous*: modifying its own beliefs with respect to experience, implies a behavioural evolution on time. It is worth noticing that learning implies *adaptation*, but not the other way around. Regardless of the internal structure, any agent can take advantage of learning to increase its own levels of autonomy. In general, there are two learning strategies that an agent can try: *tuning* its own beliefs based on a direct feedback of the executed actions and expanding its knowledge by *exploration*, that is, by executing actions leading to new experiences.

The branch of computer science focused on the study and implementation of algorithms that improve through experience is known as **machine learning**. The following section introduces ML in detail.

3.2.2   What's Machine Learning?

We have already mentioned that ML deals with algorithms that improve with experience. However, some clarifications are needed. On the one hand, *experience* refers to collecting evidence about the relation that must hold between the inputs and outputs of the algorithm. Evidence is given in the form of data samples, that is a collection of observations-outcomes pair. It is worth noticing that often the observations-outcomes pair corresponds to the inputs-outputs pair of the algorithm. However, in general, such a correspondence may depend on the problem under study and the algorithm itself (i.e. the learning strategy). Most of the ML algorithms rely on a factored representation, were both inputs and outputs are given as N-dimensional vectors of either discrete or continuous numerical values. On the other hand, *to improve* means lessening the uncertainty regarding the nature of the inputs-outputs relationship. In view of this, ML algorithms reach their objective by generalizing (or extrapolating) from specific evidence to general rules. That is, they follow an *inductive reasoning* (bottom-up paradigm). And as such, the predictions of any ML algorithm strongly depend on the evidence supplied to it: no ML algorithm is able to generalize beyond the domain of support induced by the known evidence.

Assuming that the input–output relationship can assume a functional representation, then reducing the uncertainty implies finding a better approximation, or *hypothesis*, to it. In general, different hypotheses may be consistent with the evidence, and one fundamental problem is how to select the best hypothesis among them. Based on the Ockham's razor (Mitchell 1997), the simplest consistent hypothesis should be preferred. However, in general, there should be a trade-off between the consistency and complexity of a hypothesis. Indeed, increasing the complexity reduces the *aleatoric uncertainty* (improves robustness), but at the same it increases the *epistemic uncertainty*, since generalizing becomes more difficult and requires more evidence to deal with sparsity (Hüllermeier and Waegeman 2019). Therefore, it is common to set up the quest for the best hypothesis in two steps. The first, known as *model selection*, defining the *hypothesis space*. The second, in terms of optimization to determine the best hypothesis in such a space. A learning model assuming that a finite number of parameters suffices to capture everything about the data is called *parametric*. Although, such an assumption notably restricts flexibility, the complexity of parametric models is bounded, no matter if the amount of available evidence is unbounded. In contrast, *non-parametric* assume that it is not possible to capture the data distribution in terms of a finite set of parameters. This makes such models way more flexible than the parametric ones, but their complexity increases with the amount of data provided.

Based on the information provided by the observations-outcomes samples defining the available evidence, distinct forms of learning can be identified (Bishop 2006):

1. *Supervised learning*: in this case the evidence is composed by samples of inputs-outputs pairs. Then, the learning objective is to generate the best hypothesis approximating the function that maps inputs into outputs. The best hypothesis is obtained though optimization and corresponds to the one minimizing a *loss function*, measuring the amount of utility lost between the prediction and the true output value. When the output of the algorithm corresponds to a finite number of discrete categories, or *labels*, the learning problem is called *classification*, otherwise *regression*. Many algorithms have been developed to solve this kind of problems, to name a few: decision tree learning, naive Bayes classifier, k-nearest neighbour (k-NN), metric learning, support vector machines (SVM), random forests, artificial neural network (ANN), ensembles of classifiers and Gaussian process regression.

2. *Unsupervised learning*: the evidence consists of samples containing only the inputs of the algorithm. The learning objective could be: to discover groups of samples having similar attributes (*clustering*), to project the samples into a low-dimensional space while preserving

some of their meaningful properties (*features extraction, dimensionality reduction*) or to determine the data distribution within the space (*density estimation*). Algorithm for solving unsupervised learning problems is special ANN architectures (auto-encoders, self-organizing map), k-means, DBSCAN, hierarchical clustering, principal component analysis (PCA), mixture models and Gaussian processes.

3. *Reinforcement learning* (RL): evidence is composed by a collection of samples of the form observations-reinforcements, where each observation is a state-action pair and the reinforcements can be either a *reward* or a *punishment*. The learning objective is to determine the optimal *policy* maximizing the overall total reward. In RL, a *policy* is the mapping from every possible state to the best action in that state. In practical applications, there's no prior evidence; it is obtained during the learning process by trial and error. Actions are executed based on a trade-off between *exploitation* of known state-actions pair generating high rewards and *exploration* to discover new ones. Most of the RL algorithms that can be found on literature are variants either of the policy gradient or the Q-learning methods.

It is worth mentioning that nowadays there are *semi-supervised* forms of learning dealing with evidence having a large number of data samples with uncertain or missing information about the outcomes.

In general terms, deep learning (DL) refers to the principle that learning with multiple levels of composition (hierarchy) allows to improve the learning outcomes when sufficient evidence is provided. Such a principle can be potentially applied to any ML algorithm (Deng and Yu 2014). However, in practice, due the contemporary real-world impact of deep neural network (DNN) on the fields of computer vision and natural langue processing, DL is widely understood as a synonym of DNN. From this standpoint, DL (DNN) is a special type of ANN having a very large number of hidden layers. With respect to the 1980s, today we have the enough computational power (GPGPU) and the sufficiently large datasets that such complex ANN models require to succeed: the only way to deal with the intrinsic epistemological uncertainty of a complex model is to feed it with sufficient amounts of (non-redundant) data. Moreover, although there were no significant theoretical contributions to the field of ANN since then, the use of convolutional neural networks (CNN) allows to dramatically reduce the number weights and thus to speed-up the learning algorithm. As a last remark, it is worth mentioning the technique known as *transfer learning*. In brief, the technique consists of exploiting the available knowledge for solving one task and applying it for solving a different one (Goodfellow et al. 2016). This technique is widely used on DL applications, in particular through *fine tuning*.

3.2.3 What's the Relation Between Artificial Intelligence and Machine Learning?

As a first approximation, one can say that ML *seems to be* a branch of AI. However, in analogy with the perception case, agents require learning to improve their knowledge of the environment and, consequently, to further their own goals. Therefore, learning in AI is not an end to itself, but a necessary constituent to build intelligent machines. *It follows that, although AI and ML are highly related, they pursue two different avenues.* The distinction between the two research fields can be also traced through a historical perspective.

In the early days of AI, some researchers were experimenting the ways machines can learn from data. Different approaches were developed to achieve such a goal. In particular, nowadays ANN is the most widely known. However, due to the strong emphasis that settle the AI community on the KB logical approach, by 1980 the data driven and the statistical ones were already ignored by the AI community. The latter approaches continued their way on the fields of pattern recognition and information retrieval, while the ANN enthusiast continued the research as part of the connectionism line of though. After the reinvention by them of the back-propagation algorithm, ML started to gain attention as a separate field in the 1990s. The focus of ML was no longer to achieve AI but to solve practical problems based on statistical and probabilistic methods and models.

3.3 Human–Robot Cooperation for Smart Manufacturing

Industry 4.0 foresees humans and CPS as cooperative entities. Under such a perspective, CPS need to aware not of its inner state, but also of the environmental ones, including any other entity on its surroundings. Moreover, CPS are required to smartly interact with both humans and other machines. Such a rich interaction between humans and CPS requires safe physical human-machine interaction (pHRI), unambiguous and resilient information flows, autonomous information processing and real-time decision-making capabilities. The first requirement is automatically satisfied in the context of collaborative robotics. The second deals exclusively with the Internet of Things (IoT) infrastructure. The last two are, in general, open research problems. The goal of this section is to highlight the potential of AI and ML approaches to tackle such problems in the context of human-robot cooperation in assembly.

3.3.1 CPS and Safety

Cyber-physical systems (CPS) represent one of the fundamental key enabling technologies for Industry 4.0. Although CPS are still in the making, it has been conjectured that their introduction in industry will dramatically change the way value is created along all the digitization axes of the manufacturing sector: smart product, smart manufacturing and business model. Based on the 5C architecture (Lee et al. 2015), implementing a CPS comprises the following levels:

- *Smart connection level*: is concerned with the sensing and transduction technologies and the IoT infrastructure for real-time, seamless and resilient data exchange between all parties.
- *Data-to-information conversion level*: incorporate all information retrieval methodologies aimed to understand the state of the machine and its components. In other words, this level deals with the implementation of the single machine self-awareness.

- *Cyber level*: represents a central information hub between all machines. Trough the data aggregation and subsequent analysis is could be possible to compare the performance between different machines and to predict the future behaviour of each.
- *Cognition*: includes a set of decision support systems that implement preliminary data analysis and valuable means for data visualization, aimed to transfer efficiently the inferred knowledge to the human experts.
- *Configuration level*: refers to the actuation mechanism aimed to apply any corrective or preventive decision taken at the cognition level to the physical space.

The 5C architecture is thus defined as a human-in-the-loop (HiTL) scheme were human experts, aided by decision support systems, take all decisions regarding how to improve the manufacturing process. It is worth noticing that the applicative context of this architecture is limited to classical manufacturing processes. Indeed, it doesn't account for possible interactions with the environment (safety) and it lacks of a proper design for distributed processing capabilities. Therefore, the 5C architecture is not well suited for modern robotic assembly workstations, specially for those having shared collaborative environments. Another key concept in Industry 4.0 not captured by the 5C architecture is that CPS should be able to cooperate with humans and other CPS. Cooperation implies two fundamental objectives. The first, to ensure safety; a constraint that cannot be violated by any means. The second, to conclude the assigned task; whose achievement can be only guaranteed in safe operative conditions. With regard to safety, CPS must be able to build their own knowledge not only in terms of self-awareness but also in terms of *situational-awareness*, including both, the state of the physical environment and the state of the current assembly cycle. With regard to the task completion, CPS must manifest some degrees of decision-making capabilities. In other words, they must be able to learn how to interact with the environment, including other entities, based on their beliefs about the environmental state.

With this idea in mind, let's rephrase the above considerations in AI jargon. We start by observing CPS are able to perceive and act, thus, from the very basic definition, it follows that CPS are indeed rational agents. In particular, CPS belong to the class of model-based agents: they must keep an internal representation of their physical counterpart and of their environment, including the state of the manufacturing process (self- and situational-awareness). Moreover, they should achieve multiple goals at the same time based on the current beliefs: an utility measure is required to define the proper trade-off. Consequently, CPS should plan their actions so to maximize the expected utility when averaged among all possible outcomes that can result from their actions. Furthermore, CPS must cooperate between them considering that the overall goal is to improve production; still, competing CPS willing to reach the highest performance can be desirable in a manufacturing context (paradigm defined as "self-compete" in the 5C architecture). Finally, CPS must deal with both aleatoric and epistemic uncertainty, specially on workspaces share with human beings. Nevertheless, there are some key different that makes a CPS something more *tangible* than an abstract agent. On the one hand, CPS are always associated to a physical counterpart and a concrete implementation. On the other hand, a CPS may exhibit degrees of complexity that are difficult to express or implement in terms of a single rational agent.

Based on the above considerations, we identify a structured representation for a machine or robot to be considered as a safety-aware CPS (SA-CPS), defined in terms of four interacting components (see Fig. 3.1):

1. *Safety monitor*: based on the percepts sequence and current beliefs, the aim of this block is to monitor the operative conditions of the CPS and to trigger an alarm when safety is unexpectedly lost or when it can be potentially lost in a finite time horizon. Therefore, this unit relies on an internal model to predict potential risky circumstances and to decide when to notify the other components of the CPS. This block is always active and runs in parallel with any of the other three units.

2. *Safety reflexes*: the aim of this block is to promptly react when an alarm is triggered by the safety monitor. The set of actions executed by this block seek to quickly restore the save operative conditions despite the current operative state. In terms of the AI agents taxonomy, this unit

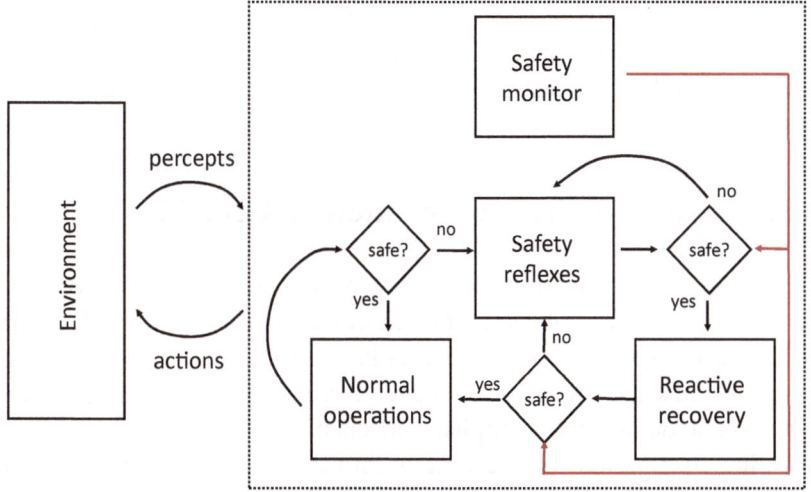

Fig. 3.1 Structured representation of the abstract safety CPS

together with the safety monitor one can be considered as a model-based reflex agent.

3. *Reactive recovery*: this block aims to restore a pre-empted operative state when the safe operative conditions are recovered again. Therefore, the goal of this component is to plan a sequence of actions allowing to ensure that the normal operations can be restarted just after a risky circumstance has been mitigated. When this component is active, normal operations are on hold.

4. *Normal operations*: this block incorporates all the functionalities required to reach the CPS goals. It can pre-empted at any time by the safety reflexes and can only restart operations after suitable recovery actions had taken place. This component can be seen as an utility-based agent, focused on the completion of the manufacturing task assigned to the CPS.

It is worth noticing that Fig. 3.1 only captures the logical relation between the four components. However, the interactions between them are in general richer and complex. As a last remark, modern collaborative robots have a similar internal structure. In particular, safety is defined in terms of physical interaction; prompt reactions imply stopping the current motion and blocking the motor actuators; and recovery actions consist of unlocking again the motors and restarting the pre-empted motion.

3.3.2 Human–Robot Cooperation in Assembly

The most dangerous risk specific to robots is the unexpected collisions between the robot and the environment (Siciliano et al. 2010). When an unexpected exertion occurs between a collaborative robot and its surrounding environment, impact forces are eased thanks to their lightweight design and compliant mechanisms and control. However, avoiding unexpected force exertions implies foreseeing dangerous situations, and thus it relies on sensing, situational awareness, planning and decision-making capabilities. *Therefore, without suitable exteroceptive sensing a collaborative robot cannot be considered as a safe companion in the context of human–robot cooperation.* Indeed, to safely interact with a human operator a collaborative robot must predict and prevent any risky circumstances based on its own situational awareness. To this end, it is required to associate to the human operator and the environment a set of meaningful spatio-temporal features that allows—with some degree of accuracy, within a finite time horizon—to model and predict the operator's behaviour and the environmental changes. In terms of safety, it is required to sense and predict the operator's motion. In terms of cooperation, it is required to understand and predict the operator's actions and intentions.

We identify three major synergic elements (see Fig. 3.2) required for a collaborative robot to be considered as a safe companion in the context of human-robot cooperation: (i) scene monitoring, (ii) tasks modelling and (iii) planning. Although these general problems can be unreasonable complex, within the context of cooperative assembly workstations where different constraints are imposed to the environment and due to the cyclic nature of the assembly process, the analysis of each element can be greatly simplified. In particular, we introduce the following simplifying assumptions:

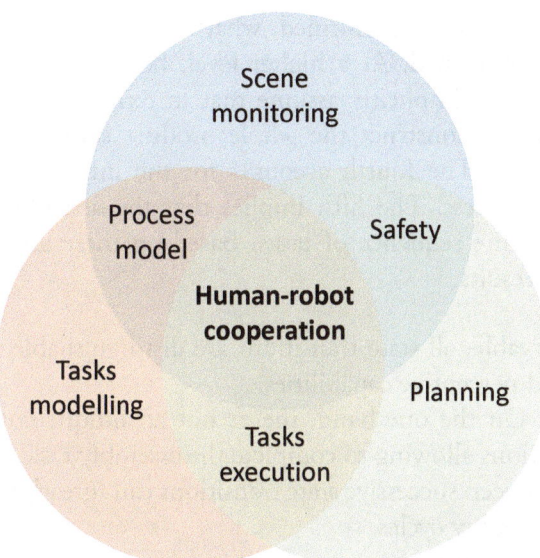

Fig. 3.2 Key enabling technologies for human-robot cooperation

i. The environment is limited to the collaborative workstation and its assembly process.

ii. There's only one human operator and one collaborative robot active a time in the workstation.

iii. The state transitions on the environment are triggered only by events.

iv. There's a finite number of sequences of state transitions that allow reaching the final state.

v. There's a finite number of environmental states.

The first and second allow us to focus on the human-robot cooperation by ignoring the interactions with the rest of the assembly line. Therefore, we assume that the inputs of the assembly process are always available and that the outputs of the assembly process are being gathered autonomously by an external entity without affecting the assembly process. The third, to limit how deep the robot's understanding of the environment should be. For example, when a human operator is finishing one part, it is not always possible to know what specific finishing touch is being performed, what are the missing ones or what were already performed. At a higher level, however, the part is being finished. Thus, we implicitly assume that in terms of cooperation it is not required to reconstruct the whole product state, but only up to the process state. The fourth accounts for the inner variability inside the assembly process. The fifth implies that the assembly process can be split in a finite sequence of tasks. Based on these assumptions, the environment results:

- Fully observable, all state transitions are distinguishable with suitable sensing and perceptive capabilities.
- Stochastic. On the one hand, there's not an unique combination of state transitions allowing to complete the assembly task. On the other, the time between successive state transitions can (greatly) vary between different assembly cycles.
- Static, mainly due to assumptions (i) and (ii). However, in the context of flexible manufacturing some clarifications are required. In case of multi-variant or multi-product lines both, the assembly cycle and the workstation layout may require some adjustments. However, such adjustments do not occur whiting the assembly cycle. Indeed, the current product under manufacture must be completed, aborted or pre-empted before switching the assembly goal. In other words, any environmental change required for a flexible assembly line will be triggered by an event (in analogy to assumption [iii]). Moreover, all possible environmental changes are necessarily countable and finite. Therefore, without loss of generality, we can assume that in the context of flexible manufacturing there exists a finite set of static environments and that each of them can be handled independently from the other.
- Discrete, by assumption (v).
- Known. All possible state transitions are well understood, in terms of the expected outcomes of the assembly process. This is also enforced by assumption (iv).
- Sequential, as the assembly process.
- Defining the environment's agency is rather ambiguous, considering that under our modelling assumptions a single CPS may be defined in terms of several interacting rational agents. However, due to the

restriction in assumption (i) and (ii), we assume that there's only one CPS, given by the collaborative robot and its associated sensing and processing capabilities.

Based on the properties of the environment, we can introduce the key enabling technologies for a safe human-robot cooperation in collaborative assembly.

Scene monitoring refers to the real-time reconstruction of the state of both the operator and the manufacturing parts and products along the whole assembly processes. Here the objective is not to reconstruct the state of the assembly process, but to increase the CPS's awareness about were the objects and operator are in physical terms (pose, motion, etc.). In other words, the goal of the scene monitoring unit is twofold. On the one hand, to extract from the percepts sequence the information required to evaluate and guarantee the safe operative conditions at every time instant. On the other hand, to extract from the percepts sequence the required information to allow further inference regarding the current and future operator's activities, and the current and future state of the ongoing assembly cycle. Therefore, the scene monitoring problem can be analysed in terms of both, the recognition and tracking of assembly parts and products, and the operator's motion tracking.

- *Object's recognition and tracking*: there are different technologies that can be used to efficiently recognize and track the pose, motion and manufacturing state of objects. To name a few, one can identify 2D/3D vision systems, RF systems, range finder, sonar, mmWave, etc. A throughout treatment of the problem of object recognition in smart manufacturing is found in (Riordan et al. 2019).
- *Operator's motion tracking*: due to the stochastic nature of the operator's body, head, arms, etc., movements while executing an assembly task, the problem of monitoring and predicting the operator's motion can be considered as a particular instance of a filtering problem (Tan and Arai 2011). That is, based on a set of past possibly noisy observations of the operator's pose determine the best estimate of the current operator's motion. Today on research and industry exist a wide range

of different sensing technologies that can be used to measure the operator's pose. The current technological trend points towards multiple networking range sensing devices or computer vision-based systems providing analogous measurements (Ferrari et al. 2018; Gkournelos et al. 2018; Agethen et al. 2016). The use of multiple sensors not only ensures a better accuracy of the estimation but also accounts for the decreasing point-density at far distances of a single sensors. Moreover, different view points are required for a reliable identification of features or markers. However, state-of-the-art deep learning models for pose estimation in RGB images (Cao et al. 2017) and 2D lidar data (Weinrich et al. 2014) allows to reach high-levels of accuracy. Indeed, the larger field of view of lidar sensors allows to track the operator beyond the field of view of the RGB-D sensor.

Tasks modelling aims to understand what are the current and future operator's activities, and the current and future state of the ongoing assembly cycle. However, considering that only the operator and robot actions can cause a process state transition, the tasks modelling problem can be restricted to the recognition and prediction of the actions executed by the operator.

- *Operator's intentions prediction*: in industrial manufacturing scenarios, the problem of task prediction is greatly simplified by the cyclic nature of the operator's work. Any manufacturing cycle is indeed defined by a finite set of atomic tasks. However, the order in which such atomic tasks are performed by the operator to conclude the cycle, in general, is not uniquely defined. Therefore, for a machine to be aware of the current state of an assembly cycle, it is required to recognize any atomic task executed by the operator and to model the transitions between them (Alati et al. 2019b). Identifying a task implies understanding the actions being performed by the operator, while understanding the transitions between tasks implies predicting the operator's intentions. Based on this idea, the prediction of intentions problem can be analysed in terms of two distinct processes: (i) action recognition, and (ii) action prediction. Action recognition refers to the prompt identification of the current task executed by the operator. The aim of this process is to continuously monitoring the operator's actions on real-time. The action identification can be driven by different cues, like gestures (Carrasco and Clady 2010), scene objects being manipulated (Koppula and Saxena 2015) or environmental information (Casalino et al. 2018). Action recognition has been also extensively studied in terms of whole body motion tracking and segmentation (Natola et al. 2015; Tome et al. 2017). Although there are no specific manufacturing datasets for the evaluation of action recognition models, in recent years different deep network architectures had been demonstrated high levels of accuracy on totally unrelated but similar manipulation tasks, like the one proposed by the Epic Kitchens challenge (Damen, et al. 2018; Wang et al. 2018). Action prediction refers to the total or partial reconstruction of the possible sequence of actions that the operator would execute just after concluding the current task. Consequently, this process implies the generation and constant refinement of an action transition model (Zanchettin and Rocco 2017; Zanchettin et al. 2018). In general, it can be also assumed that all operator's states and actions are fully observable and that the operator can only execute one action a time.

In general, **planning** actions in collaborative workstations requires finding a suitable and safe plan to complete a manipulation or a mobility task assigned to the CPS. However, we will restrict our attention to the manipulation case, since in most collaborative workstation the CPS is defined on top of a robot manipulator with a fixed inertial base. The objective here is to superimpose the robot's state on top of the assembly process model, such that to allow the real-time analysis and generation of the robot plan. In other words, the robot's collaborative behaviour is achieved by dynamically allocating its tasks, in terms of the predicted operator's actions and the relative action transition model (Alati et al. 2019a). As a result, the objective of the action planning is to reach a designated assembly process goal state. This implies that any goal state includes both the operator's and robot's states. Therefore, it is expected that one particular goal can be reached from a finite set of initial candidate states, each one depending on the particular sequence of actions performed by the operator. Consequently, in human–robot collaborative environments, the action planning process deals with the robot behaviour adaptation (Mitsunaga et al. 2008) to the time-varying set of constraints imposed the operators' actions. In turns, imposed by the customer requirements, diversity of the available manufacturing variants and operator's task execution preferences (Munzer et al. 2017). Therefore, the robot adaptation should provide a proactive (anticipatory) collaborative behaviour driven by the different forms of human-robot interaction associated to each target goal (Mason and Lopes 2011). In other words, robots working alongside humans should model how to anticipate a belief about possible future human actions (Koppula et al. 2016). In complete analogy to the operator's case, the cyclic nature of the assembly process implies that there's a predefined number of goals that the robot can reach, a finite set of deterministic actions that it can perform and a finite set of states that it can have. Moreover, it can be also assumed that the robot can only execute one action a time. However, in general, the execution time of any planned cannot be defined in advance since it also depends on the current state of the assembly process. Specially in the cases when the action execution requires explicit synchronization with the operator.

We observe that each key enabling technology comprises different perceptive or cognitive processes that, based on the structured representation of the SA-CPS, can be mapped to one or more of its building blocks. In particular, the scene monitoring greatly overlaps with the safety monitor. However, the scope of the former is not only to evaluate risks but also to understand the current process state and its evolution in the near future, which belongs to the normal operations block. Planning is required for safety reflexes, reactive recovery and normal operations. Finally, task modelling belongs mainly to the normal operations blocks. However, understanding the assembly sequence provide useful hints on the prediction of risky circumstances.

3.4   Conclusions

Industry 4.0 foresees humans and CPS as cooperative entities. Under such a perspective, CPS need to aware not of its inner state, but also of the environmental ones, including any other entity on its surroundings. To smartly interact with both humans and other machines, CPS must be endowed with real-time decision-making capabilities. Although still today there are many open problems on the field, different AI and ML techniques can be combined together to provide feasible solutions to real-world problems, specially on the fields of HRC and automated adaptation of a multi-variant collaborative manufacturing system.

Within these applicative context, it is required to provide a strong emphasis on safety, concept that to our knowledge has not being taken into account on any formalization of the concept of CPS. A safety-aware CPS is composed at least by four fundamental blocks:

- A constantly running safety monitor system to evaluate the safety status independently of any other functionalities of the CPS.
- A safety reflexes block to be activated when a risky circumstance has been detected.
- A reactive recovery unit to restore safe operative conditions just after the safety has been guarantee by the prompt actions of the safety reflexes unit.
- A normal operations module, which normally runs unless pre-empted due to safety issues.

HRC can be effectively implemented through the exploitation of three key enabling technologies, namely: scene monitoring, task modelling and planning. Different state-of-the-art AI and ML algorithms can deal with deferent aspect of one or more of these technologies. The research in this area is still in an early stage, so this contribution aims to motivate other researchers to do further research and practitioners to collaborate with research institutions for conducting tests on practical applications in real case studies.

Acknowledgements This project has received funding from the European Union's Horizon 2020 research and innovation programme under the Marie Skłodowska-Curie grant agreement No 734713.

References

Agethen, P., M. Otto, S. Mengel, and E. Rukzio. 2016. Using marker-less motion capture systems for walk path analysis in paced assembly flow lines. *Procedia CIRP* 54: 152–157. https://doi.org/10.1016/j.procir.2016.04.125.

Alati, E., L. Mauro, V. Ntouskos, and F. Pirri. 2019a. Anticipating next goal for robot plan prediction. *SAI Intelligent Systems Conference*, 792–809. https://doi.org/10.1007/978-3-030-29516-5_60.

Alati, E., L. Mauro, V. Ntouskos, and F. Pirri. 2019b. Help by predicting what to do. *IEEE International Conference on Image Processing*, 1930–1934. https://doi.org/10.1109/ICIP.2019.8803155.

Bishop, C.M. 2006. *Pattern recognition and machine learning*. Berlin: Springer-Verlag.

Cao, Z., T. Simon, S.E. Wei, and Y. Sheikh. 2017. Realtime multi-person 2d pose estimation using part affinity fields. *IEEE Conference on Computer Vision and Pattern Recognition*, 7291–7299. https://doi.org/10.1109/CVPR.2017.143.

Carrasco, M., and X. Clady. 2010. Prediction of user's grasping intentions based on eye-hand coordination. *IEEE/RSJ International Conference on Intelligent Robots and System*, 4631–4637. https://doi.org/10.1109/IROS.2010.5650024.

Casalino, A., C. Messeri, M. Pozzi, A.M. Zanchettin, P. Rocco, and D. Prattichizzo. 2018. Operator awareness in human-robot collaboration through wearable vibrotactile feedback. *IEEE Robotics and Automation Letters* 3 (4): 4289–4296. https://doi.org/10.1109/LRA.2018.2865034.

Damen, D. et al. 2018. Scaling egocentric vision: The Epic-Kitchens dataset. *European Conference on Computer Vision*. arXiv:1804.02748 [cs.CV]. Available at https://arxiv.org/abs/1804.02748.

Deng, L., and D. Yu. 2014. Deep learning: Methods and applications. *Foundations and Trends in Signal Processing* 7 (3–4): 1–199. https://doi.org/10.1561/2000000039.

Ferrari, E., M. Gamberi, F. Pilati, and A. Regattieri. 2018. Motion analysis system for the digitalization and assessment of manual manufacturing and assembly processes. *IFAC-PapersOnLine* 51 (11): 411–416. https://doi.org/10.1016/j.ifacol.2018.08.329.

Gkournelos, C., P. Karagiannis, N. Kousi, G. Michalos, S. Koukas, and S. Makris. 2018. Application of wearable devices for supporting operators in human-robot cooperative assembly tasks. *Procedia CIRP* 76: 177–182. https://doi.org/10.1016/j.procir.2018.01.019.

Goodfellow, I., Y. Bengio, and A. Courville. 2016. *Deep learning*. MIT Press.

Hüllermeier, E., and W. Waegeman. 2019. Aleatoric and epistemic uncertainty in machine learning: A tutorial introduction to concepts and methods. ArXiv:1910.09457 [cs.LG]. Available at https://arxiv.org/abs/1910.09457.

Koppula, H.S., and A. Saxena. 2015. Anticipating human activities using object affordances for reactive robotic response. *IEEE Transactions on Pattern Analysis and Machine Intelligence* 38 (1): 14–29. https://doi.org/10.1109/TPAMI.2015.2430335.

Koppula, H.S., A. Jain, and A. Saxena. 2016. Anticipatory planning for human-robot teams. In *Experimental robotics. Springer tracts in advanced robotics*, ed. M. Hsieh, O. Khatib, and V. Kumar, 109. https://doi.org/10.1007/978-3-319-23778-7_30.

Lee, J., B. Bagheri, and H. Kao. 2015. A cyber-physical systems architecture for Industry 4.0-based manufacturing systems. *Manufacturing Letters* 3: 18–23. https://doi.org/10.1016/j.mfglet.2014.12.001.

Mason, M., and M. Lopes. 2011. Robot self-initiative and personalization by learning through repeated interactions. *ACM/IEEE International Conference on Human-Robot Interaction*, 433–440. https://doi.org/10.1145/1957656.1957814.

Mitchell, T.M. 1997. *Machine Learning*. New York: McGraw-Hill.

Mitsunaga, N., C. Smith, T. Kanda, H. Ishiguro, and N. Hagita. 2008. Adapting robot behavior for human-robot interaction. *IEEE Transactions on Robotics* 24 (4): 911–916. https://doi.org/10.1109/TRO.2008.926867.

Munzer, T., M. Toussaint, and M. Lopes. 2017. Preference learning on the execution of collaborative human-robot tasks. *IEEE International Conference* on *Robotics and Automation*, 879–885. https://doi.org/10.1109/ICRA.2017. 7989108.

Natola, F., V. Ntouskos, M. Sanzari, and F. Pirri. 2015. Bayesian non-parametric inference for manifold based MoCap representation. *IEEE International Conference on Computer Vision*, 4606–4614. https://doi.org/ 10.1109/ICCV.2015.523.

Poole, D., and A. Mackworth. 2017. *Artificial intelligence: Foundations of computational agents*. New York: Cambridge University Press.

Riordan, A., D. Toal, T. Newe, and G. Dooly. 2019. Object recognition within smart manufacturing. *Procedia Manufacturing* 38: 408–414. https://doi.org/ 10.1016/j.promfg.2020.01.052.

Ruiz Garcia, M.A., R.A. Rojas, L. Gualtieri, E. Rauch, and D. Matt. 2019. A human-in-the-loop cyber-physical system for collaborative assembly in smart manufacturing. *Procedia CIRP* 81: 600–605. https://doi.org/10.1016/j.pro cir.2019.03.162.

Russell, S., and P. Norvig. 2010. *Artificial intelligence: A modern approach*. Prentice Hall.

Siciliano, B., and O. Khatib. 2016. *Springer handbook of robotics*. Springer International Publishing.

Siciliano, B., L. Sciavicco, L. Villani, and G. Oriolo. 2010. *Robotics: Modelling, planning and control*. London: Springer.

Tan, J.T.C., and T. Arai. 2011. Triple stereo vision system for safety monitoring of human-robot collaboration in cellular manufacturing. *International Symposium on Assembly and Manufacturing*, 1–6. https://doi.org/10.1109/ ISAM.2011.5942335.

Tome, D., C. Russell, and L. Agapito. 2017. Lifting from the deep: Convolutional 3d pose estimation from a single image. *IEEE Conference on Computer Vision and Pattern Recognition*, 2500–2509. https://doi.org/10.1109/CVPR. 2017.603.

Tseng, M.M., J. Jiao, and M.E. Merchant. 1996. Design for mass customization. *CIRP Annals* 45 (1): 153–156. https://doi.org/10.1016/S0007-850 6(07)63036-4.

Wang, L., et al. 2018. Temporal segment networks for action recognition in Videos. *IEEE Transactions on Pattern Analysis and Machine Intelligence* 41 (11): 2740–2755. https://doi.org/10.1109/TPAMI.2018.2868668.

Weinrich, C., T. Wengefeld, C. Schroeter, and H. Gross. 2014. People detection and distinction of their walking aids in 2D laser range data based on generic distance-invariant features. *IEEE International Conference on Robot and Human Interactive Communication,* 767–773. https://doi.org/10.1109/ROMAN.2014.6926346.

Zanchettin, A.M., A. Casalino, L. Piroddi, and P. Rocco. 2018. Prediction of human activity patterns for human–robot collaborative assembly tasks. *IEEE Transactions on Industrial Informatics* 15 (7): 3934–3942. https://doi.org/10.1109/TII.2018.2882741.

Zanchettin, A.M., and P. Rocco. 2017. Probabilistic inference of human arm reaching target for effective human-robot collaboration. *IEEE/RSJ International Conference on Intelligent Robots and Systems,* 6595–6600. https://doi.org/10.1109/IROS.2017.8206572.

4

Industrial Assistance Systems to Enhance Human–Machine Interaction and Operator's Capabilities in Assembly

Benedikt G. Mark, Erwin Rauch, and Dominik T. Matt

4.1 Introduction

This chapter discusses industrial assistance systems, which can be used to enhance the operator's capabilities and the human-machine interaction (HMI) during production processes. It presents solutions for HMI and automation, and delivers insights into different possibilities

B. G. Mark (✉) · E. Rauch · D. T. Matt
Industrial Engineering and Automation (IEA), Faculty of Science and Technology, Free University of Bozen-Bolzano, Piazza Università 1, 39100 Bolzano, Italy
e-mail: benediktgregor.mark@unibz.it

E. Rauch
e-mail: erwin.rauch@unibz.it

D. T. Matt
e-mail: dominik.matt@unibz.it; dominik.matt@fraunhofer.it

D. T. Matt
Innovation Engineering Center (IEC), Fraunhofer Research Italia S.C.a.R.L, Via A.-Volta 13a, 39100 Bolzano, Italy

© The Author(s) 2021
D. T. Matt et al. (eds.), *Implementing Industry 4.0 in SMEs*,
https://doi.org/10.1007/978-3-030-70516-9_4

129

to enhance the various types of operators' skills in industrial assembly. In the course of the fourth industrial revolution, also called Industry 4.0 (I4.0), that was introduced in the year 2011 at the Hannover fair (Deutscher Bundestag 2016), the world of work is changing comprehensively. Smaller lot sizes, an increasing variability of products and increasing complexity in the modern industrial production present new challenges for operators working in manual assembly (Matt 2007, 2009; Rauch et al 2017). Industrial assistance systems help the worker during these production tasks to enhance their capabilities. The development of these systems is not only characterized by questions of the potential feasibility of new technical systems, but also by the possibilities of closer cooperation between humans and machines. Furthermore, it aims to synergize the outstanding abilities of humans with the special features of machines to bring together the best from both worlds. With the knowledge that is given in this chapter, each worker can be individually equipped with suitable supporting systems in order to be best prepared for future challenges in the daily production.

This chapter is structured as follows: It is divided into five sections. After a short introduction in the first section, Sect. 2 gives an overview of the theoretical background. In this section, industrial assistance systems are introduced in general and user groups of assistance systems are presented. Further, the importance of human-machine interaction in production is outlined and a brief analysis of the relevance of assistance systems in literature is given. Based on the three categories of aid systems (sensorial, cognitive, and physical), Sect. 3 explains and presents individual assistance systems within each category. Section 4 discusses risks, challenges, and potential in the context of industrial assistance systems. Finally, in Sect. 5, a short conclusion summarizes the main findings of this chapter. The content of this chapter should not only be of interest for researchers, but especially also for practitioners from SME companies in the field of assembly and manufacturing.

4.2 Theoretical Background

4.2.1 Industrial Assistance Systems

The demographic development in most of the highly industrialized countries in Europe and many other emerging challenges increase the need for assistance systems (Mueller et al. 2018). Assistance or aid systems should help the operator to conduct his/her daily work appropriately. Within these systems, it can be differentiated between technological systems that support and systems that substitute humans. The following four aspects characterize a technical system that can be seen as a worker assistance system (Weidner et al. 2015):

1. The technical system supports the operator and does not replace him,
2. The technical system can always be overruled by the operator,
3. The technical system is used by the operator,
4. The technical system does not provoke any hazard for the operator.

Assistance systems can have diverse functionalities and advantages that support the operator's daily work. The following list points out the most relevant features that can be increased, decreased, and enabled by the usage of these systems (Mark et al. 2019a):

- *Increase of:* physical support, cognitive support, speed and productivity, quality control, comfort and convenience, ergonomics, worker capacity, worker safety, worker integration, location independence.
- *Decrease of:* mental stress, language barrier, search times.
- *Enabling of:* health control.

In the field of technological change, the range of tasks and requirements of people in manufacturing companies will change dramatically. Assistance systems are an important element on the way to the smart factory. Visualized work instructions increase the skills of the employees, the quality of the products and ultimately the productivity and competitiveness of a company (Rothenberger 2020). Nonetheless, the human being with his characteristic of complex perception, his ability to see, touch, grasp, and hear as well as his cognitive abilities offers suitable conditions to react flexibly and quickly to production conditions (Böhle 2005; Arnold and Furmans 2019). The frequent changes in manual work processes, which are caused by the changes on the market, on the one hand lead to more interesting and varied work content for the operator. On the other hand, the worker's requirements regarding performance, concentration, and stamina increase, which may cause a greater susceptibility to errors (Reinhart and Zäh 2014). Due to the high variance of products, a high qualification of the employees is required to ensure a consistently high quality of the manual activities (Bächler et al. 2018). This can be reached by increasing the operator's capabilities through worker assistance systems.

4.2.2   User Groups in Production

With the introduction of the fourth industrial revolution, the production moves towards a factory in which robots and workers interact and collaborate with each other (Gualtieri et al. 2020). In addition, employees are supported by web-based technologies and diverse assistance systems (Gorecky et al. 2014). When looking at different user groups working in nowadays production, one realizes that there are many different types of operators. Although it looks like this in reality, in most of scientific literature, only one general type of employee or worker is mentioned and most of the papers do not address different types of user groups. The term "Operator 4.0" has been introduced by Romero to describe the change of the industrial worker in production and the "operator of the future", a smart and skilled worker, who undertakes work with the help of machines and digital/technical systems (Romero et al. 2016a). The term "Operator 4.0" announces the fourth generation of worker. These four different operator generations can be seen in Fig. 4.1.

As already mentioned, in literature, only few papers address specific user groups. Mostly, in order to deduce specifications for the aid system, a primary aim is addressed, such as to support a new employee (Hallewell et al. 2018). For the help of elderly and impaired operators, "gamification" can be used. Motivation mechanics, which originally come from

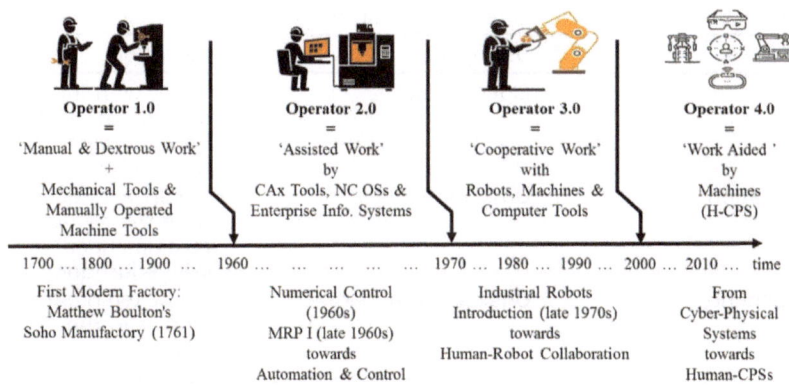

Fig. 4.1 The four operator generations (Romero et al. 2016a)

the game design, can be implemented with gamification to the industrial environment (Korn 2012; Hinrichsen et al. 2016). When looking at different types of user groups, the authors propose a distinction into nine groups (Mark et al. 2019a, 2020a). These different user groups of assistance systems in production can be seen in Table 4.1. The approach for classifying different workers with regard to their needs and limitations, due to, e.g. their origin or age, helps to adequately equip them with appropriate assistance systems.

4.2.3   Importance of Human–Machine Interaction in Production

HMI is nondescript and at the same time occurs rather naturally. Therefore, it is often forgotten how important a user-friendly interaction is. HMI occurs through human–machine interfaces, which can be described as the interface between machine and operator. It is the part of a machine used by people to interact and through which they can intervene. This ranges from simple everyday objects such as the steering wheel (to control a car on the street) to high-performance and complex systems (Juschkat 2019). Nonetheless, this always entails the risk that the complexity of HMI grows along with the performance and capability of industrial

Table 4.1 User groups of industrial assistance systems (Mark et al. 2019a)

Variable	User group	Description
Age	Elder worker	Worker with increasing age, which might have an impact on the task performance
Education	Unskilled worker	Worker, who does not have the required or recommended skills/education
Experience	Unexperienced worker	New or temporary worker in the company, department, or the specific workplace
Variety of work content	Flexible worker	Worker, who switches often between different types of work (or products) within a company (e.g. "Jolly")
Occupational Health and Safety (OHS)	Worker with safety risk	Worker with work conditions that might have an impact on the safety
	Worker with health risk	Worker with work conditions that might have an impact on the health and ergonomics
Handicap presence	Physically handicapped worker	Worker with physical disability that might have an impact on the task performance
	Mentally handicapped worker	Worker with mental disability that might have an impact on the task performance
Migration	Migrant worker	Worker who usually has a different background in terms of culture and language

machines. A noticeable change is the increasing technical and informative complexity and hence also the increasing specialist knowledge that the user needs to operate with an interface. It has become increasingly complex over the years, and in most of the cases, modern user interfaces can no longer be understood by non-specialists (Juschkat 2019).

The evolution of human–machine interfaces has passed many milestones over the years starting from the beginning of automation. These stages can be seen in Fig. 4.2. The fastest step could be noticed right after the first industrial revolution. In the course of the first industrial revolution, machines were brought on the market and hence in industrial production. Workers regulated machines directly through levers, cranks, or pedals. With the electricity, that was brought in the course of the second industrial revolution, employees started to control machines via switches to turn power on and off and lights for the purpose of signalization. This can be seen as the origin of modern human–machine interfaces, the so-called HMI 0.0 (Papcun et al. 2018). In the following and in Fig. 4.2, the different evolutionary steps towards modern human–machine interfaces are shown:

- *Human–Machine Interface 1.0.* The first level of human–machine interfaces starts with the third industrial revolution in the 1960s in which buttons, displays, and lights were used to facilitate the work for the employees and to interact with the machines in a better way.
- *Human–Machine Interface 2.0.* The second level started in the 1990s and uses desktop visualization and touch panels.

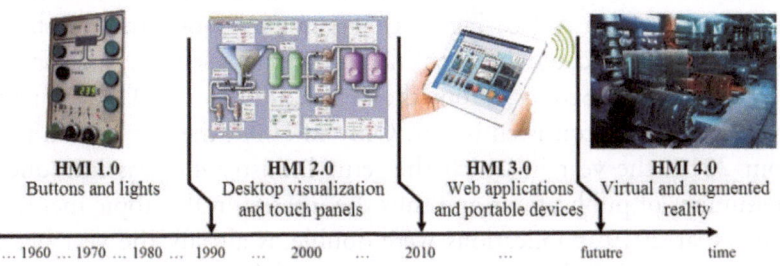

Fig. 4.2 Evolution of human–machine interfaces (Papcun et al. 2018)

- *Human–Machine Interface 3.0.* The third level works with wireless visualization boards and portable devices with diverse web applications.
- *Human–Machine Interface 4.0.* The fourth level of human–machine interfaces uses augmented and virtual reality on mobile terminals with a camera or corresponding glasses to expand the real world with additional elements.

It was already possible in early stages to trigger any effects with devices that were placed far away by just a few hand movements. Therefore, human–machine interfaces have already made a decisive contribution to machine safety. A human–machine interface consists of four components: (i) human (operator), (ii) display, (iii) input box, and (iv) machine. The user approaches the display with the input box. The action that had been conducted in the input box triggers certain actions on the machine. The purpose of today's interfaces is to show the reaction of the machine in its essential points on the display. This gives the human operator feedback on the effect of his action on the input box and not necessarily on the machine itself (Juschkat 2019).

4.2.4 Relevance of Assistance Systems in Literature

For an extended literature review, the database Scopus was used. A selection of relevant keywords was established with which the search was carried out. When looking at the scientific literature about assistance systems in production, it becomes evident that the number of publications increases (see Fig. 4.4). Figure 4.4 shows the number of publications that could be identified with the scientific citation and abstract database Scopus. From the years 2003 to 2010, the number of publications is consistently low. Only in 2008, there is a small peak. From 2011, the year in which the term Industry 4.0 was introduced, the number of publications and thus the interest in this topic increased. In the year 2019; publications were double as already the year before. This confirms that there is a growing interest in technologies that help the operator during production tasks. A total of 171 scientific publications could be identified between the years 2003 and 2019, out of which 87.7 per cent are conference articles, 10 per cent journal articles, and 2.3 per cent book chapters (see Fig. 4.3). Figure 4.5 illustrates the countries from which the research publications originate. Germany, the country in which the term "Industry 4.0" was born, is leading and claims 66.1 per cent of the publications, followed by Austria with 7.0 per cent and the USA with 5.3 per cent.

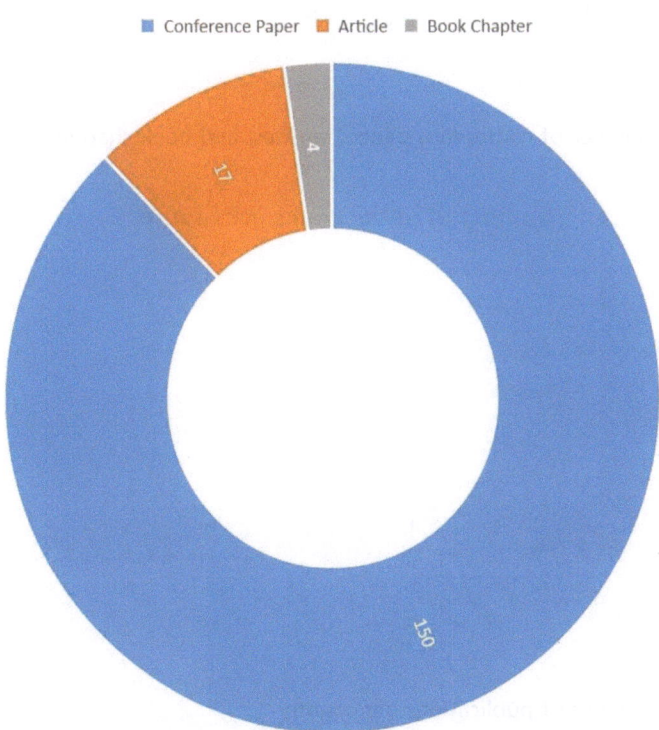

Number of conference papers, articles, and book chapters

Fig. 4.3 Number of publications per year

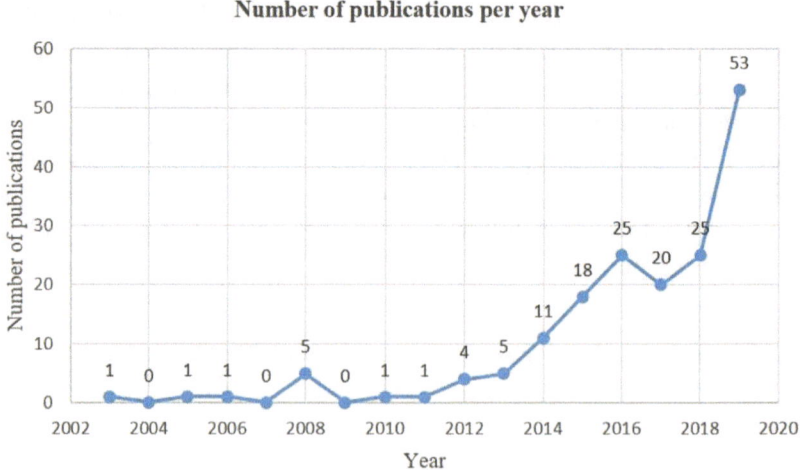

Fig. 4.4 Number of conference papers, articles, and book chapters

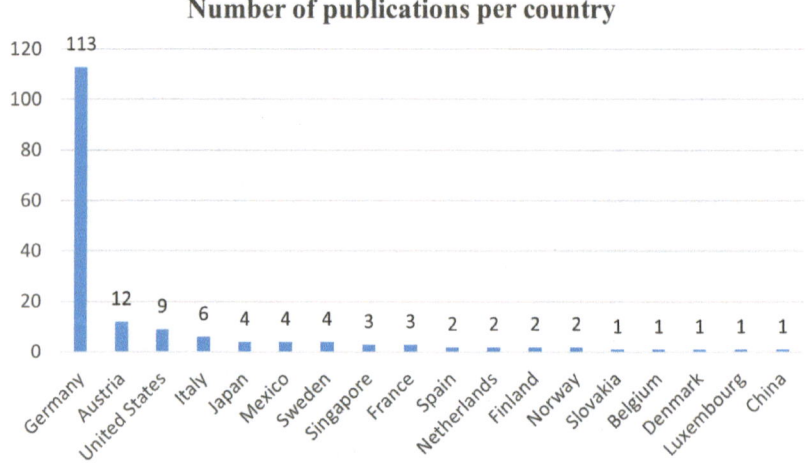

Fig. 4.5 Number of publications per country

4.3   Overview of Industrial Assistance Systems in Production

After the introduction and the presentation of the theoretical background in industrial assistance systems, we want to give a clear overview of categories of such aid systems and a presentation of the different systems as well as exemplary applications in industrial assembly. To present the three main categories of industrial assistance systems, we need to introduce the term "capability". The definition of the capability is the "measure of the ability of an entity (e.g. department organisation, person, system) to achieve its objectives, especially in relation to its overall mission" (Business Dictionary 2016). This means, related to humans/workers, that they have the ability and assets to apply the capabilities for a certain objective (Romero et al. 2016a). Based on this concept of capabilities, in literature, industrial assistance systems are subdivided into three categories: (i) sensorial, (ii) cognitive, and (iii) physical assistance systems based on their ability to extend the capabilities of a worker (see Fig. 4.6).

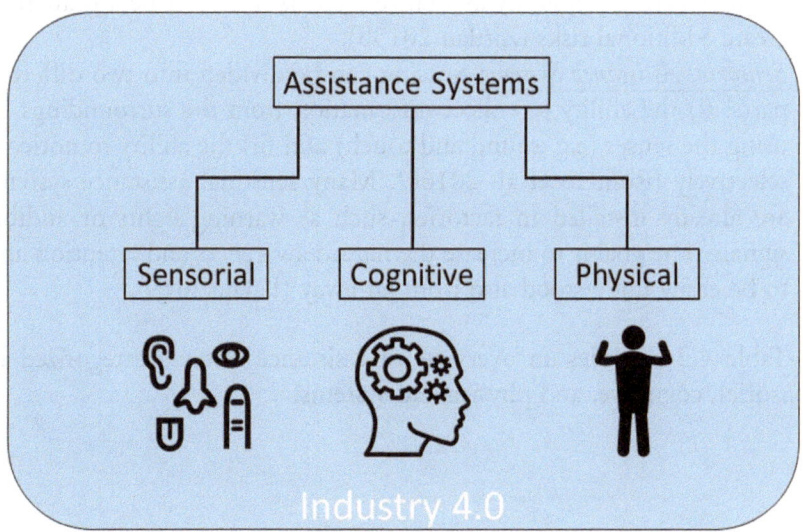

Fig. 4.6   Categorization of industrial assistance systems

- *Cognitive Assistance Systems* Cognitive assistance systems provide information and support learning in real time. They help the user with information processing and with the execution of his/her work. Examples of cognitive assistance system are smartphones, virtual reality, augmented reality, tablets, smartwatches, and wearables. Cognitively supportive assistance systems can provide information, generate guidelines for actions, steps, and processes, and provide rated feedback. Furthermore, they can collect data for use, e.g. work progress, working speed, and times (Zittlau 2015a).
- *Physical Assistance Systems* Physical assistance systems can support the physical performance of an operator. Often, exoskeletons are mentioned in this context. They can be worn as "robot suits" or support individual limbs, such as the arms or legs. Exoskeletons especially help with carrying, lifting, walking, and special systems even sitting without a chair. The aim of physical assistance systems is to combine the advantages of humans (e.g. flexibility) and technology (e.g. endurance), hence facilitating movements of the user and preventing health risks. It is important to adapt the assistance systems to the individual operator needs and the corresponding situation. Wearing physically assistance systems must not be a burden or create additional risks (Zittlau 2015b).
- *Sensorial Assistance Systems* Sensing can be divided into two different parts: (i) the ability to collect information from the surroundings by using the senses (e.g. sound and touch) and (ii) the ability to notice it selectively (Romero et al. 2016b). Many sensorial assistance systems are already installed in factories, such as warning lights or audible signals. The goal is to increase the hazard awareness and attention and to be easily understood also from far away (Layne 2019).

Table 4.2 provides an overview of assistance systems categorized in sensorial, cognitive, and physical aid systems.

Table 4.2 Overview of industrial assistance systems in production

Category	Assistance systems
Sensorial (extend sensing capabilities)	Eye Tracking
	Galvanic Skin Response (GSR)
	Physiological Sensor - Heart Rate (HR)
	Intelligent Hand Tracking
	Motion Tracking and Gesture Recognition Device
	Smartwatch
	Wearable Tracker
	Haptic Glove
	Infrared Camera
	Position Tracking System
Physical (extend physical capabilities)	Exoskeleton
	Arm Support
	Leg Support
	Back Support
	Flexible Assembly Assist Robot
	Robots/Automats
	Telemanipulator/ Balancer/Lifting Aid
	Wearable Lifting/Holding Aid
	Ergonomic Manual Workplaces
	Robot Assistance System with ToF Camera
	Collaborative Robot
Cognitive (extend cognitive capabilities like "orient" or "decide")	Augmented Reality (AR)
	Virtual Reality (VR)
	Mixed Reality (MR)
	Tablet
	Visual Computing System
	Projection-Based Assistance System
	Head-Mounted Display (HMD)
	Smart Scan Glove
	Smartphone
	Laser Projection System

(continued)

Table 4.2 (continued)

Category	Assistance systems
	Computer Assisted Instructions (CAI)
	Projector
	Monitor
	Pictorial Instruction
	Voice Control
	AI-Based Intelligent Personal Assistant

4.3.1 Sensorial Worker Assistance Systems

Figure 4.7 visualizes and summarizes possible sensorial assistance systems. These systems will be presented in detail in the following.

Eye Tracking: The eye tracking technology refers to the procedure of tracking the absolute Point of Gaze (POG) or the eye movements referring to the user's gaze point in the visual place (Mark et al. 2019c). It is useful in many different applications that range from medical and

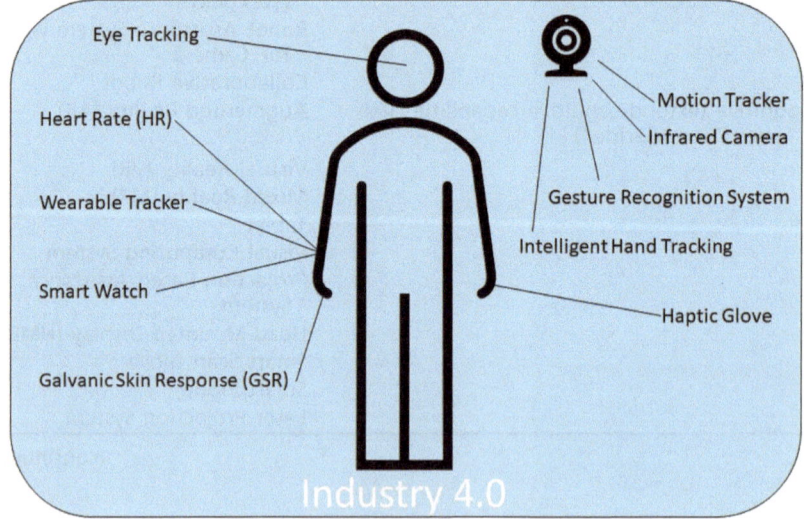

Fig. 4.7 Overview of sensorial worker assistance systems

psychological research diagnostics to gaze-controlled, interactive applications and usability studies (Majaranta and Bulling 2014). The variety and quantity of publications using eye tracking for their research purpose show that there has been a drastic increase in employing this technology (Reingold 2014). Lusic et al. use eye tracking to draw conclusions on the efficiency of the communication between the actual operator's behaviour, the object to be assembled, and the medium that provides information (Lusic et al. 2016).

Galvanic Skin Response (GSR): Galvanic skin response has been interesting for academic research since the 1900s. The human skin is an organ, which is entirely innervated by the nervous system (Gollan et al. 2018). The GSR is able to analyse and measure the skin's electrodermal activity (EDA) which embodies a reflection of synaptic excitation as the increased conductance of the skin appears in considerable correlations with neuronal actions (Frith and Allen 1983). Therefore, GSR can be used as an indicator of excitation increases with attention during the performance of a task (Kahneman 1973).

Physiological Sensor – Heart Rate (HR): The heart rate which is an example of cardiac functions stands for another fundamental indicator of excitation and arousal and hence of the activation of attention as a direction to changes in the nervous system (Graham 1992). Heart Rate Response (HRR), Heart Rate Variability (HRV), and T-Wave amplitude analysis represent the most expressive indicators of excitation (Suriya-Prakash et al. 2015). The mobile and stationary evaluation of cardiac data can be done both, with medical and customer products through diverse sensors (Gollan et al. 2018).

Intelligent Hand Tracking: An intelligent hand tracking system uses depth cameras in order to track the operator's hands and make him aware in case of wrong actions or errors during the assembly tasks of workers in manual assembly. The system can use different hand tracking algorithms (Büttner et al. 2017).

Motion Tracking and Gesture Recognition Device: Motion tracking or motion capturing is the process of collecting data of people's and object's movements. It is not only used in robotics and in the validation of computer vision, but also in military and entertainment (Noonan et al. 2009; Yamane and Hodgins 2009). A motion sensing and gesture recognition device is a technology that includes different hardware, such as infrared projectors, RGB cameras, detectors, and microphones. In literature, it is used for experiments about visual feedback that guides the worker from one assembly step to the next (Funk et al. 2018) and for tracking the assembly process (Gupta et al. 2012).

Smartwatch: A smartwatch is a wearable, small computer that can be worn on the wrist. It usually consists of a touchscreen interface while it is connected to the smartphone app that provides telemetry and management, such as biomonitoring. In industrial literature, it is used to include it in daily production and support the worker with gesture and voice control (Müller et al. 2018b). In addition, it can support step by step instructions of assembly tasks (Aehnelt et al. 2014).

Wearable Tracker: Wearable tracker is the generic term for devices that measure and record stress, Global Positioning System (GPS) location, activity, heart rate, and additional health-related metrics as well as other data, such as biometrics. Nowadays, there is a large selection of such systems. Currently, there are first steps in tracking the complex human brain during tasks, which brings it to the next level. This might take some time to be implemented also in industry but gives an idea of what is already possible (Romero et al. 2016b).

Haptic Glove: A haptic glove can be a kind of glove that makes the operator get in touch with a computer device through haptic technology. It can provide tactile feedback of virtually based objects. When it is activated, it gives feedback about the sensation in terms of the sense of touch of having a virtual object in the hands. In industrial experiments, it is also a glove equipped with diverse sensors to get feedback from real objects in the operator's hand (Otten et al. 2016).

Infrared Camera: Infrared camera, or also thermographic camera, is a device, which can create images by infrared radiations. In literature, operators can express their order and intention by gestures and a system recognizes them via infrared cameras (Oka et al. 2002). An enhanced work desk in production might therefore consist of a desk, projector, infrared camera, and a display (Sugi et al. 2005).

Position Tracking System: Position tracking systems track the position of body parts or systems that are worn by the operator, e.g. a head-mounted display. The definition of the precise orientation and position of an object can be done by special markers or trackers (Müller et al. 2018a).

4.3.2 Physical Worker Assistance Systems

Figure 4.8 visualizes and summarizes possible physical assistance systems. These systems will be presented in detail in the following.

Exoskeleton: An exoskeleton is a wearable device that is placed on the operator's body and restores, reinforces, or augments the performance. They can be made out of different materials and optionally being equipped with actuators and sensors. Even though they can be seen as universal technical concept, they must be individually adapted to the activity and operator (Weidner et al. 2018).

Arm/Leg/Back Support: There are different types of exoskeleton support systems. According to the individual desire, there also exist arm, leg, back, and overall supports, which are based on soft controls, flexible structures, and textile components (Otten et al. 2016).

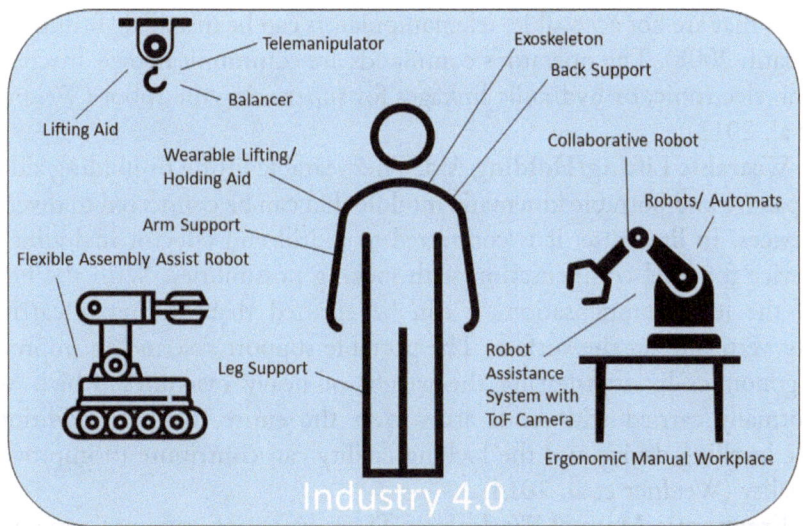

Fig. 4.8 Overview of physical worker assistance systems

Flexible Assembly Assist Robot: Flexible assembly assist robots can have high potential for assembly and manufacturing. They can improve the cooperation between machine and human (and thus the human-machine interaction) when considering the safety issues and workplace conditions (Drust et al. 2013).

Human–Robot (machine) Collaboration: Human-machine collaboration (HRC) is the general term for a model in which the operator works together with machines, robots, or other (intelligent) systems and not only using them as an instrument. The goal of such a relationship is to help each other with diverse abilities and to use strengths to fill the other's weaknesses (Rouse 2017).

Robots/Automats: Technical systems for robotic systems and automated solutions for industrial use are free programmable and standardized robots, with parallel and/or serial chains. They can be used together with other robots (or humans), automated machine tools, and alone (Weidner et al. 2013).

Telemanipulator/Balancer/Lifting Aid: Balancers are used to transport an assembly part or workpiece from one workplace to the other. Different from industrial robots, the motion that is given to a balancer is initiated by the operator himself. In dangerous working environments or areas that are not accessible, telemanipulators can be installed (Bruno and Khatib 2008). The operator's commands are communicated by mechanical, electronic, or hydraulic linkages for supervising the robot (Weidner et al. 2013).

Wearable Lifting/Holding Aid: The wearable lifting or holding aid is a passive and portable kinematic module that can be connected to diverse devices. In literature, it is connected to a drill end effector including a device for level compensation with locking possibilities. With the help of the level compensation, it can be ensured that drilling is carried out vertically on the surface. The portable support system can improve ergonomics by transferring the weight of heavy machines, which are normally carried with both arms, over the entire torso. In addition, the levelling device and the locking facility can contribute to improved quality (Weidner et al. 2014).

Ergonomic Manual Workplace: The purpose of ergonomic manual workplaces is that the operator can perform all tasks and gets support by pneumatic, electrical, and mechanical tools, such as a screwdriver. The design of an ergonomic workplace seeks to optimize it regarding work-flow and organization (Buch et al. 2008) and also to make the design age-differentiated, e.g. with touchscreens (Vetter et al. 2010).

Robot Assistance System with Time-of-flight (ToF) Camera: The technology constitutes the basis for collision-free collaboration between industrial robots and humans. With the help of the real-time depth infor-mation, dynamic and static objects are detected which is necessary to install industrial robots as assistants of the operator (Ramer and Franke 2014).

Collaborative Robot: Collaborative robots, also Cobots, are indus-trial robots that can perform a variety of non-ergonomic and repetitive tasks. They are designed for collaborating directly with the operator while ensuring safety regarding collision and force sensing (Romero et al. 2016b; Gualtieri et al 2019).

4.3.3 Cognitive Worker Assistance Systems

Figure 4.9 visualizes and summarizes possible cognitive assistance systems. These systems will be presented in detail in the following.

Augmented Reality (AR): Augmented reality refers to any computer-aided expansion of the real world. A use case of the augmented reality technology is supporting operators in search activities by leading their attention towards the important targets (Renner et al. 2018).

Virtual Reality (VR): Virtual reality is a simulated world, which can be totally different from the normal world. This technology can be of great benefit for applications in industry by creating simulations of main-tenance tasks, design review, and prototyping. It is used for the training of workers of different equipment, which might reduce problems, risks, and expenses (Wolfartsberger et al. 2018).

Mixed Reality (MR): In mixed reality, the virtual and real world are merged to create new visualizations and environments. The digital and physical objects interact and co-exist in real time. Regarding mixed reality, experiments are conducted to provide virtual support to the worker in form of projected instructions onto a workplace environment.

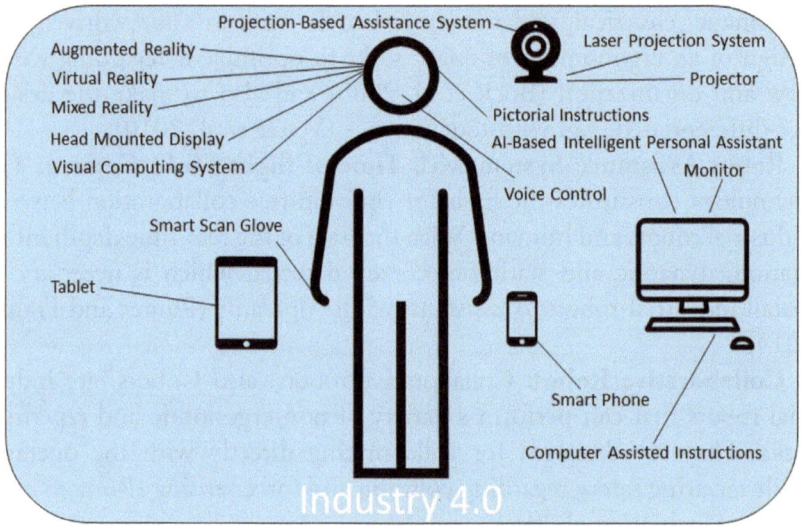

Fig. 4.9 Overview of cognitive worker assistance systems

This could be in form of a mounted projector over the assembly station (Rodriguez et al. 2015).

Tablet: A tablet is a portable, thin, small, and lightweight computer with a touch screen. It has speakers, microphone, and a mechanical or virtual keyboard. In literature, it is used as assistance system at workplaces in assembly lines to show key steps in assembly production (Hallewell Haslwanter and Blazevski 2018).

Visual Computing System: Visual computing is a term for computer science disciplines which deal with 3D models and image information. According to Posada et al. (2015), visual computing is a key technology within Industry 4.0 and contains technologies such as virtual and augmented reality, visual analytics, HMI interfaces, and collaborative robotics interaction (Segura et al. 2018).

Projection-Based Assistance System: A projection-based assistance system is a system, which projects information, e.g. instructions, directly on the operator's field of vision on the workplace (Mark et al. 2020b). They can be flexibly designed in order to be moved to any workplace (Hinrichsen et al. 2018).

Head-Mounted Display (HMD): A head-mounted display is a device equipped with a display that can be separately worn on the head or be part of a helmet. In most applications mentioned in literature, HMD uses augmented reality, which can be used for training applications and industrial education (Besbes et al. 2012).

Smart Scan Glove: The smart scan glove is a small and light barcode scanner that can be connected to information within the industrial Internet of things. Experiments showed that this type of barcode scanner is better compared to the traditional handheld variant in terms of user acceptance, physical fatigue, and support to humans (Scheuermann et al. 2016).

Smartphone: A smartphone is a device that can be carried by the operator during daily work. It can be connected to the company's internal network. Speaking about Industry 4.0, wearables have great potential although they are still in their infancy. For instance, in literature, it is used to share expert knowledge via push notifications (Scheuermann et al. 2015).

Laser Projection System: Laser projection systems are visual assistance systems that can ensure the quality and efficiency during industrial production and assembly processes. They are used to generate laser lines that can display outlines on objects and surfaces. The shapes can be created from Computer-Aided Design (CAD) files (Müller et al. 2016).

Computer Assisted Instructions (CAI): Computer assisted instructions are usually used in complex tasks that might have a complex set of assembly instructions. The worker has here the possibility to select the appropriate instruction manual online when needed (Tang et al. 2003).

Projector: Projectors are often used in combination with monitors and diverse cameras in order to project instructions or assistance on the workplace of the operator. The price of this technology has been sinking over the last years (Hinrichsen et al. 2018) which makes it more attractive to the companies and researchers.

Monitor: A monitor, similar to the tablet, can be mounted on a workstation and combined with other technologies such as a camera, laser projector, and tracking system (Müller et al. 2018a).

Pictorial Instructions: Pictorial instructions communicate information in form of drawings whereat text is only used occasionally to clarify the information. Pictures are usually easier to remember and understand. Unlike reading, researchers think that humans learn through exposure and experience how to follow and understand pictorial instructions (Paul 2012). In industrial production, pictorial instructions can be put above boxes with assembly pieces to show how to perform the assembly (Funk et al. 2015).

Voice Control: A voice control system can be combined with collaborative robots in order to work hands-free and enhance the human–machine interaction. It can raise the efficiency, quality, and automation of the process. The implementation of a voice client enables the interaction between operator and technical system, e.g. projection system, and makes it possible to adapt the voice control specifically to the industrial working environment (Müller et al. 2018b).

AI-Based Intelligent Personal Assistant: An intelligent personal assistant is artificial intelligence or a software agent that can assist the worker in interfacing with databases, computers, machines, and other information systems. A main feature is the voice interaction to the operator, which enhances operational efficiency and productivity (Romero et al. 2016b).

4.4   Discussion of Risks, Challenges, and Potential

Already in the year 2014 Creighton writes in the EU Factpack that "Europe is going grey" (Creighton 2014). While at the moment mostly Western and Northern countries in Europe have the oldest populations, according to the EU Factpack this will change by the year 2060 and the countries in the East will have the highest number of inhabitants older than 65 (Creighton 2014). This shows that the ageing of society in Europe and thus also of workers in industrial production will be a major challenge for manufacturing companies in the future. According to Thun et al., this phenomenon was already observable in the year 2007, and the process of aging workforce will accelerate (Thun et al. 2007).

Assistance systems could therefore counteract this process and support older workers in production. Another aspect is the user-friendliness of assistance systems. An assistance system, to be accepted by employees, must have a practical and supportive effect on the worker and not hinder him/her in any way in daily work. This aspect plays a leading role in this respect. Therefore, when designing and developing these systems, not only the technical aspect and benefits must be considered, but also the relevance and applicability to the worker himself.

One point that is often forgotten is the legal regulation. It has already been mentioned that due to demographic change, a shortage of workers will also occur in manufacturing companies. In this respect, the firms should also be interested in the inclusion of workers with disabilities into the daily work routine. Meanwhile, there is a wide range of different assistance systems that can not only support the worker himself, but also monitor and control the work steps. In literature, scientists are also increasingly concerned about the integration of people with disabilities into everyday working life through suitable assistance systems. Often, however, the practical implementation is difficult due to legal regulations (Mark et al. 2019b).

Due to the increasing complexity of machines and systems within the company, also the human–machine interfaces must be continuously improved. Research is currently developing a way to make HMI safer and more durable with the help of contactless input commands. There are three possible ways of contactless input: (i) input via facial expressions, (ii) input via gestures, and (iii) input via languages (Juschkat 2019). From a research perspective, much progress remains to be made in this area. An example of a HMI discussed, are headsets that measure brain waves (Beigh and Beigh 2018) and could thus in the future directly convert the thoughts of the user into interactions of a machine—e.g. to control robotic arms. Today, this technology is primarily intended for people with robotic prostheses or paraplegics. However, the new technology could also be used for machine control (Juschkat 2019). In addition, technical opportunities, such as machine learning, artificial intelligence, and automation, can play an important role in the future of work (Wang and Siau 2019).

When looking at the analysed publications, it can be denoted that the area of assistance systems in production is growing due to the increasing number of literature and industrial applications. Many technologies are getting cheaper over the years, which makes them more affordable not only for research centres of universities and large companies but also for research departments of small and medium-sized enterprises (SME).

In conclusion, the following main risks/challenges and potential in the context of worker assistance systems can be pointed out:

Risks/challenges:

- Demographic change,
- User-friendliness of assistance systems,
- Legal foundations.

Potential:

- Inclusion of workers with disabilities,
- Advanced human–machine interfaces,
- Affordability of systems.

4.5 Conclusion

This chapter examines the opportunity and potential of assistance systems in industrial production to enhance the operator's capabilities and human-machine interaction. The introduction of assistance systems in assembly changes the daily production comprehensively. On the one hand, such systems can give benefit in form of increased operator's capabilities (upgrading) and on the other hand support the operators with mental or physical limitations (compensation). Having in mind that a typical worker of a factory can be classified into different kinds of user groups, with diverse capabilities and limitations, makes it easier to find adequate support. The list of available assistance systems together with an explanation helps the reader to consider each system and estimate their individual potential. The important thing is that these systems provide the employee significant form of support and do not cause any additional stress or work. This is the only way to ensure that it can be used in practice sustainably. In addition, the increasing importance of human–machine interface is pointed out. This is mainly due to the high complexity of the systems and machines with which the worker has to work in the company today. In the past, many important technologies have been developed, but most of them only improved the productivity of the company and were generally always company oriented. This is different with assistance systems. They follow the approach of putting the worker in the centre of attention and to improve the company as a whole by equipping the operator with adequate support.

Acknowledgements This project has received funding from the European Union's Horizon 2020 research and innovation programme under the Marie Skłodowska-Curie grant agreement No 734713.

References

Aehnelt, M., and B. Urban. 2014. Follow-me: Smartwatch assistance on the shop floor. In Lecture Notes in Computer Science (including subseries Lecture Notes in Artificial Intelligence and Lecture Notes in Bioinformatics): 279–287. https://doi.org/10.1007/978-3-319-07293-7_27.

Arnold, D., and K. Furmans. 2019. *Materialfluss in Logistiksystemen*. Berlin: Springer. https://doi.org/10.1007/978-3-642-01405-5.

Bächler, A., L. Bächler, S. Autenrieth, H. Behrendt, M. Funk, G. Krüll, T. Hörz, T. Heidenreich, C. Misselhorn, and A. Schmidt. 2018. Systeme zur Assistenz und Effizienzsteigerung in manuellenProduktionsprozessen der Industrie auf Basis von Projektion und Tiefendatenerkennung. In: Wischmann S., and E. Hartmann (eds) Zukunft der Arbeit – EinepraxisnaheBetrachtung. Springer Vieweg, Berlin, Heidelberg. https://doi.org/10.1007/978-3-662-49266-6_3.

Beigh, N.T., and F.T. Beigh. 2018. A review on brain wave signal appliance control. International Journal for Research in Applied Science and Engineering Technology 6. https://doi.org/10.22214/ijraset.2018.4403.

Besbes, B., S.N. Collette, M. Tamaazousti, S. Bourgeois, and V. Gay-Bellile. 2012. An interactive augmented reality system: A prototype for industrial maintenance training applications. In IEEE International Symposium on Mixed and Augmented Reality (ISMAR): 269–270. https://doi.org/10.1109/ISMAR.2012.6402568.

Böhle, F. 2005. Erfahrungswissen – die verborgenenSeitenprofessionellenHandelns. EineHerausforderungfür die beruflicheBildung. Bonn: Fachtagung des BundesinstitutsfürBerufsbildung.

Bruno, S., and O. Khatib. 2008. Springer handbook of robotics. Berlin: Business Media.

Buch, M., J. Weichel, and E. Frieling. 2008. Analyse und Gestaltung von Montagearbeitsplätzen in der Automobilindustrie - EinBeitrag zur GenerierungaltersgerechterArbeitssysteme. GFA (Hrsg.) Produkt- und Produktions-Ergonomie—AufgabefürEntwickler und Planer. GFA Press: 411–414.

Büttner, S., O. Sand, and C. Röcker. 2017. Exploring design opportunities for intelligent worker assistance: a new approach using projetion-based AR and a novel hand-tracking algorithm. In Lecture Notes in Computer Science (including subseries Lecture Notes in Artificial Intelligence and Lecture Notes in Bioinformatics): 33–45. https://doi.org/10.1007/978-3-319-569 97-0_3.

Capability (General) business dictionary. 2016. https://www.businessdiction ary.com.

Creighton, H. 2014. Europe's ageing demography. ILC-UK2014 EU Factpack. www.ilcuk.org.uk. Published in November 2014. Registered Charity Number: 1080496.

Deutscher Bundestag. 2016. WissenschaftlicheDienste: AktuellerBegriffIndustrie 4.0. Available online: https://www.bundestag.de/resource/blob/474528/cae2bfac57f1bf797c8a6e13394b5e70/industrie-4-0-data.pdf. Nr. 23/16 (article from September 26, 2016). Accessed on June 22, 2020.

Drust, M., T. Dietz, A. Pott, and A. Verl. 2013. Production assistants: The rob@work family. In 44th International Symposium on Robotics (ISR): 6695746. https://doi.org/10.1109/ISR.2013.6695746.

Frith, C.D., and H.A. Allen. 1983. The skin conductance orienting response as an index of attention. Biological Psychology 17 (1): 27–39. https://doi.org/10.1016/0301-0511(83)90064-9.

Funk, M., L. Lischke, S. Mayer, A.S. Shirazi, and A. Schmidt. 2018. Teach me how! Interactive assembly instructions using demonstration and in-situ projection. *Cognitive Science and Technology* 49–73. https://doi.org/10. 1007/978-981-10-6404-3_4.

Funk, M., S. Mayer, and A. Schmidt. 2015. Using in-situ projection to support cognitively impaired workers at the workplace. In Proceedings of the 17th International ACMSIGACCESS Conference on Computers and Accessibility: 185–192. https://doi.org/10.1145/2700648.2809853.

Gollan, B., M. Haslgruebler, A. Ferscha, and J. Heftberger. 2018. Making sense: Experiences with multi-sensor fusion in industrial assistance systems. In Proceedings of the 5th International Conference on Physiological Computing Systems: 64–74. https://doi.org/10.5220/0007227600640074.

Gorecky, D., M. Schmitt, M. Loskyll, and D. Zuhlke. 2014. Human-machine-interaction in the industry 4.0 era. In Proceedings of the 12th IEEE International Conference on Industrial Informatics: 289–294. https://doi. org/10.1109/INDIN.2014.6945523.

Graham, F.K. 1992. Attention: The heartbeat, the blink, and the brain. Attention and information processing in infants and adults: Perspectives from human and animal research. In B.A. Campbell, H. Hayne, and R. Richardson (Eds.), Attention and information processing in infants and adults: Perspectives from human and animal research: 3–29.

Gualtieri, L., E. Rauch, R. Vidoni, and D.T. Matt. 2019. An evaluation methodology for the conversion of manual assembly systems into human-robot collaborative workcells. *Procedia Manufacturing* 38: 358–366. https:// doi.org/10.1016/j.promfg.2020.01.046.

Gualtieri, L., I. Palomba, F.A. Merati, E. Rauch, and R. Vidoni. 2020. Design of human-centered collaborative assembly workstations for the improvement of operators' physical ergonomics and production efficiency: A case study. *Sustainability* 12 (9): 3606. https://doi.org/10.3390/su12093606.

Gupta, A., D. Fox, B. Curless, and M. Cohan. 2012. DuploTrack: A real-time system for authoring and guiding duplo block assembly. In Proceedings of the 25th Annual ACM Symposium on User Interface Software and Technology: 389–401. https://doi.org/10.1145/2380116.2380167.

Hallewell Haslwanter, J.D., and B. Blazevski. 2018. Experiences with an assistive system for manual assembly. *ACM International Conference Proceeding Series* 46–49. https://doi.org/10.1145/3197768.3203173.

Hinrichsen, S., D. Riediger, and A. Unrau. 2016. Assistance systems in manual assembly. Production Engineering and Management. In Proceedings 6th International Conference. Publication Series in Direct Digital Manufacturing, eds. F. J. Villmer, E. Padoano: Volume: 6

Hinrichsen, S., D. Riediger, and A. Unrau. 2018. Development of a projection-based assistance system for maintaining injection molding tools. *IEEE International Conference on Industrial Engineering and Engineering Management* 1571–1575. https://doi.org/10.1109/IEEM.2017.8290157.

Juschkat, K. 2019. ElektrotechnikAutomatisierung. Was it human machine interface? Definition, Geschichte & Beispiele. Available online: https://www.elektrotechnik.vogel.de/was-ist-human-machine-interface-definition-geschichte-beispiele-a-718202/ (article from April 10, 2019). Accessed on June 22, 2020.

Kahneman, D. 1973. *Attention and effort*, vol. 1063. NJ: Prentice-Hall Enlegwood Cliffs.

Korn, O. 2012. Industrial playgrounds: How gamification helps to enrich work for elderly or impaired persons in production. In Proceedings of the 2012 ACMSIGCHI Symposium on Engineering Interactive Computing Systems: 313–316. https://doi.org/10.1145/2305484.2305539.

Layne, E. 2019. Small business: How are pictographs used in a workplace? Available online: https://smallbusiness.chron.com/pictographs-used-workplace-38856.html. Accessed on 15 September 2019.

Lusic, M., C. Fischer, K.S. Braz, M. Alam, R. Hornfeck, and J. Franke. 2016. Static versus dynamic provision of worker information in manual assembly: A comparative study using eye tracking to investigate the impact on productivity and added value based on industrial case examples. *ProcediaCIRP* 57: 504–509. https://doi.org/10.1016/j.procir.2016.11.087.

Majaranta, P., and A. Bulling. 2014. Chapter 3: Eye tracking and eye-based human-computer interaction. In Advances in Physiological Computing. Human-Computer Interaction Series (S.H. Fairclough, K. Gilleade). Springer-Verlag London. https://doi.org/10.1007/978-1-4471-6392-3_3.

Mark, B.M., L. Gualtieri, E. Rauch, R. Rojas, D. Buakum, and D.T. Matt. 2019a. Analysis of user groups for assistance systems in production 4.0. In International Conference on Industrial Engineering and Engineering Management (IEEM). https://doi.org/10.1109/IEEM44572.2019.8978907.

Mark, B.G., S. Hofmayer, E. Rauch, and D.T. Matt. 2019b. Inclusion of workers with disabilities in production 4.0: Legal foundations in Europe and potentials through worker assistance systems. Sustainability (Switzerland) 11(21): 5978. https://doi.org/10.3390/su11215978.

Mark, B. G., E. Rauch, Y. Borgianni, and D.T. Matt. 2019c. Eye tracking in der produktion 4.0: Eye tracking alsnützliche Technologie zur Optimierung der Produktionsprozesse im Zeitalter von Industrie 4.0. ZWF Zeitschrift fürwirtschaftlichenFabrikbetrieb, 114(1–2): 72–75. https://doi.org/10.3139/104.112032.

Mark, B.G., L. Gualtieri, M. De Marchi, E. Rauch, and D.T. Matt. 2020a. Function-based mapping of industrial assistance systems to user groups in production. (Accepted at CIRPe 2020 – 8th CIRP Global Web Conference – Flexible Mass Customisation).

Mark, B. G., E. Rauch, and D.T. Matt. 2020b. Study of the impact of projection-based assistance systems for improving the learning curve in assembly processes. In ProcediaCIRP, 88: 98–103. https://doi.org/10.1016/j.procir.2020.05.018.

Matt, D.T. 2007. Reducing the structural complexity of growing organizational systems by means of axiomatic designed networks of core competence cells. Journal of Manufacturing Systems, 26(3–4): 178–187. https://doi.org/10.1016/j.jmsy.2008.02.001.

Matt, D.T. 2009. Design of lean manufacturing support systems in make-to-order production. *Key Engineering Materials* 410: 151–158. https://doi.org/10.4028/www.scientific.net/KEM.410-411.151.

Müller, R., M. Vette, M. Scholer, and J. Ball. 2016. Assembly assistance and position data feedback by means of projection lasers. SAE Technical Papers.

Mueller, R., M. Vette-Steinkamp, L. Hoerauf, C. Speicher, and A. Bashir. 2018. Intelligent and flexible worker assistance systems assembly assistance platform for planning assisted assembly and rework as well as execution of a worker-centered assistance. *VISIGRAPP* 2018: 77–85. https://doi.org/10.5220/0006613900770085.

Müller, R., M. Vette-Steinkamp, L. Hörauf, C. Speicher, and A. Bashir. 2018a. Worker centered cognitive assistance for dynamically created repairing jobs in rework area. *ProcediaCIRP* 72: 141–146. https://doi.org/10.1016/j.procir.2018.03.137.

Müller, R., R. Müller-Polyzou, L. Hörauf, A. Bashir, M. Karkowski, D. Vesper, and S. Gärtner. 2018b. Intuitive control of laser based assembly assistance [Intuitive BedienunglaserbasierterMontageassistenz]. ZWF Zeitschrift fuer-WirtschaftlichenFabrikbetrieb, 113(6): 363–368. https://doi.org/ https://doi.org/10.3139/104.111922.

Noonan, D., P. Mountney, D. Elson, A. Darzi, and G.Z. Yang. 2009. A stereoscopic fibroscope for camera motion and 3D depth recovery during minimally invasive surgery. *ProcediaICRA* 2009: 4463–4468. https://doi.org/10.1109/ROBOT.2009.5152698.

Oka, K., Y. Sato, and H. Koike. 2002. Real-time fingertip tracking and gesture recognition. In IEEE Computer Graphics and Applications, 22(6): 64–71. https://doi.org/10.1109/MCG.2002.1046630.

Otten, B., P. Stelzer, R. Weidner, A. Argubi-Wollesen, and J.P. Wulfsberg. 2016. A novel concept for wearable, modular and soft support systems used in industrial environments. *Proceedings of the Annual Hawaii International Conference on System Sciences* 542–550. https://doi.org/10.1109/HICSS.2016.74.

Papcun, P., E. Kajati, and J. Koziorek. 2018. Human machine interface in concept of industry 4.0. In World Symposium on Digital Intelligence for Systems and Machines (DISA). https://doi.org/10.1109/DISA.2018.8490603.

Paul, Y., Techwhirl. Pictorial instructions—What are they good for? Available online: https://techwhirl.com/pictorial-instructions-what-good/ (Article form December 5, 2012). Accessed on June 30, 2020.

Posada, J., C. Toro, I. Barandiaran, D. Oyarzun, D. Stricker, R. De Amicis, E.B. Pinto, P. Eisert, J. Dollner, and I. Vallarino. 2015. Visual computing as a key enabling technology for industries 4.0 and industrial internet. In IEEE Computer Graphics and Applications 35 (2): 26–40. https://doi.org/10.1109/MCG.2015.45.

Ramer, C., and J. Franke. 2014. Work space surveillance of a robot assistance system using a ToF camera. *Advanced Materials Research* 907: 291–298. https://doi.org/10.4028/www.scientific.net/AMR.907.291.

Rauch, E., P. Dallasega, and D.T. Matt. 2017. Critical factors for introducing lean product development to small and medium sized enterprises in Italy. *ProcediaCIRP* 60: 362–367. https://doi.org/10.1016/j.procir.2017.01.031.

Reingold, E. 2014. Eye tracking research and technology: Towards objective measurement of data quality. *Visual Cognition* 22 (3): 635–652. https://doi.org/10.1080/13506285.2013.876481.

Reinhart, G., and M. Zäh. 2014. *Assistenzsysteme in Der Produktion. WtWerkstattstechnik-Online* 104 (9): 516.

Renner, P., J. Blattgerste, and T. Pfeiffer. 2018. A path-based attention guiding technique for assembly environments with target occlusions. In IEEE Conference on Virtual Reality and 3D User Interfaces (VR) – Proceedings, 8446127: 671–672. https://doi.org/10.1109/VR.2018.8446127.

Rodriguez, L., F. Quint, D. Gorecky, D. Romero, and H.R. Siller. 2015. Developing a mixed reality assistance system based on projection mapping technology for manual operations at assembly workstations. *Procedia Computer Science* 75: 327–333. https://doi.org/10.1016/j.procs.2015.12.254.

Romero, D., P. Bernus, O. Noran, J. Stahre, and A. Fast-Berglund. 2016a. The operator 4.0: Human cyber-physical systems & adaptive automation towards human-automation symbiosis work systems. In IFIP International Conference on Advances in Production Management Systems: 677–686. https://doi.org/10.1007/978-3-319-51133-7_80.

Romero, D., J. Stahre, T. Wüst, O. Noran, P. Bernus, A. Fast-Berglund, and D. Gorecky. 2016b. Towards an operator 4.0 typology: A human-centric perspective on the fourth industrial revolution technologies. In 46th International Conferences on Computers and Industrial Engineering: 1–11.

Rothenberger, R. IT Production online – Das Industrie 4.0-MagazinfürerfolgreicheProduktion. DigitaleAssistenzsysteme in der Produktion. Available online: https://www.it-production.com/fertigung snahe-it/unterstuetzung-fuer-werker/#:~:text=Digitale%20Assistenzsyst eme%20sind%20ein%20wichtiger,Produktivit%C3%A4t%20und%20W ettbewerbsf%C3%A4higkeit%20eines%20Unternehmens. (Article from January 23, 2020). Accessed on June 23, 2020.

Rouse, M., Definition: Machine-human collaboration. Available online: https://whatis.techtarget.com /definition/machine-human-collaboration#. (Article from January, 2017). Accessed on June 29, 2020.

Scheuermann, C., B. Bruegge, J. Folmer, and S. Verclas. 2015. Incident localization and assistance system: A case study of a cyber-physical human system. *IEEE/CIC International Conference on Communications in China - Workshops, CIC/ICCC* 57–61. https://doi.org/10.1109/ICCChinaW.2015.7961580.

Scheuermann, C., M. Strobel, B. Bruegge, and S. Verclas. 2016. Increasing the support to humans in factory environments using a smart glove: An evaluation. In IEEE Conferences on Ubiquitous Intelligence & Computing, Advanced and Trusted Computing, Scalable Computing and Communications, Cloud and Big Data Computing, Internet of People, and Smart World Congress: 847–854. https://doi.org/ https://doi.org/10.1109/UIC-ATC-ScalCom-CBDCom-IoP-SmartWorld.2016.0134.

Segura, Á., H.V. Diez, I. Barandiaran, A. Arbelaiz, H. Álvarez, B. Simoes, J. Posada, A. García-Alonso, and R. Ugarte. 2018. Visual computing technologies to support the Operator 4.0. In Computers and Industrial Engineering. https://doi.org/10.1016/j.cie.2018.11.060.

Sugi, M., M. Nikaido, Y. Tamura, J. Ota, T. Arai, K. Kotani, K. Takamasu, S. Shin, H. Suzuki, and Y. Sato. 2005. Motion control of self-moving trays for human supporting production cell "attentive workbench". *Proceedings - IEEE International Conference on Robotics and Automation* 4080–4085. https://doi.org/10.1109/ROBOT.2005.1570746.

Suriya-Prakash, M., G. John-Preetham, and R. Sharma. 2015. Is heart rate variability related to cognitive performance in visuospatial working memory. *Psychology.* https://doi.org/10.7287/PEERJ.PREPRINTS.1377V1.

Tang, A., C. Owen, F. Biocca, and W. Mou. 2003. Comparative effectiveness of augmented reality in object assembly. *Conference on Human Factors in Computing Systems* 73–80. https://doi.org/10.1145/642611.642626.

Thun, J.H., A. Größler, and S. Miczka. 2007. The impact of the demographic transition on manufacturing: Effects of an ageing workforce in German industrial firms. *Journal of Manufacturing Technology Management.* https://doi.org/10.1108/17410380710828299.

Vetter, S., J. Bützler, and N. Jochems. 2010. Ergonomic workplace design for the elderly: Empirical analysis and biomechanical simulation of information input on large touch screens. In W. Karwowski, G. Salvendy (Hrsg.) Conference proceedings of the 3rd international conference on applied human factors and ergonomics (AHFE): 1–8. ISBN: 0–9796435–4–6, 978–0–9796435–4–5.

Wang, W., and K. Siau. 2019. Artificial intelligence, machine learning, automation, robotics, future of work and future of humanity: A review and research agenda. Journal of Database Management (JDM). https://doi.org/10.4018/JDM.2019010104.

Weidner, R., N. Kong, and J.P. Wulfsberg. 2013. Human hybrid robot: A new concept for supporting manual assembly tasks. *Production Engineering* 7 (6): 675–684. https://doi.org/10.1007/s11740-013-0487-x.

Weidner, R., T. Redlich, and J.P. Wulfsberg. 2014. Passive and active support systems for production processes [Passive und aktiveUnterstützungssystemefür die Produktion]. *WTWerkstattstechnik* 104 (9): 561–566.

Weidner, R., T. Redlich, and J.P. Wulfsberg. 2015. *TechnischeUnterstützungssysteme.* Berlin: Springer.

Weidner, R., B. Otten, A. Argubi-Wollesen, and Z. Yao. 2018. Support technologies for industrial production. *Biosystems and Biorobotics* 23: 149–156. https://doi.org/10.1007/978-3-030-01836-8_14.

Wolfartsberger, J., R. Lindorfer, J.D. Hallewell Haslwanter, M. Jungwirth, R. Froschauer, and D. Wahlmüller. 2018. Industrial perspectives on assistive systems for manual assembly tasks. *ACM International Conference Proceeding Series* 289–291. https://doi.org/10.1145/3197768.3201552.

Yamane, K., and J.K. Hodgins. 2009. Simultaneous tracking and balancing of humanoid robots for imitating human motion capture data. In Intelligent Robots and Systems (IROS) IEEE/RSJ International Conference on Intelligent robots and systems: 2510–2517. https://doi.org/10.1109/IROS.2009.5354750.

Zittlau, K. 2015a. HaufeArbeitsschutz Office. Assistenzsyteme in der Arbeitswelt 4.0 / 1.1 KognitivunterstützendeAssistanzsysteme. Available online: https://www.haufe.de/arbeitsschutz/arbeitsschutz-office/assistenzsys teme-in-der-arbeitswelt-40-11-kognitiv-unterstuetzende-assistenzsysteme_i desk_PI957_HI11359363.html. Accessed on June 22, 2020.

Zittlau, K. 2015b. HaufeArbeitsschutz Office. Assistenzsyteme in der Arbeitswelt 4.0 / 1.2 PhysischunterstützendeAssistenzsystem. Available online: https://www.haufe.de/arbeitsschutz/arbeitsschutz-office/assistenzsys teme-in-der-arbeitswelt-40-12-physisch-unterstuetzende-assistenzsysteme_i desk_PI957_HI11359364.html. Accessed on June 22, 2020.

Part II

Implementing Industry 4.0 for Smart Logistics in SMEs

5

Investigation of the Potential to Use Real-Time Data in Production Planning and Control of Make to Order (MTO) Manufacturing Companies

Manuel Woschank, Patrick Dallasega, and Johannes A. Kapeller

5.1   Introduction

In an explorative research study Woschank et al. (2020) the authors stated that the usage of real-time data in logistics and operations management is an important principle of Industry 4.0. This becomes especially important because manufacturing processes are constantly generating a large volume of data in the fourth industrial revolution. Up to now, only a limited amount of data is used for production planning and control (PPC) strategies. Real-time-orientated PPC strategies enable highly responsive, reconfigurable, and time-efficient production systems

M. Woschank (✉) · J. A. Kapeller
Montanuniversitaet Leoben, Erzherzog-Johann-Strasse 3/1, 8700 Leoben, Austria
e-mail: manuel.woschank@unileoben.ac.at

P. Dallasega
Industrial Engineering and Automation (IEA), Faculty of Science and Technology, Free University of Bozen-Bolzano, Bolzano, Italy
e-mail: patrick.dallasega@unibz.it

© The Author(s) 2021
D. T. Matt et al. (eds.), *Implementing Industry 4.0 in SMEs*,
https://doi.org/10.1007/978-3-030-70516-9_5

(Arica and Powell 2014; Dallasega et al. 2019a; Dallasega et al. 2019b; Dallasega et al. 2020) based on the concept of mass customization (Bednar and Modrak 2014; Matt and Rauch 2016). Moreover, the integration of modern information and communication technology, interconnected networks, and physical processes is named Cyber-Physical Systems (CPS). CPS capture data of the physical world via sensors, use the Internet, and cloud computing to communicate between the connectors, and interact with the physical world utilizing mechatronic actuators (Lee 2008; Zsifkovits and Woschank 2019). This enables autonomous control systems, which can satisfy customer demands in real-time (Spath et al. 2013; Dallasega et al. 2017). CPS, as well as the Internet of Things (IoT), allow enterprises to sense deviations from the production plan as soon as they appear and identify delays in real-time (Magoutas et al. 2014; Chaopaisarn and Woschank 2019).

This chapter further investigates the postulated impact of different levels of planning periods on logistics performance indicators in combination with the application of three PPC strategies. Therefore, the initial simulation model from Woschank et al. (2020) was updated, reconfigured, and subsequentially validated in three additional simulation experiments by using a discrete event simulation study that was configured based on the data from an industrial case study in the field of electronics manufacturing.

5.2   Problem Formulation

Basically, in the area of PPC strategies, most Industry 4.0-related approaches are focusing on the principles of decentralization and/or aim to integrate real-time data for the ongoing improvement of the overall logistics performance in terms of promised delivery dates, work in progress, capacity utilization, and lead-times.

In a first step, the authors aim to evaluate significant differences regarding the impact of a material requirement planning (MRP), a KANBAN, or a constant work in process (CONWIP) strategy on logistics performance indicators in an make to order production system. In the second step, the authors further investigate the impact of real-time data usage by simulating a monthly, a two-week, and a weekly planning period, within the MRP, KANBAN, and CONWIP strategy on logistics performance indicators in an make to order production system.

Consequently, the two research questions of this chapter will be formulated as follows:

- RQ1: There is a significant difference in lead-time (LT) and work in progress (WIP) between the MRP, KANBAN, and CONWIP strategy.
- RQ2: There is a significant difference in LT and WIP between a monthly plan, a two-week plan, and a weekly plan within the MRP, KANBAN, and CONWIP strategies.

5.3 Related Work

By reviewing the recent literature on PPC approaches in an MTO environment, the Woschank et al. (2020) identified only a handful of relevant research studies that will be presented within the next paragraphs.

- Cadavid et al. (2020) present a systematic literature review analyzing the state of the art of Machine Learning (ML) approaches applied to PPC. According to their results, scientific literature rarely considers customer, environmental, and human-in-the-loop aspects when linking ML to PPC. Moreover, applications rarely link PPC to product and process design as well as to the logistics processes.
- Cadavid et al. (2020) suggest using IoT technologies in future research to collect data and update the ML model to adapt it to manufacturing system changes.
- Panetto et al. (2019) summarize the challenges for Cyber-Physical Production Systems (CPPS) as studied by the IFAC research community. According to their results, an infrastructure is needed that supports the adaptation of models according to the changing environment over time to support modification and (self-) adaptation.
- Similary, Bendul and Blunck (2019) present the vision of Industry 4.0 to assign tasks of production control to "smart" objects, such as

machines, parts, and products, to reach distributed control architectures with higher flexibility, higher adaptability, and, as such, a higher logistics performance.

- According to Ivanov et al. (2018), Industry 4.0 technology enables data interchange between the product and workstations, flexible stations able to execute various technological operations, and real-time capacity utilization control. However, modern production and supply chains are challenged by increasing uncertainty and risks as well as multiple feedback cycles where control theory could contribute to gain further insights regarding the management of these challenges.
- Gräßler and Pöhler (2018) describe the change of a milling workstation by adding different sensors and computers to reach a self-controlling cyber-physical device in a laboratory environment. Specifically, the system is distributed and decentralized whereby, through the negotiation of resources; a common planning schedule for all orders is reached. The real-time measurement of data is used to improve assessments in the planning process, improvements of process execution as well as for the identification of consequences of disturbances. However, no quantitative improvement of logistics performance indicators compared to a conventional workstation was reported.
- Similarly, Choi et al. (2017) state that, in a smart manufacturing environment, procurement, production, logistics, service, and the product itself are connected to the network and controlled in real-time based on CPS. According to them, to establish CPS manufacturing systems, real-time information exchanges from the shop-floor level to the business level need to be enabled. They argue that data acquisition from non-equipment factors, like human operators, is much more difficult to obtain than from machines because of issues like non-standardized working environments and data protection regulations.
- Hortskemper and Hellingrath (2016) present the concept of Order Allocation Flexibility and the potential of CPS in implementing and empowering the concept leading to a further increase of flexibility in the production system. However, they argue that the concept might introduce further complexity into the PPC system and the costs for such a system might not be worthwhile for all companies.

- According to Strandhagen et al. (2017), moving toward real-time control requires new conceptual models for planning and control. They state that real-time control is, today, mostly applied on machine and production line level, while, on the planning levels, existing concepts are based on conventional concepts like cyclic data processing and re-planning. However, according to them, Industry 4.0 technologies have the potential to enable real-time planning and control of all planning activities. They argue that real-time planning and control is easier to be applied in repetitive production environments because collecting data may be easier, enabling higher volumes and quality of production data.
- Similarly, Ruiz Zúñiga et al. (2017) state that, even in more advanced Industry 4.0 manufacturing companies, real-time data gathered at the shop-floor level are mostly used for monitoring different machines and work centers (e.g., processing times, failures, waiting and blocking times) and not for optimizing PPC processes.

The systematic literature review confirms the fact that there is little knowledge regarding the cause-effect relationships between different centralized and/or decentralized PPC strategies and logistics performance indicators in a make to order environment. Moreover, in contrast to the Industry 4.0 philosophy of real-time data usage, most of the data generated in production systems are frequently not completely exploited for PPC purposes. This further undermines the importance to answer the proposed research questions.

5.4   Research Design/Methodology

This chapter aims to investigate and revalidate the impact of different PPC strategies and real-time data usage in production systems on logistics performance indicators by conducting simulation with data from an industrial case study. Thereby, the two research questions assume different PPC strategies and that different levels of planning periods and will have a significant effect on the logistics performance of a production system in an MTO environment.

In this context, the authors focus on the investigation of MRP, KANBAN, and CONWIP as the most important PPC strategies in industrial enterprises (Kapeller 2018). Thereby, in the centralized MRP approach, the material is pushed to the subsequent machine after processing the order. KANBAN, as a decentralized pull system, uses a control system based on cards as a trigger for the transport of material from the outbound storage. CONWIP, as a hybrid system, pursues the goal of regulated order release procedures based on the current WIP (Kapeller 2017; Kumar et al. 2007; Gstettner 1998; Jodlbauer and Huber 2008, Dolgui and Proth 2010). Moreover, according to Unver (2013), we derived three levels of planning periods from the ISA 95 framework leading to three test groups: (1) a monthly plan, (2) a two-week plan, and (3) a weekly plan (Woschank et al. 2020). Figure 5.1 displays the basic concept of our research.

In sum the two research questions were formulated as follows:

- RQ1: There is a significant difference in lead-time (LT) and work in progress (WIP) between the MRP, KANBAN, and CONWIP approach.
- RQ2: There is a significant difference in LT and WIP between a monthly plan, a two-week plan, and a weekly plan within the MRP, KANBAN, and CONWIP strategies.

Based on the formulation of the basic research questions, the research process furthermore includes the following phases:

- Phase 1: Data collection based on the production process of an electronics manufacturer

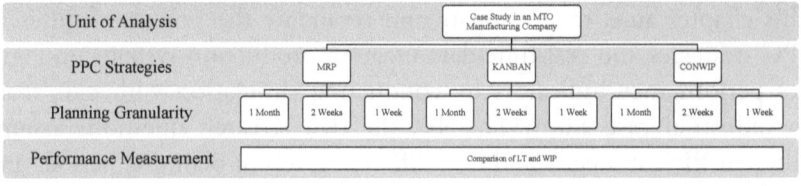

Fig. 5.1 Concept of research

- Phase 2: Programming of the simulation model
- Phase 3: Conducting a set of simulation runs
- Phase 4: Systematic evaluation of the calculated results

According to Woschank et al. (2020), in phase 1, the data collection was based on the production process of an electronics manufacturer working as an SME in a make to order environment. Therefore, we used the secondary data from a Value Stream Mapping (VSM) as a method for the data gathering, where we focused on one specific product group. Based on the VSM, we identified the following four value-adding production processes: Step 1: production process 1 (PP1: raw printing, solver paste printing, printing check), step 2: production process 2 (PP2: picking and placing of components, soldering), step 3: production process 3 (PP3: programming and function control and step 4: production process 4 (PP4: final assembly). The supporting processes SP1 and SP2 are used for a temporary storage of material. Moreover, we recorded the following parameters for every process step: change over time, cycle time, lot size, availability, meantime to repair, LT, pieces per shift, and number of shifts. In sum, we identified the following problems: (1) The productivity of production and logistics department is quite low, (2) frequently, the customer demand cannot be satisfied, (3) high cost due to high stock-levels within the production, and (4) the planning data in the production planning system are not up to date.

Figure 5.2 displays the production process based on the initial process analysis.

Phase 2 focused on the programming of the simulation model as a tool for the systematic evaluation of research question 1 and research question 2. Thereby, we used discrete event simulation as a research method because this approach offers a high internal validity, high reliability, and the possibility to systematically isolate potential confounding variables because of the pre-defined modeling procedures (Cooper and Schindler 2014; Rabe et al. 2008; März et al. 2011; Woschank et al. 2020). Also, the transferability of the established research findings, respectively, the external validity of the simulation procedures will be ensured by using the data from the conducted VSM analysis (Bortz and Döring 2007; Woschank et al. 2020).

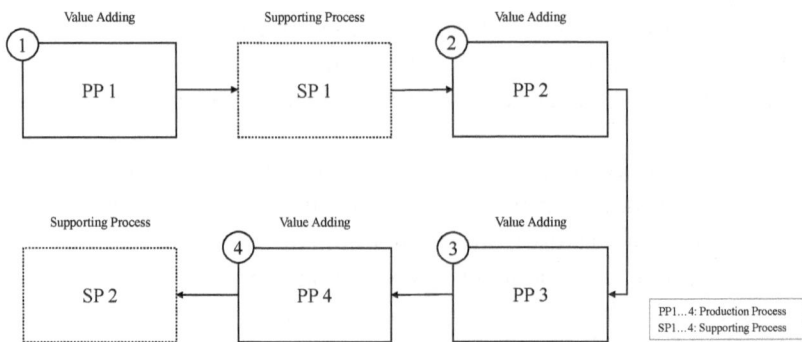

Fig. 5.2 Production process

Moreover, the research design of this paper is based on the VDI 3633 guidelines for simulation of logistics systems which consider the steps of preparation, simulation, and evaluation (März et al. 2011). In the preparation phase, we used the software Tecnomatix Plant Simulation 15.1 by Siemens PLM, which is a tool package for discrete event simulation (Woschank et al. 2020). The simulation approach is displayed in Fig. 5.3.

The final simulation approach includes three PPC strategies (MRP, KANBAN, and CONWIP), three different levels of planning periods (1 month, two weeks, and a weekly plan), and two indicators for the measurement of the logistics performance (lead-time (LT) and work in progress (WIP)). The simulation includes five machines which represent

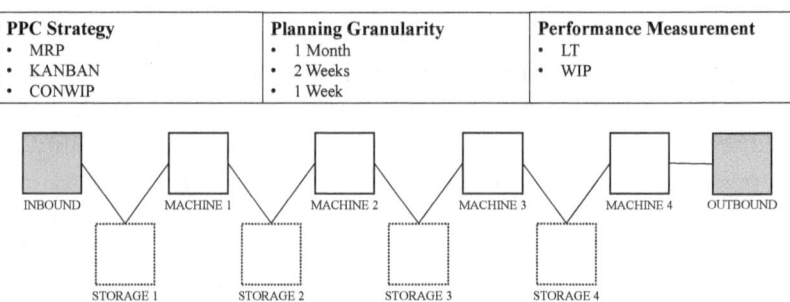

Fig. 5.3 Simulation approach

the identified value-adding production processes (PP1-PP2), four inter-mediate storages in the production system as well as an inbound and an outbound warehouse.

In phase 3, the author conducted a set of simulation runs by applying the MRP, KANBAN, and CONWIP approach to the production system. Thereby, in line with Woschank et al. (2020), the different levels of planning periods were simulated by generating three different test groups. Test group 1 is based on the usage of a monthly plan, test group 2 uses a two-week plan, and test group 3 uses a weekly plan for the PPC process. The logistics performance was operationalized by using a set of manifest indicators. In this case, we focused on the measurement of LT and the measurement of WIP in the production system.

In sum, the authors conducted three simulation experiments with nine simulation models, leading to 27,818 simulation runs within the first simulation experiment. Moreover, within a second simulation experi-ment, the authors furthermore computed 83,454 simulation runs, which did not show any significant difference ($p < 0.05$) in comparison with the initially computed results. Finally, the authors conducted a third simulation experiment with 24,559 simulation runs leading to 49,118 performance indicators for the subsequent statistical analysis.

5.5 Results and Discussion

Figure 5.4 displays an overview of the conceptualized PPC strategies in the present case study.

In phase 4, the authors systematically evaluate the calculated results from the third simulation experiment to answer the following research questions:

- RQ1: There is a significant difference in lead-time (LT) and work in progress (WIP) between the MRP, KANBAN, and CONWIP strategy.
- RQ2: There is a significant difference in LT and WIP between a monthly plan, a two-week plan, and a weekly plan within the MRP, KANBAN, and CONWIP strategies.

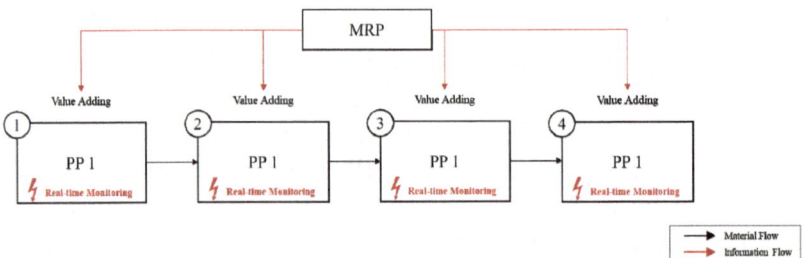

PPC strategy 1: MRP concept coupled with real-time progress measurement

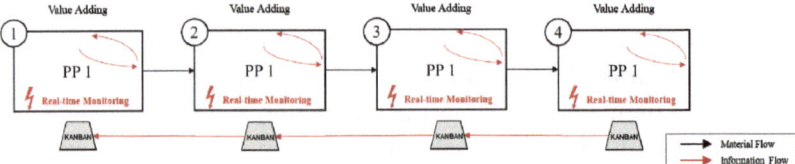

PPC strategy 2: KANBAN concept coupled with real-time progress measurement

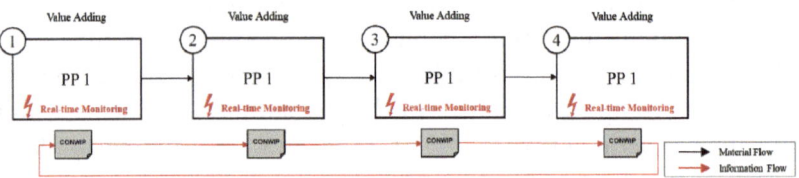

PPC strategy 3: CONWIP concept coupled with real-time progress measurement

Fig. 5.4 PPC strategies

5.5.1 Research Question 1 (RQ1): Comparative Evaluation of PPC Strategies

To answer research question 1, the authors conducted a comparative evaluation of the three pre-defined PPC strategies. Therefore, a one-way analysis of variance (ANOVA) was used to evaluate significant differences in the two logistics performance indicators lead-time (LT) and (WIP) between the MRP, KANBAN, and CONWIP strategies. Therefore, the authors used the software package IBM SPSS Statistics 26 for the computation of the statistical procedures. The summarized results of the conducted ANOVA analyses are displayed in Fig. 5.5 and Fig. 5.6.

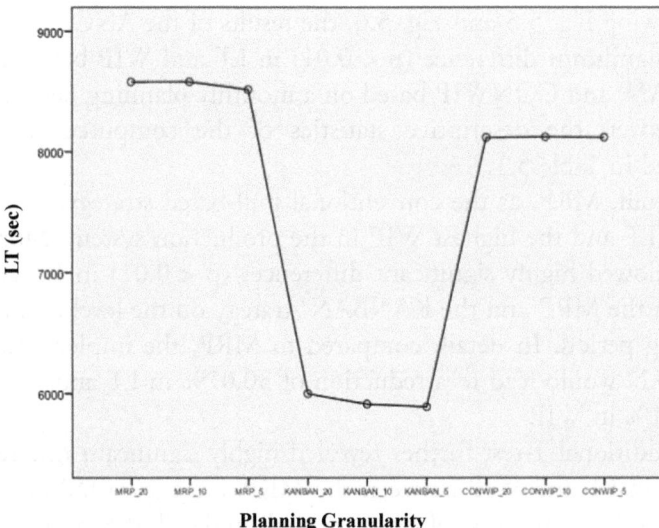

Fig. 5.5 Comparison of PPC strategies—lead-time (LT)

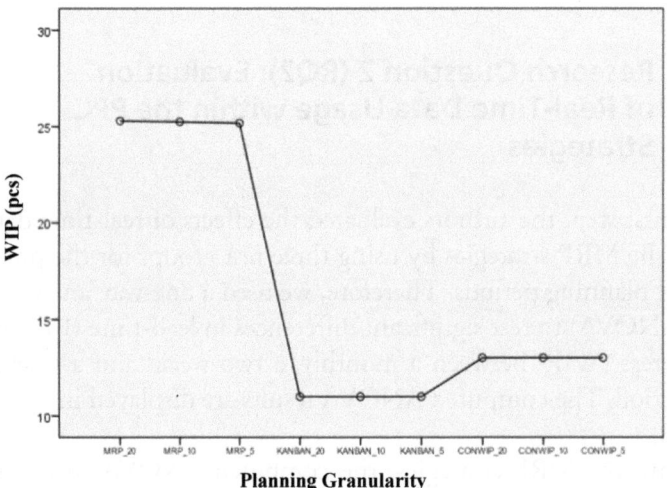

Fig. 5.6 Comparison of PPC strategies—work in progress (WIP)

Following Fig. 5.5 and Fig. 5.6, the results of the ANOVA revealed a highly significant difference ($p < 0.01$) in LT and WIP between MRP, KANBAN, and CONWIP based on a monthly planning period.

Moreover, the descriptive statistics of the computed sample are displayed in Table 5.1.

In detail, MRP, as the conventional pull-based strategy, leads to the longest LT and the highest WIP in the production system. Moreover, a T-test showed highly significant differences ($p < 0.01$) in LT and WIP between the MRP and the KANBAN strategy on the level of a monthly planning period. In detail, compared to MRP, the implementation of KANBAN would lead to a reduction of 30.07% in LT and a reduction of 56.51% in WIP.

An additional T-test further revealed highly significant differences ($p < 0.01$) in LT and WIP between the MRP and the CONWIP strategy on the level of a monthly planning period. In detail, the implementation of CONWIP would lead to a reduction of 5.37% in LT and 48.60% in WIP, in comparison with the MRP approach.

5.5.2   Research Question 2 (RQ2): Evaluation of Real-Time Data Usage within the PPC Strategies

In the first step, the authors evaluated the effects of real-time data usage within the MRP strategies by using three test groups for the pre-defined levels of planning periods. Therefore, we used a one-way analysis of variance (ANOVA) to test significant differences in lead-time (LT) and work in progress (WIP) between a monthly, a two-week, and a weekly planning period. The computed ANOVA results are displayed in Fig. 5.7 and Fig. 5.8.

Within the MRP strategies, the computed ANOVA revealed highly significant differences ($p < 0.01$) in lead-time (LT) and work in progress (WIP) between a monthly, a two-week, and a weekly planning period. The detailed statistical analysis revealed, that, within the MRP strategies, a higher level of planning will lead to better LT and a lower WIP, but only with a relatively low effect. In detail, a reduction in planning from

Table 5.1 Descriptive statistics

		N	Mean	Std. Deviation	Std. Error	95% Confidence Interval for Mean		Minimum	Maximum
						Lower Bound	Upper Bound		
LT	1 MRP_20	6756	8579.55	275.314	3.350	8572.99	8586.12	3719	9483
	2 MRP_10	3349	8580.69	313.914	5.424	8570.05	8591.32	3705	9393
	3 MRP_5	1111	8516.85	506.351	15.191	8487.05	8546.66	3565	9229
	4 KANBAN_20	4290	5999.76	636.984	9.725	5980.69	6018.83	336	8253
	5 KANBAN_10	2128	5914.68	656.433	14.230	5886.77	5942.58	336	8253
	6 KANBAN_5	711	5892.07	606.584	22.749	5847.40	5936.73	1515	7467
	7 CONWIP_20	3740	8118.98	294.056	4.808	8109.55	8128.41	3735	10,184
	8 CONWIP_10	1863	8124.12	320.379	7.423	8109.56	8138.68	3807	11,003
	9 CONWIP_5	611	8120.83	462.261	18.701	8084.10	8157.56	3730	10,384
	Total	24,559	7701.42	1208.442	7.711	7686.30	7716.53	336	11,003
WIP	1 MPR_20	6756	25.29	0.498	0.006	25.28	25.30	22	27
	2 MRP_10	3349	25.24	0.510	0.009	25.23	25.26	21	27
	3 MRP_5	1111	25.18	0.509	0.015	25.15	25.21	22	26
	4 KANBAN_20	4290	11.00	0.000	0.000	11.00	11.00	11	11
	5 KANBAN_10	2128	11.00	0.000	0.000	11.00	11.00	11	11
	6 KANBAN_5	711	11.00	0.000	0.000	11.00	11.00	11	11
	7 CONWIP_20	3740	13.00	0.000	0.000	13.00	13.00	13	13
	8 CONWIP_10	1863	13.00	0.000	0.000	13.00	13.00	13	13
	9 CONWIP_5	611	13.00	0.000	0.000	13.00	13.00	13	13
	Total	24,559	18.02	6.692	0.043	17.94	18.11	11	27

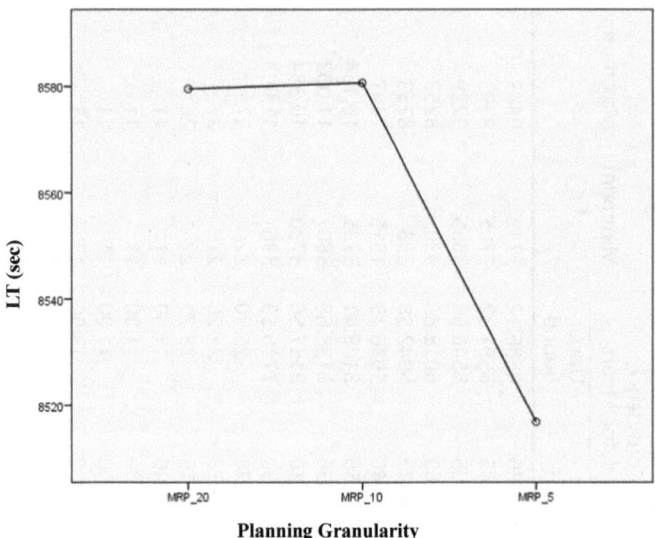

Fig. 5.7 Real-time data usage within the MRP strategy—lead-time (LT)

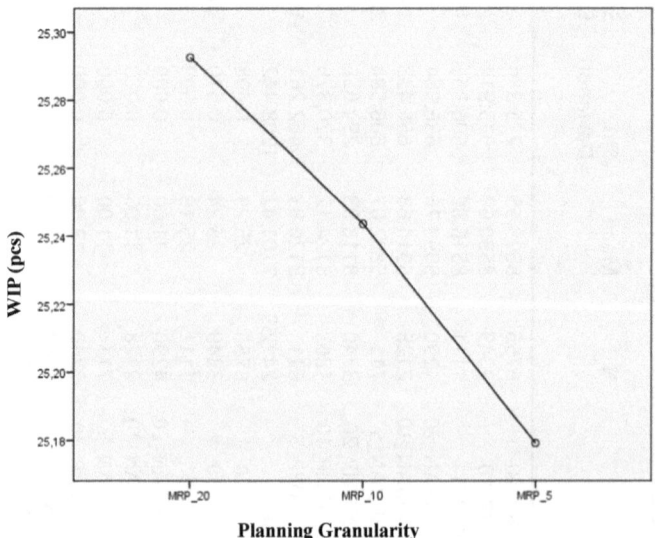

Fig. 5.8 Real-time data usage within the MRP strategy—work in progress (WIP)

a monthly plan to a weekly plan would result in a reduction of 0.73% in LT and 0.45% in WIP.

In the second step, the authors evaluated the effects of real-time data usage within the KANBAN strategies by using a one-way analysis of variance (ANOVA) to test significant differences in lead-time (LT) and work in progress (WIP) between a monthly, a two-week, and a weekly planning period. The computed ANOVA results are displayed in Fig. 5.9.

Within the KANBAN strategies, the computed ANOVA revealed highly significant differences ($p < 0.01$) in lead-time (LT), but no significant differences ($p > 0.05$) work in progress (WIP) between a monthly, a two week, and a weekly planning period. Thereby, the detailed statistical analysis revealed, that, within the KANBAN strategies, a higher level of planning will lead to a reduction of 1.79% in LT.

In the third step, the authors evaluated the effects of real-time data usage within the CONWIP strategies by using three test groups for the pre-defined levels of planning periods. Therefore, we used a one-way analysis of variance (ANOVA) to test significant differences in lead-time

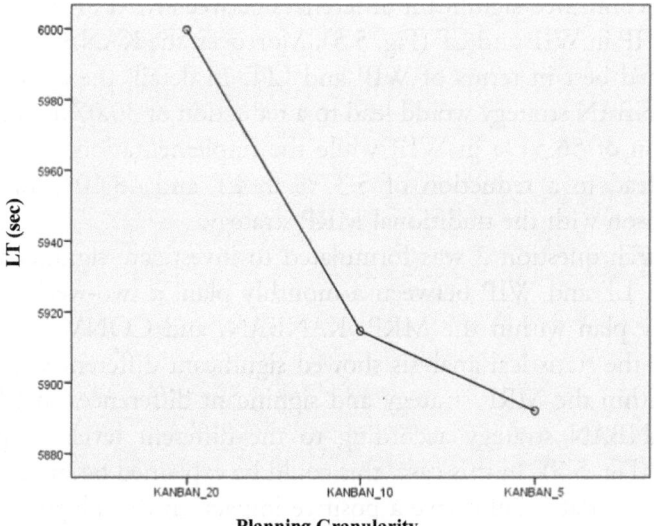

Fig. 5.9 Real-time data usage within the KANBAN strategy—lead-time (LT)

(LT) and work in progress (WIP) between a monthly, a two-week, and a weekly planning period. However, within the CONWIP strategies, the computed ANOVA results showed no significant differences ($p < 0.010$) in LT and no significant differences in WIP between a monthly plan, a two-week plan, and a weekly plan. Therefore, the authors conclude that different planning periods have no significant effect on the LT and WIP within the CONWIP approach.

5.6   Conclusions and Outlook

In this chapter, the authors have updated and reconfigured the simulation model by Woschank et al. (2020) to investigate and subsequently validate and the impact of different PPC strategies and different levels of planning periods on logistics performance indicators by using a discrete event simulation based on an industrial case study.

Research question 1 aimed to investigate significant differences in lead-time (LT) and work in progress (WIP) between the MRP, KANBAN, and CONWIP strategy. Thereby, the results of the statistical analysis confirmed significant differences between MRP, KANBAN, and CONWIP in WIP and LT (Fig. 5.5). Moreover, the KANBAN approach performed best in terms of WIP and LIT. In detail, the application of the KANBAN strategy would lead to a reduction of 30.07% in LT and a reduction of 56.51% in WIP, while the implementation of CONWIP would lead to a reduction of 5.37% in LT and 48.60% in WIP, in comparison with the traditional MRP strategy.

Research question 2 was formulated to investigate significant differences in LT and WIP between a monthly plan, a two-week plan, and a weekly plan within the MRP, KANBAN, and CONWIP strategies. Hereby, the statistical analysis showed significant differences in LT and WIP within the MRP strategy and significant differences in LT within the KANBAN strategy according to the different levels of planning periods (Fig. 5.7). In this case, this could be explained because the usage of real-time data could have a positive impact on the adaptation of the system to changes, thus making the system more responsive. Within the CONWIP strategy, no significant differences in LT and WIP were indicated between the different levels of planning periods.

In general, this chapter contributes to a better understanding of PPC strategies and the usage of real-time data in production systems. The updated and reconfigured simulation model from Woschank et al. (2020) further increases the transferability and, therefore, the external validity of the computed statistical results and the established research findings. Future research should focus on the further development of the proposed model by transferring it to other industrial branches, by incorporating additional PPC strategies, or by using a different set of products and/or machines within the simulation model.

Acknowledgements This research is part of the project "SME 4.0 – Industry 4.0 for SMEs", which has received funding from the European Union's Horizon 2020 research and innovation program under the Marie Skłodowska-Curie grant agreement No. 734713.

References

Arica, E., and D.J. Powell. 2014. A framework for ICT-enabled real-time production planning and control. *Advances in Manufacturing* 2 (2): 158–164. https://doi.org/10.1007/s40436-014-0070-5.

Bednar, S., and V. Modrak. 2014. Mass customization and its impact on assembly process' complexity. *International Journal for Quality Research* 8(3): 417–430. https://www.ijqr.net/journal/v8-n3/10.pdf.

Bendul, J.C., and H. Blunck. 2019. The design space of production planning and control for industry 4.0. *Computers in Industry* 105: 260–272. https://doi.org/10.1016/j.compind.2018.10.010.

Bortz, J., and N. Döring. 2007. *Forschungsmethoden und Evaluation. Für Human- und Sozialwissenschaftler*, 4th ed. Berlin et al.: Springer.

Cadavid, J.P.U., S. Lamouri, B. Grabot, R. Pellerin, and A. Fortin. 2020. Machine learning applied in production planning and control: A state-of-the-art in the era of industry 4.0. *Journal of Intelligent Manufacturing* 31: 1531–1558. https://doi.org/10.1007/s10845-019-01531-7.

Chaopaisarn, P., and M. Woschank. 2019. Requirement analysis for SMART supply chain management for SMEs. In *Proceedings of the International Conference on Industrial Engineering and Operations Management*, 3715–3725. Bangkok, Thailand.

Choi, S., G. Kang, C. Jun, J.Y. Lee, and S. Han. 2017. Cyber-physical systems: A case study of development for manufacturing industry. *International Journal of Computer Applications in Technology* 55 (4): 289–297. https://doi.org/10.1504/IJCAT.2017.086018.

Cooper, D.R., and P.S. Schindler. 2014. *Business research methods*, 12th ed. New York: McGraw-Hill.

Dallasega, P., R.A. Rojas, E. Rauch, and D.T. Matt. 2017. Simulation based validation of supply chain effects through ICT enabled real-time-capability in ETO production planning. *Procedia Manufacturing* 11: 846–853. https://doi.org/10.1016/j.promfg.2017.07.187.

Dallasega, P., R.A. Rojas, G. Bruno, E. Rauch, and E. 2019. An agile scheduling and control approach in ETO construction supply chains. *Computers in Industry* 112: 103–122. https://doi.org/10.1016/j.compind.2019.08.003.

Dallasega, P., M. Woschank, S. Ramingwong, K. Tippayawong, and N. Chonsawat. 2019b. Field study to identify requirements for smart logistics of European, US and Asian SMEs. In *Proceedings of the International Conference on Industrial Engineering and Operations Management*, 844–855. Bangkok, Thailand.

Dallasega, P., M. Woschank, H. Zsifkovits, K.Y. Tippayawong, and C.A. Brown. 2020. Requirement analysis for the design of smart logistics in SMEs. In Matt, D.T., V. Modrák, and H. Zsifkovits (eds.). *Industry 4.0 for SMEs*, 147–162. Cham: Springer.

Dolgui, A., and J.P. Proth. 2010. *Supply chain engineering. Useful methods and techniques*. London et al.: Spinger.

Gräßer, I., and A. Pöhler. 2018. Intelligent devices in a decentralized production system concept. *Procedia CIRP* 67: 116–121. https://doi.org/10.1016/j.procir.2017.12.186.

Gstettner, S. 1998. *Leistungsanalyse von Produktionssystemen*. Heidelberg: Physica.

Horstkemper, D., and B. Hellingrath. 2016. Employing order allocation flexibility in cyber-physical production systems. *Procedia CIRP* 57: 345–350. https://doi.org/10.1016/j.procir.2016.11.060.

Ivanov, D., S. Sethi, A. Dolgui, and B. Sokolov. 2018. A survey on control theory applications to operational systems, supply chain management, and Industry 4.0. *Annual Reviews in Control* 46: 134–147. https://doi.org/10.1016/j.arcontrol.2018.10.014.

Jodlbauer, J., and A. Huber. 2008. Service level performance of MRP, Kanban, CONWIP and DBR due to parameter stability and environmental robustness. *International Journal of Production Research* 46 (8): 2179–2195. https://doi.org/10.1080/00207540600609297.

Kapeller, J.A. 2017. Explorative Potentialevaluierung der Kombination von Fertigungssteuerungs-strategien. Ein simulationsbasierter Ansatz zur Evaluierung von Produktions- und Fertigungsverfahren im Bereich der Linienfertigung. In Zsifkovits, H., and S. Altendorfer-Kaiser (eds.). *Logistisches Produktionsmanagement,4. Wissenschaftlicher Industrielogistik-Dialog in Leoben (WiLD)*, 71–82. Berlin: epubli.

Kapeller, J.A. 2018. *EoPaMS. Systematische Potentialevaluierung der sequentiellen Kombination von Fertigungssteuerungsstrategien für den Bereich der Linienfertigung. Ein simulationsbasierter Ansatz.* Dissertation. Montanuniversitaet Leoben: Leoben.

Kumar, C.S., and R. Panneerselvam. 2007. Literature review of JIT-KANBAN system. *the International Journal of Advanced Manufacturing Technology* 32 (3): 393–408. https://doi.org/10.1007/s00170-005-0340-2.

Lee, E.A. 2008. Cyber physical systems: Design challenges. In *Proceedings of the 11th IEEE International Symposium on Object and Component-Oriented Real-Time Distributed Computing*, 363–369.

Magoutas, N., N. Stojanovic, A. Bousdekis, D. Apostolou, G. Mentzas, and L. Stojanovic. 2014. Anticipation-driven architecture for proactive enterprise decision making. *Proceedings of the CAiSE* 2014: 121–128.

März L., W. Krug, O. Rose, and G. Weigert. 2011. *Simulation und Optimierung in Produktion und Logistik. Praxisorientierter Leitfaden mit Fallbeispielen.* Berlin: Springer.

Matt, D.T., and E. Rauch. 2016. Designing assembly lines for mass customization production systems. In Modrak, V. (ed.). *Mass customized manufacturing. Theoretical concepts and practical approaches*, 15–36. Boca Raton: CRC Press.

Panetto, H., B. Iung, D. Ivanov, G. Weichhart, X. Wang, and X. 2019. Challenges for the cyber-physical manufacturing enterprises of the future. *Annual*

Reviews in Control 47: 200–213. https://doi.org/10.1016/j.arcontrol.2019. 02.002.

Rabe, M., S. Spieckermann, and S. Wenzel. 2008. *Verifikation und Validierung für die Simulation in Produktion und Logistik. Vorgehensmodelle und Techniken.* Berlin, Heidelberg: Springer.

Ruiz Zúñiga, E., A. Syberfeldt, and M. Urenda Moris. 2017, The Internet of Things, Factory of Things and Industry 4.0 in Manufacturing: Current and Future Implementations. In *Proceedings of the 15th International Conference on Manufacturing Research*, 221–226. Greenwich, United Kingdom. https://doi.org/10.3233/978-1-61499-792-4-221.

Spath, D., S. Gerlach, M. Hämmerle, S. Schlund, and T. Strolin. 2013. Cyber-physical system for self-organised and flexible labour utilization. In *22nd International Conference on Production Research*, 1–6.

Strandhagen, J.W., E. Alfnes, J.O. Strandhagen, and L.R. Vallandingham. 2017. The fit of Industry 4.0 applications in manufacturing logistics: A multiple case study. *Advances in Manufacturing* 5: 344–358. https://doi.org/10.1007/s40436-017-0200-y.

Unver, H.O. 2013. An ISA-95-based manufacturing intelligence system in support of lean initiatives. *The International Journal of Advanced Manufacturing Technology* 65 (5): 853–866. https://doi.org/10.1007/s00170-012-4223-z.

Woschank, M., P. Dallasega, and J.A. Kapeller. 2020. The impact of planning granularity on production planning and control strategies in MTO: A discrete event simulation study. *Procedia Manufacturing*. In press 1–6.

Zsifkovits, H.E., and M. Woschank. 2019. Smart Logistics – Technologiekonzepte und Potentiale. *BHM Berg- Und Hüttenmännische Monatshefte* 164 (1): 42–45. https://doi.org/10.1007/s00501-018-0806-9.

6

Readiness Model for Integration of ICT and CPS for SMEs Smart Logistics

Sakgasit Ramingwong, Lachana Ramingwong,
Trasapong Thaiupathump, and Rungchat Chompu-inwai

6.1   Introduction

Industry 4.0 has broadened the gap between modern and traditional small- and medium-sized enterprises (SMEs). Industry 4.0 involves application of a number of the most advanced technologies (Lasi et al. 2014) in all areas, including logistics. Logistics management involves

S. Ramingwong (✉) · L. Ramingwong · T. Thaiupathump
Department of Computer Engineering, Chiang Mai University, 239 Huay
Kaew Road, Muang District, 50200 Chiang Mai, Thailand
e-mail: sakgasit@eng.cmu.ac.th

L. Ramingwong
e-mail: lachana.ramingwong@cmu.ac.th

T. Thaiupathump
e-mail: trasapong@eng.cmu.ac.th

S. Ramingwong
Center of Excellence in Logistics and Supply Chain Management, Chiang
Mai University, 239 Huay Kaew Road, Muang District, 50200 Chiang Mai,
Thailand

© The Author(s) 2021
D. T. Matt et al. (eds.), *Implementing Industry 4.0 in SMEs*,
https://doi.org/10.1007/978-3-030-70516-9_6

planning, implementing, and controlling the efficient, effective forward and reverse flow and storage of goods, services, and related information from one point to another according to customers' requirements (CSCMP 2020). It is not just a process of transportation, but many activities are also involved in the logistics process. It is an integrating function, which coordinates and optimizes all logistics activities, as well as integrates logistics activities with other activities (CSCMP 2020).

Due to constantly changing customer needs, traditional logistics may not be able to meet more complicated customer requirements. Barreto et al. (2017) defined "Smart Logistics" as a logistics system that can increase flexibility and make adjustments to meet the changing needs of customers. Smart logistics is technology-driven. Flexibility, adaptability, and proactivity are more important than before and can only be achieved by integration of new technologies (Uckelmann 2008).

Information and Communication Technology (ICT) and Cyber-Physical Systems (CPS) have been integrated for smart logistics (Schrauf and Berttram 2016). ICT refers to the integrations of devices, networking components, applications, and systems that allow people and organizations to exchange the information. CPS can be generally defined as the integrations of computation, networking, and physical processes that enables a physical system to be closely monitored, controlled by a computer algorithm (Lu 2017). In Industry 4.0, numerous industrial CPS-based applications have been developed and deployed. Building efficient connectivity of the CPS and external systems is quite challenging. A conceptual framework for enabling the connectivity of cyber-physical production systems inside smart factories is proposed by Rojas et al. (2017). Lee et al. (2015) proposed a functional model called 5C architecture for CPS. This model defines 5 levels of CPS functions and then provides a guideline for construction of a CPS. Similar to 5C, Porter and

R. Chompu-inwai
Department of Industrial Engineering, Chiang Mai University, 239 Huay Kaew Road, Muang District, 50200 Chiang Mai, Thailand
e-mail: rungchat@eng.cmu.ac.th

Heppelmann (2014) suggested building a connectivity-based infrastructure. The capability model advised by Hernández and Reiff-Marganiec (2014) suggests how CPS can be improved.

"Logistics 4.0" is focused on the specific applications of Industry 4.0 in logistics (Amr et al. 2019; Facchini et al., 2020). In addition, Barreto et al. (2017) also defined Logistics 4.0 as the combination of using logistics with innovations and applications of CPS. Winkelhaus and Grosse (2019) defined Logistics 4.0 as *"the logistical system that enables the sustainable satisfaction of individualized customer demands without an increase in costs and supports this development in industry and trade using digital technologies."*

Although smart logistics is one of the essential elements which are critical to survive in this highly competitive system, investing in technologies can be a double-edged sword for SMEs. It is important to keep all business dimensions balanced. Self-evaluation of an organization's own readiness for integrating technologies for smart logistics is important. From previous research studies, it was found that there have been several proposed readiness models based on different perspectives and different levels of implementation (Schumacher et al. 2016; Maasouman and Demirli 2015; Akdil et al. 2018; Gökalp et al. 2017). This research proposes a novel readiness model which can be used to evaluate readiness for integration of ICT and CPS for smart logistics. The model proposed in this chapter can be used as a guideline for organization assessment which aims to evaluate their technological readiness for smart logistics, particularly for SMEs. Therefore, they can accordingly make decisions on further investment.

6.2 Related Works

Numerous Industry 4.0 maturity and readiness models on different perspectives have been proposed since the term Industry 4.0 was introduced in 2015. Some of the well-recognized models are IMPULS (2015), Industry 4.0/Digital Operations Self-Assessment by PwC (2016), the Connected Enterprise Maturity Model by Rockwell Automation (2016), and the Industry 4.0 Maturity Model by Schumacher et al. (2016).

IMPULS—Industrie 4.0 readiness model, proposed by Lichtblau et al. in 2015, consists of six levels of Industry 4.0 readiness: 0-outsider, 1-beginner, 2-intermediate, 3-experienced, 4-expert, and 5-top performer, and six Industry 4.0 dimensions: strategy and organization, smart factory, smart operations, smart products, data-driven services, and employees. This model is an example of models focusing on technological aspects. The IMPULS model considers logistics operations within the context of the vertical and horizontal integration involving several company's departments and hierarchical levels (Lichtblau et al. 2015).

The digital operations self-assessment for Industry 4.0 was proposed by PwC in 2016. PwC provided a maturity model for companies to assess their capabilities. PwC's maturity model was organized in four stages and seven dimensions. Four stages are (1) digital novice, (2) vertical integrator, (3) horizontal collaborator, and (4) digital champion. Vertical integration is the integration of IT systems at multiple hierarchical manufacturing and production levels into one complete solution, whereas horizontal integration is the integration of IT systems across the multiple business planning and production processes. This model assesses companies' level of maturity with seven dimensions: digital business models and customer access, digitization of product and service offerings, digitization and integration of vertical and horizontal value chains, data and analytics as core capability, agile IT architecture, compliance, security, legal and tax, organization, employees, and digital culture. PwC provides an online self-assessment tool that enables companies to assess their level of Industry 4.0 maturity (Geissbauer et al. 2016).

The connected enterprise maturity model was proposed by Rockwell Automation in 2014. This model consists of five stages and four technology-oriented dimensions. Maturity stages in this model are (1) assessment; (2) secure and upgraded network and controls; (3) defined and organized working data capital (WDC); (4) analytics, and (5) collaboration. This model primarily focuses on four dimensions of OT/IT (Operations Technology/Information Technology) network: information infrastructure (hardware and software), controls and devices (sensors, actuators, motor controls, switches, etc.) that feed and receive data, networks that move all of this information and security policies (understanding, organization, enforcement) (Rockwell Automation 2014).

The Industry 4.0 maturity model proposed by Schumacher et al. (2016) has nine dimensions and sixty-two maturity items in assessing companies Industry 4.0 maturity levels. Nine dimensions are strategy, leadership, customers, products, operations, culture, people, governance, and technology. Maturity levels are investigated under five levels, from level 1 to 5.

Since the term Logistics 4.0 was coined in 2016, several maturity models for Logistics 4.0 have been proposed (Sternad et al. 2018; Oleśków-Szłapka and Stachowiak 2018; Facchini et al. 2020). Sternad et al. (2018) proposed the maturity levels for Logistics 4.0 based on NRW's Industry 4.0 maturity model (Kompetenzzentrum Mittelstand NRW 2020). The NRW's levels were defined based on the definition of automation in the Industry 4.0 concept. These levels also represent the evolution from separated software systems to networked systems. Sternad et al. (2018) used NRW's Industry 4.0 maturity model levels in identifying the maturity levels in four subsystems of logistics: purchase, production, distribution, and after-sales. Oleśków-Szłapka and Stachowiak (2018) defined three dimensions of Logistics 4.0: management, flow of material, and flow of information and also identified the areas of evaluation for each dimension. The five maturity levels of this model are ignoring, defining, adopting, managing, and integrated. Facchini et al. (2020) applied Oleśków-Szłapka's Logistics 4.0 maturity model and verified the feasibility of this model in a real case study.

Compared to large-size companies, SMEs generally have less resources in adoption of Industry 4.0 technologies and concepts which is considered to be one of the key obstacles. Chatzoglou and Chatzoudes (2016) identified various factors affecting the adoption of Industry 4.0 technologies in SMEs. These factors involved adoption costs, competitive pressure, firm size, firm scope, CEO's knowledge, Internet skills, and IT infrastructure. There have been various studies focusing on the development of readiness/maturity models for SMEs and related issues (Ganzarain and Errasti 2016; Trotta and Garengo 2019; Rauch et al. 2020). Rauch et al. (2020) proposed a maturity level-based assessment tool to enhance the implementation of Industry 4.0 in SMEs. Some works focused on the logistics process in SMEs, such as a self-assessment of Industry 4.0 technologies focusing on the internal logistics

Table 6.1 Readiness pillars and areas of evaluation

Readiness pillars	Areas of evaluation
Production	Products, facilities, operations, quality, real-time status, automation, agility
Technology	Implementation, security, networking
People	People, stakeholders, culture
Logistics	Internal logistics, transportation
Strategy	Governance, strategy

for SMEs by Schiffer et al. (2019) and a requirement analysis for implementing smart logistics in SMEs by Dallasega et al. (2020). The results from this study involved: lean and agility; real-time status; digitization, connectivity, and network; tracking, PPC and WMS; culture, people, and implementation; security and safety; ease of use; transportation, and automation. This chapter expands this concept and systematically constructs a model for self-investigating for readiness in emerging SMEs.

6.3   Readiness Model for Integration of ICT and CPS for Smart Logistics

Based on the aforementioned research, all factors are analyzed and revised into 5 pillars, i.e. production, technology, people, logistics, and strategy. Each of them is further classified into a specific process area. Each readiness pillar and dedicated evaluation areas are depicted in Table 6.1. Detailed explanations of each process area as well as their objectives and passing criteria are subsequently described in the next section.

6.4   Stages of Readiness

Five stages of readiness, i.e. initial, trial, organized, automated, and optimized, are proposed in this chapter as shown in Table 6.2.

Table 6.2 illustrates the level of readiness and relevant process areas. There is only one process area on the product for the first level of

Table 6.2 Readiness model for implementing ICT and CPS

Readiness Level	Process Areas				
	A-Production	B-Technology	C-People	D-Logistics	E-Strategy
5-Optimized	A-5-1 Agility	B-5-1 Intelligence			E-5-1 Strategy
4-Automated	A-4-1 Quality A-4-2 Real-time Status A-4-3 Automation	B-4-1 Digitization	C-4-1 Culture		
3-Organized	A-3-1 Operations	B-3-1 Security B-3-2 Networking	C-3-1 Stakeholders	D-3-1 Internal Logistics D-3-2 Transportation	E-3-1 Governance
2-Trial	A-2-1 Facilities	B-2-1 Implementation	C-2-1 People		
1-Initial	A-1-1 Products				

readiness. Then, basic resources of the organization are the core maturity of the second level. The third level of readiness explores the overall production, technology, people, logistics, and strategy. More advanced implementation of technology is evaluated in the fourth level. Finally, the fifth readiness level depicts the optimized utilization of technology.

The organization needs to satisfy every criterion of each process area to secure its readiness level. The first stage, initial, is essentially a preparation stage for organizations who are interested in implementing ICT and CPS.

1-Initial

Only one process area, i.e. A-1-1 Products, is identified at this level. An organization needs to at least understand basic requirements and purposes of the ICT and CPS implementation. Indeed, this is the first stage of investing and implementing technology in general. Although noting concrete is found in this early level, it projects a critical direction for the organizations.

Readiness: The organizations on this readiness level are not yet ready for advanced investment in technology. They should initially investigate their own needs and consult experts accordingly.

2-Trial

The second stage of the readiness model highlights an organization beginning its investment in ICT and CPS. Three major resources, infrastructure, implementation of technology, and people, are investigated in this stage. At this stage, technology is widely used in the organizations. However, numbers of deficiencies and repeated trials and errors are expected. The organizations which satisfy all requirements in this level are recognized as the beginners who are ready to excel in smart technologies in the near future.

Readiness: The organizations on this readiness level have basic experience on utilizing technology. However, they are still in an initial stage. Investment can slowly be made on the areas which they are familiar with. Consultancy from experts is still needed.

3-Organized

After a period of implementing technologies, the organizations begin to seek standardization of their processes. This level of readiness encapsulates 7 process areas which cover all of the 5 pillars. The organizations which achieve this level of readiness are familiar with the technologies and begin to expand their benefits to a larger scope of operations. Logistics begins to be a core strength. Production of these organizations has become stable. Deficiency can be found occasionally.

Readiness: The organizations are equipped with experiences and expertise on utilizing technology in several areas. Computer systems have become an important part in their core business. They are ready for more advanced technology.

4-Automated

The manufacturing lines in these organizations are stable. Production can be accurately forecasted. Higher technologies such as robotics and automation are fully utilized. Errors and defects are less likely to occur. The organizations also begin to build up their own learning cultures. Data begins to be one of important assets, although it has not yet been fully utilized.

Readiness: The organizations utilize automated systems in several areas of core competencies. They also efficiently utilize their knowledge base and are ready to upgrade themselves for intelligent technologies.

5-Optimized

The highest level of readiness depicts organizations which fully utilize technology. With strong culture and proactive organizational strategy, they are ready to implement any technologies for Industry 4.0 and are likely to succeed systematically. The organization also becomes heavily data-oriented. Standards are continuously evaluated and improved. The production lines are fully optimized. Very few errors or deficiencies can be found, and irregularly.

Readiness: The organizations have matured in their use of technology. The organizations are heavily data- and knowledge-driven. Intelligent systems have been implemented and efficiently utilized. They are ready for implementation of any unfamiliar technologies.

6.5   Readiness Process Areas

This section explains more detailed information on each process area following respective readiness pillars, as shown in Table 6.2.

A. Production

Production is the backbone of this readiness model. It evaluates the entire life cycle of manufacturing by determining the maturity of relevant elements. A-1-1 Products, A-2-1 Facilities, A-3-1 Operations, A-4-1 Quality, A-4-2 Real-time Status, A-4-3 Automation, and A-5-1 Agility are specific process areas of this pillar. The advancement in this readiness level clearly depicts advancement in implementation of technologies.

A-1-1 Products

Products is the single starting point of readiness proposed by this chapter. This process area studies the feasibility as well as investigates the potential improvement of products and production lines using ICT and CPS (Schumacher et al. 2016). The organization needs to conduct a background inspection on the product life cycle and determine whether implementing ICT and CPS could be beneficial. Although the implementation of ICT and CPS is generally beneficial for most products, its effectiveness depends on many factors.

The organization needs to analyze what, when, where, who, and how ICT and CPS can improve its products and product lines. Different products may receive different benefits from the same approach. For example, implementing ICT and CPS may improve the quality of product A while the same application cannot provide significant changes to product B. Additionally, areas of improvement should also be identified with clear and achievable indicators. Examples of these indicators are increased production speed, increased quantity, improved product quality, reduced defects, and improved visibility of production processes.

Although technical and financial feasibility plays a more important role for this process area. Other feasibility dimensions such as strategic, environmental, legal, schedule, safety, and resources should not be overlooked.

Objectives: To investigate the feasibility of implementation ICT and CPS.

Passing Criteria: Products and production lines which are suitable for and will be beneficial from implementation of ICT and CPS.

A-2-1 Facilities

Maturity of manufacturing infrastructure is a critical foundation of Industry 4.0. Even though the organizations have a strategic plan and suitable products for implementation, without a suitable infrastructure, implementing advanced technology can be extremely challenging. This process area involves availability of computer equipment, appropriateness of technology, and adequate spacing for automation (Maasouman and Demirli 2015). Network connectivity is not considered in this process area. Therefore, these facilities can be disconnected from each other. Inspection for adequateness should be conducted from strategic and technical perspectives.

A facility management process is also required for this process area. The main role of facility management is to monitor and maintain routine activities. A reporting mechanism and facility knowledge base are favorable for providing and communicating with stakeholders in case that undesirable events surface. Data from facility management can subsequently be used for strategic planning.

Objectives: To ensure that the organization has adequate facilities and facility management for integration of ICT and CPS.

Passing Criteria: Adequate and appropriate facilities and facility management for implementing ICT and CPS.

A-3-1 Operations

As the level of readiness increases, standards of operations need to be clearly defined. These standards should be developed and documented. Technologies are actively introduced at this stage. Although the utilization of ICT and CPS may be far from maturity, long-term plans or roadmaps for automating the operations are obviously preferred.

This process area focuses on all core business functions of the organizations, regardless of their relevance to ICT and CPS (Dallasega et al. 2020; Maasouman and Demirli 2015). Examples of these functions include accounting, researching, servicing, training, etc. In order to satisfy this process area's requirements, all relevant business functions need to be organized and visible. Their input, output, measurement, and responsibilities need to be systematically defined. This information will later assist in progressing to higher levels of readiness.

Objectives: To define core business operations.

Passing Criteria: Adequate documentation on core business operations.

A-4-1 Quality

At the fourth level of readiness, production processes are largely organized and standardized. Consequently, the focus of the pillar turns to quality. It is undeniable that quality of the products usually increases after the implementation of ICT and CPS systems (Maasouman and Demirli 2015). However, the organizations still need to define the desired level of quality for both products and production line. Indeed, needs for higher quality usually require higher investment in technology in both short and long terms.

In order to achieve desirable quality, ICT and CPS should be used as the major mechanisms to collect and analyze data from the production lines. This can be used to identify and subsequently resolve production problems.

Objectives: To investigate the desired level of quality of products and product line.

Passing Criteria: A balance in quality expectation and investment on ICT and CPS.

A-4-2 Real-Time Status

With a computerized system, monitoring and collecting data from the production lines as well as other business operations in real time become critical options to increase business competitiveness (Dallasega et al. 2020). These digital feedbacks can not only be used to improve the overall performance of the manufacturing, but also to make strategic decisions. Indeed, achieving real-time collection of data involves a certain level of investment. Therefore, the organizations may need to clearly prioritize on areas which have the most potential to gain benefit from this system.

Utilizing real-time status at this level of technological readiness spans throughout the entire supply chain. This means not only internal logistics, but information from suppliers and customers can also play important roles in this process area. Appropriate usage of real-time status can help the organizations to gain massive competitive advantages over their competitors, at the same time that it can help reduce unnecessary expenses.

Objectives: To make use of real-time data in improving operational efficiency.

Passing Criteria: A process of analyzing real-time data and examples of utilizing them in strategic decision-making or proactive planning.

A-4-3 Automation

Robotics and automation usually involve major investments (Dallasega et al. 2020). However, although they generally provide benefits to the organization, maximizing performances from these technologies is challenging. Implementing automation requires various areas of expertise and experience. The Integration of ICT and CPS is a backbone to the core production line in the organization at this stage. With this process area, the organizations are expected to have higher productivity, higher accuracy, and lower defects.

Due to the high cost of investment, the organizations may begin the implementation of automation on a smaller scale for the core business competencies. This can help them to steadily manage changes as well as expectations from stakeholders.

Disruption from the implementation of automation is also expected, especially from employees. This ranges from personal stress to major resistance. The organization needs to prepare and manage these negative forces. The balance between technology and humanity is key for a sustainable future.

Also in this stage, the organizations should begin to collect data from automated processes. These data can be used for forecasting productivity, improving overall performance, and making decisions. They are also essential for the next level of technological maturity.

Objectives: To gain the benefits of automation in terms of productivity, visibility, and reduced deficiencies.

Passing Criteria: An automated production line with adequate data collecting mechanism.

A-5-1 Agility

At the final stage of maturity, the organizations expand their focus toward optimization, agility, and leanness. Unnecessary activities are restructured (Dallasega et al. 2020; Maasouman and Demirli 2015). All resources are used at their maximum capacity. Knowledge bases are maintained. The organizations also have adequate flexibility to handle any unexpected scenarios and requirements.

Agility also encourages innovations and continuous improvement. These are essential characteristics which improve long-term business competitiveness.

Objectives: To maximize the use of organizational resources and capital.
Passing Criteria: A prototype of an efficient logistics system in a production line.

B. Technology

Technology pillar highlights the readiness of infrastructure, equipment, as well as technological protocols implemented in the organizations. It is involved in the second through the final stage of this readiness model.

B-2-1 Implementation

As for the second level of the readiness, technology is in the early introductory stage. However, the organizations need to adequately prepare relevant basic infrastructure and equipment. This usually includes a computer system and specialists (Dallasega et al. 2020). Implementation is at least done in offline mode and a number of trials and errors are expected.

A basic computer system includes hardware, software, and network equipment. Other relevant machinery such as robotics, conveyor belts, sensors, and other CPS can also be counted toward technological assets. The organizations need to learn to utilize this equipment in real production environments.

At this stage, the implementation of technology is in an introductory phase. However, the organizations need to inspect and cultivate lessons learnt from the process. With this knowledge, the organization will be able to efficiently identify and plan for long-term utilization of technology.

Objectives: To learn to utilize technology in a production line.
Passing Criteria: A production line which has technology successfully implemented.

B-3-1 Security

With more mature implementation of technology, more critical information is digitized. As a result, the organizations need to establish standards to secure their information (Dallasega et al. 2020).

Appropriate and adequate security and safety mechanisms need to be implemented to the system. The organizations should seek the latest and most secure standards to date. Additionally, an update interval needs to be scheduled to make sure that all information is secured. Backup systems are required for critical digital units.

Physical safety is also another important facet of this process area. All computer equipment needs to be secured in an appropriate space which is shielded from theft, fire, flood, or other disasters.

Objectives: To ensure that adequate security mechanisms have been implemented and maintained.

Passing Criteria: Appropriate and adequate security measures are implemented and maintained.

B-3-2 Networking

Networking of computer systems is an essential step toward digitization (Dallasega et al. 2020). At the third stage of readiness, all critical systems must be networked. This allows more efficient and accurate data collection which will be used in higher levels of technological readiness.

Redundancy systems are also preferable for this process area. In scenarios which online activities are crucial, the organizations may consider subscribing to more than one Internet or technology providers. This therefore provides redundancy to the core business functions.

Objectives: To ensure the stability and redundancy of networking systems.

Passing Criteria: A network system which has appropriate structure and redundancy.

B-4-1 Digitization

At the fourth stage of readiness, the organizations have become largely digitized. Old-fashioned paper-based documentation is converted to computerized systems (Dallasega et al. 2020; Chatzoglou and Chatzoudes 2016). Information is visible, secured, and accurate. The organizations begin to use this information in strategic planning and decision-making.

Objectives: To utilize the digitized core business function.

Passing Criteria: A core business function is digitized and its data is used in strategic planning.

B-5-1 Intelligence

At the final stage of readiness, the manufacturing becomes intelligent. All core processes are data-driven. Machine learning and other computational intelligence become the backbone of the organization.

This further increases the efficiency and effectiveness of logistics and other core business functions.

Objectives: To utilize business intelligence to continuously improve core business functions.

Passing Criteria: An appropriate utilization of intelligence on core business functions.

C. People

People is undeniably one of the most crucial capitals in modern businesses. Managing people can be extremely difficult especially when unfamiliar scenarios are introduced. This readiness pillar highlights the relevant people and their required quality in implementation of digital systems.

C-2-1 People

This process area focuses on internal staff. Indeed, in order to successfully implement advanced technology, people with the appropriate experience and expertise are required (Dallasega et al. 2020; Maasouman and Demirli 2015; Schumacher et al. 2016). Although external experts can be sought, the organizations need to appoint internal staff on at least basic solutions and maintenance of computer systems. This is to ensure that these systems can be regularly used for supporting relevant business functions.

Training is an important investment for this process area. This is specifically true when new technologies are introduced or procured. The organizations should consider both short- and long-term training for their employees. Although training can be costly, it usually provides greater benefits in the long run.

Another challenging aspect of implementing technology is a method to sustain staff's morale. It is obvious that staff, especially older and less technologically experienced employees, are likely to fear technology. Minor resistance to major disruptions can be a result of this. The organizations need certain mechanisms to lessen these potential struggles.

Objectives: To ensure that adequate staffing is allocated for technology division.

Passing Criteria: Adequacy in numbers and quality of staff for technology-related tasks.

C-3-1 Stakeholders

The focus of people in the third stage of readiness expands to external stakeholders (Schumacher et al. 2016). Due to the more implementation of technology, the organizations need to improve their communication channels in which they can cope up with digital information more appropriately. In the same way, their major stakeholders are expected to use the same level of technology. Similar to other process areas, this helps coordination between stakeholders to become more effective, efficient, accurate, and visible.

An important key to establishing effective communication between the organizations and their stakeholders is standards. There are a number of standards which are widely implemented by leading organizations. This needs to be agreed upon by their counterparts.

Objectives: To ensure that there is a standard for communicating with stakeholders.

Passing Criteria: A standard for communication between stakeholders.

C-4-1 Culture

The highest level process area of people pillar is culture. It is important for the organization to establish a culture which encourages and embraces changes, especially for technology (Dallasega et al., 2020; Schumacher et al. 2016). Indeed, the speed of change in technology is gradually increasing. Lagging behind can have the cost of a major reduction in business competitiveness. To prevent this, the organization needs to continuously update its digital capital as well as educate its human resources.

Knowledge is a critical element for maintaining culture in the organization. It can be used as the main media to share and transfer culture between staff and stakeholders.

Objectives: To build a culture which embraces implementation of technology.

Passing Criteria: A mechanism which lays the foundations for knowledge sharing and culture sustaining.

D. Logistics

The logistics pillar highlights all essential business processes regarding logistics and supply chain management.

D-3-1 Internal Logistics

At the third stage of technological maturity, all internal logistics need to be structured and manageable. The visibility of core business functions is crucial for expanding to higher levels of readiness (Dallasega et al. 2020). Obviously, implementation of computer systems in the line of logistics can satisfy this need.

The use of ICT and CPS can immensely introduce a number of novel perspectives to the manufacturing. For example, the use of RFID can make the entire logistics become thoroughly visible. Moreover, the processes are faster and more manageable. Information from this can be used for more complex logistics functions such as strategic planning and forecasting.

Objectives: To digitize internal logistics function as a prototype.

Passing Criteria: A production line which has ICT and CPS implemented and has data collected.

D-3-2 Transportation

Technology and transportation have been intensively interrelated (Dallasega et al. 2020). The use of technology to improve the efficiency of transportation is required for this readiness level. Unlike the previous process area, which has internal focus, this process area highlights external logistics and the transportation of products.

Integrating ICT and CPS in transportation can noticeably improve visibility and manageability of the logistics. Similar to internal logistics, the transportation becomes more efficient and accurate. Information from this process area can be another critical input for a higher level of technological advancement.

Objectives: To digitize external logistics function as a prototype.

Passing Criteria: A transportation function which has ICT and CPS implemented and has data collected.

E. Strategy

The final pillar in this proposed readiness model involves strategy. This includes top-level management policy and strategy to build competitiveness and sustainability for the implementation of ICT and CPS.

E-3-1 Governance

At the third stage of readiness, the organizations need to have a certain level of governance. Governance includes policies on authorities, accountabilities, roles, and responsibilities of stakeholders involved by the organizations (Chatzoglou and Chatzoudes 2016; Maasouman and Demirli 2015; Schumacher et al. 2016). An effective governance policy can also help the organizations to be flexible and competitive.

A guideline for evaluating performance of technology utilization can be identified in the governance policy. This further improves the overall manageability of the organizations. It also helps in measuring, evaluating, and judging events which may surface later.

Objectives: To ensure the availability and implementation of a digital governance.

Passing Criteria: An adequate digital governance policy.

E-5-1 Strategy

This process area highlights a long-term plan on integration of ICT and CPS for the organization. Similar to business strategy, technology strategy leads the direction of development of digital transformation (Chatzoglou and Chatzoudes 2016; Schumacher et al. 2016). Relevant technologies, governance framework, and implementation policy should be continuously reviewed and revised on a timely basis.

Strategy has to be based on facts from data cultivated from all other process areas in the lower level of readiness model. With actual data, planning is likely to be more accurate and the implementation is likely to be more efficient. Several levels of strategy, e.g. short-term, medium-term and long-term, may be needed according to business needs and competition.

Objectives: To establish a strategy for integration of ICT and CPS.

Passing Criteria: An adequate strategy based on actual data.

6.6   Conclusion and Outlook

The proposed model can be used for assessing the readiness of integration of ICT and CPS for organizations. Five essential pillars of readiness have been identified, which are production, technology, people, logistics, and strategy. Each pillar encapsulates several process areas. The organizations need to satisfy the objectives and passing criteria in order to accredit each process area. The readiness model is classified into 5 levels.

The first level is the initial stage of technology implementation. In order to achieve this level, the organizations need to understand the potential of their products and conduct a basic feasibility study. As for the second level of readiness, the organizations begin their pilot integration. This includes several ad hoc processes which may result in errors and defects. Yet, it is an important stage which builds the foundations for more advanced technologies to be implemented later. The third stage of readiness defines the organizations which have already established the utilization of ICT and CPS in production lines. Although deficiencies can be regularly found, the organizations appear to realize the benefits of technology and aim to further utilize it in the future. As for the fourth stage of the readiness, the organizations build up their knowledge base and partly automate core business processes. Finally, the final stage identifies the organizations which optimize their technological usage and continuously improve their perspectives. In order to acquire a level of readiness, all process areas in that and lower levels need to be achieved.

This readiness model is suitable to be adopted by SMEs or larger organizations. Depending on the size, focus, and market, SMEs can use the objective and passing criteria identified in each process area to evaluate their readiness. The model will help SMEs to investigate their incompetencies, which they can address later.

References

Akdil, K.Y., Ustundag, A., and Cevikcan, E. 2018. Maturity and Readiness Model for Industry 4.0 Strategy. Industry 4.0: Managing The Digital Transformation, Springer Series in Advanced Manufacturing. Springer, Cham. https://doi.org/10.1007/978-3-319-57870-5_4.

Amr, M., M. Ezzat, and S. Kassem. 2019. Logistics 4.0: Definition and historical background. *Novel Intelligent and Leading Emerging Sciences Conference (NILES)* 2019: 46–49. https://doi.org/10.1109/NILES.2019.8909314.

Barreto, L., A. Amaral, and T. Pereira. 2017. Industry 4.0 implications in logistics: An overview. *Procedia Manufacturing* 13: 1245–1252. https://doi.org/10.1016/j.promfg.2017.09.045.

Chatzoglou, P., and D. Chatzoudes. 2016. Factors affecting e-business adoption in SMEs: An empirical research. *Journal of Enterprise Information Management* 29 (3): 327–358. https://doi.org/10.1108/JEIM-03-2014-0033.

Council of Supply Chain Management Professionals (CSCMP). 2020. CSCMP's Definition of logistics management. https://cscmp.org Accessed on August 13, 2020.

Dallasega, P., Woschank, M., Zsifkovits, H., Tippayawong, K., Brown, C.A. 2020. Requirement analysis for the design of smart logistics in SMEs. In Matt D., Modrák V., Zsifkovits H. (eds.) Industry 4.0 for SMEs. Palgrave Macmillan, Cham. https://doi.org/10.1007/978-3-030-25425-4_5.

Facchini, F., J. Oleśków-Szłapka, L. Ranieri, and A. Urbinati. 2020. A maturity model for logistics 4.0: An empirical analysis and a roadmap for future research. *Sustainability* 12 (1): 86. https://doi.org/10.3390/su12010086.

Ganzarain, J., and N. Errasti. 2016. Three stage maturity model in SME's toward industry 4.0. *Journal of Industrial Engineering and Management (JIEM)* 2016–9 (5): 1119–1128. https://doi.org/10.3926/jiem.2073.

Geissbauer, R., Vedso, J., Schrauf, S. 2016. Industry 4.0: Building the digital enterprise. https://www.pwc.com/gx/en/industries/industries-4.0/landing-page/industry-4.0-building-your-digital-enterprise-april-2016.pdf. Accessed on July 30, 2020.

Gökalp, E. Sener, U., and Eren, P. 2017. Development of an assessment model for industry 4.0: Industry 4.0-MM. International Conference on Software Process Improvement and Capability Determination. https://doi.org/10.1007/978-3-319-67383-7_10.

Hernández, M.E.P. and Reiff-Marganiec, S. 2014. Classifying smart objects using capabilities. 2014 International Conference on Smart Computing, Hong Kong, 309–316. https://doi.org/10.1109/SMARTCOMP.2014.7043873.

Kompetenzzentrum Mittelstand NRW. 2020. Quick check industrie 4.0 Reifegrad. https://indivsurvey.de/umfrage/53106/uHW7XM. Accessed on June 25, 2020.

Lasi, H., Fettke, P., Feld, T. and Hoffmann, M. 2014. Industry 4.0. Business & Information Systems Engineering, 6(4), 239–242. https://aisel.aisnet.org/bise/vol6/iss4/5.

Lee, J., B. Bagheri, and H.A. Kao. 2015. A cyber-physical systems architecture for industry 4.0-based manufacturing systems. *Manufacturing Letters* 3: 18–23. https://doi.org/10.1016/j.mfglet.2014.12.001.

Lichtblau, K., Stich, V., Bertenrath, R., Blum, M., Bleider, M., Millack, A., Schmitt, K., Schmitz, E., Schröter, M. 2015. IMPULS-Industrie 4.0-Readiness. Impuls-Stiftung des VDMA, Aachen-Köln.

Lu, Y. 2017. Cyber Physical System (CPS)-based industry 4.0: A survey. *Journal of Industrial Integration and Management.* 2 (3): 1750014. https://doi.org/10.1142/S2424862217500142.

Maasouman, M.A., and K. Demirli. 2015. Assessment of lean maturity level in manufacturing cells. *IFAC-PapersOnLine* 48 (3): 1876–1881. https://doi.org/10.1016/j.ifacol.2015.06.360.

Oleśków-Szłapka, J., and A. Stachowiak. 2018. The framework of logistics 4.0 maturity model. *Intelligent Systems in Production Engineering and Maintenance* 771–781. https://doi.org/10.1007/978-3-319-97490-3_73.

Porter, M.E., and J.E. Heppelmann. 2014. How smart, connected products are transforming competition. *Harvard Business Review* 92 (11): 64–88.

Rauch, E., M. Unterhofer, R. Rojas, L. Gualtieri, M. Woschank, and D. Matt. 2020. A maturity level-based assessment tool to enhance the implementation of industry 4.0 in small and medium-sized enterprises. *Sustainability.* 12: 1–18. https://doi.org/10.3390/su12093559.

Rockwell Automation. 2014. The Connected Enterprise Maturity Model. https://literature.rockwellautomation.com/idc/groups/literature/documents/wp/cie-wp002_-en-p.pdf. Accessed on June 30, 2020.

Rojas, R., E. Rauch, R. Vidoni, and D.T. Matt. 2017. Enabling connectivity of cyber-physical production systems: A conceptual framework. *Procedia Manufacturing* 11: 822–829. https://doi.org/10.1016/j.promfg.2017.07.184.

Schiffer, M., H.H. Wiendahl, and B. Saretz. 2019. Self-assessment of industry 4.0 technologies in intralogistics for SME's. *IFIP International Conference on Advances in Production Management Systems (APMS)* 339–346. https://doi.org/10.1007/978-3-030-29996-5_39.

Schrauf, S., and Berttram, P. 2016. Industry 4.0: How digitization makes the supply chain more efficient, agile, and customer-focused. https://www.strategyand.pwc.com/gx/en/insights/2016/digitization-more-efficient.html. Accessed on June 30, 2020.

Schumacher, A., S. Erol, and W. Sihn. 2016. A maturity model for assessing Industry 4.0 readiness and maturity of manufacturing enterprises. *Procedia CIRP* 52: 161–166. https://doi.org/10.1016/j.procir.2016.07.040.

Sternad M., Lerher T., and Gajšek B. 2018. Maturity Levels for Logistics 4.0 Based on NRW's Industry 4.0 Maturity Model. 18th international scientific conference Business Logistics in Modern Management. 695–708.

Trotta, D., and Garengo, P. 2019. Assessing industry 4.0 maturity: An essential scale for SMEs. 2019 8th International Conference on Industrial Technology and Management (ICITM). https://doi.org/10.1109/ICITM.2019.8710716.

Winkelhaus, S., and E.H. Grosse. 2019. Logistics 4.0: A systematic review towards a new logistics system. *International Journal of Production Research* 58 (1): 18–43. https://doi.org/10.1080/00207543.2019.1612964.

Uckelmann D. 2008. A definition approach to smart logistics. In Balandin S., Moltchanov D., Koucheryavy Y. (eds.) Next generation teletraffic and wired/wireless advanced networking. NEW2AN 2008. Lecture Notes in Computer Science, 5174. Springer, Berlin, Heidelberg. https://doi.org/10.1007/978-3-540-85500-2_28.

7

Automated Performance Measurement in Internal Logistics Systems

Chiara Raith, Manuel Woschank, and Helmut Zsifkovits

7.1 Introduction

With ongoing digitalization, requirements regarding performance, efficiency, and adaptability of logistics systems are steadily increasing. In terms of this digital transformation, demand for an increase of performance among transparency, cost efficiency, and innovation capability of companies is growing. Studies show that industrial companies of all sectors have already recognized the relevance of automation and digitalization for planning and strategy development (Dallasega et al. 2019b, 2020; Staufen AG, Staufen Digital Neonex GmbH 2018).

C. Raith (✉) · M. Woschank · H. Zsifkovits
Chair of Industrial Logistics, Montanuniversitaet Leoben, Leoben, Austria
e-mail: chiara.raith@unileoben.ac.at

M. Woschank
e-mail: manuel.woschank@unileoben.ac.at

H. Zsifkovits
e-mail: helmut.zsifkovits@unileoben.ac.at

© The Author(s) 2021
D. T. Matt et al. (eds.), *Implementing Industry 4.0 in SMEs*,
https://doi.org/10.1007/978-3-030-70516-9_7

In that context, Industry 4.0 concepts, more specific automation and robotics together with cloud computing and blockchain technology have been ranked as megatrends to meet the high requests for efficiency and adaptability (SCI Verkehr 2019). Especially, in internal logistics systems approaches for implementing automation are regarded as relevant tools for increasing efficiency (Bundesvereinigung Logistik 2018).

Examples of practice show the combination of machine, storage and picking systems, which are connected by conveyor and transport systems and linked via information and communication systems. Automated high-bay warehouses and automated small-parts warehouses are often used for this purpose. These are commonly supplied by automated loading and unloading systems or industrial robots and are connected to picking stations and manual workstations via a combination of various types of carousels, conveyor belts and conveyor-based sorting systems, Automated Guided Vehicles (AGVs), automatic forklift trucks, etc. Such networked systems operate via high-performance control technology and are assisted by information and communication systems, Transport Management Systems (TMS), Warehouse Management Systems (WMS), Enterprise Resource Planning Systems (ERP), Product Lifecycle Management (PLM) solutions, Inventory Management Software, etc. (Inboundlogistics 2018).

7.2   Problem Formulation

Thus, these networked systems can be regarded as integrated internal logistics systems of high complexity. These complexity and variety emerge from the fact that the system cannot be comprehensively described by a single performance indicator (e.g., throughput). Moreover, an exhaustive characterization of the system properties concerning performance and availability is only possible employing complementary parameters (Follert and Nagel 2006).

Despite a large number of possible benefits from implementing automation concepts in internal logistics processes, Granlund and Wiktorsson (2014) highlight among other factors the need for an automation strategy and the lack of performance measurement of internal logistics as challenges to automation in this sector. Availability and performance are important parameters for monitoring and controlling of these automated systems. To track them over the life cycle of the plant and to be able to make transparent statements about the condition of the plant, it is necessary to collect selected key performance indicators (KPIs) for systematic monitoring (Gottmann 2016).

Currently, the determination and calculation of performance and availability parameters within the site acceptance tests of automated logistics systems is plant-specific and involves a great amount of manual effort. In this chapter, the authors develop a concept for the automated determination of performance indicators for storage and conveying systems. The remainder is organized as follows. In Sect. 7.3, the contribution of constant monitoring and controlling as an enabler for high-level responsiveness and systematic planning is discussed. Section 7.4 will review state-of-the-art approaches, give an overview of applied standards, and outline their limitations in practical application. In Sect. 7.5, the results of the conducted expert interviews are reported and analyzed. Taking these findings into consideration, the authors develop a concept for an exhaustive evaluation regarding the performance and availability of automated systems.

7.3 Monitoring and Controlling—Enablers for High-Level Responsiveness and Systematic Planning

The term 'Controlling' was strongly influenced by business practice and is therefore used in various fields of activity. It describes the interaction of planning, control, and information supply (Weber and Wallenburg 2010). There are different levels of the view of controlling from a pure information system to the integration of a personnel management and organization system (Arnold et al. 2008). For this reason, a large number of definitions can be found in the literature. Koch (2012) defines controlling as the task of providing information to support the decision-making process. Arnold et al. (2008) assigns controlling not only the task of processing and preparing information but also the development and support of operative and strategic planning. According to Klaus et al. (2012), controlling has the task of ensuring the rationality of management. Controlling has to ensure that management decisions are made in a ratio of intuition and reflection that is appropriate to the respective problem. A study conducted by the International Controller Association (ICV) evaluated the understanding of controlling functions based on a large-scale field study (Weber and Wallenburg 2010). The study shows that controlling is not only assigned a transparency function but also the task of ensuring the rationality of managerial activities. Rational decisions can only be made based on a comprehensive knowledge of the action alternatives and their effects on the set objectives.

Based on this definition of controlling and the described lack in performance measurement, greater relevance can be accorded to the monitoring and controlling of internal logistics systems. Within the concept of controlling, monitoring is intended to take over tasks to provide support via regular reports. This includes the rapid availability of key figures and graphical preparation and visualization. Standardized monitoring is realized with the help of precisely defined key figures to record processes and document their development (Wagner and Patzak 2015). However, the comprehensive performance measurement enables the early detection of deviations from set objectives. Thus, non-value-adding activities and rationalization potentials can be identified (Werner 2014).

In the control of internal logistics systems, the economic aim is the optimal and efficient operation of the system. In this context, automation offers possibilities to optimize material availability and material flow coordination as well as to gain error reduction and to improve machine utilization. Automated material flow systems are based on information and control technology which are linked via suitable communication technology (Jünemann and Beyer 1998). According to that, the described systems also have the task to continuously record the movements of storage objects to enable value- and quantity-based reporting. The control technology of these material flow systems is designed to enable the mapping of control functions at different system levels, to provide a high level of data security and availability (Schulte 2013). At all these system levels, the data required for monitoring and controlling are collected to help to quickly detect and eliminate faults and malfunctions in the system and keep downtimes low (Jünemann and Beyer 1998; ten Hompel et al. 2008).

7.4   State-of-the-Art and Literature Review

In the context of monitoring the condition of internal logistics systems, respectively, plants in general, literature frequently refers to the term 'reliability of technical systems' (Eberlin and Hock 2014; vom Bovert and Jünemann 2001; Gudehus 1976). Thereby, the term reliability covers the technical availability of a system and describes it as the expected value for a plant component to be in a functional state at a certain time under given circumstances (VDI 4001).

The performance of internal logistics systems in line with defined requirements represents a crucial success factor for plant operators as well as for plant suppliers. Typically, within the site acceptance tests, the two main performance measurement indicators, throughput and technical availability are calculated (Maier et al. 2011). Since the 1970s the term technical availability has been continuously developed and several standards have been created to regulate the definition and calculation (Fig. 7.1).

The chronological classification of the development of the valid set of standards ranges from basic thoughts on availability, as Timm Gudehus introduces them, to current considerations on the term performance availability. In practice, the elicitation of the two factors performance and availability is regulated by the following set of defined standards (Table 7.1).

FEM 9.851 describes a procedure for calculating the cycle time and the related handling performance of storage and retrieval machines (SRM) with automatic control and pick-up of a loading unit. Despite simplifications, such as the definition of typical movements and average cycle times, a good approximation to the exact average value can be achieved.

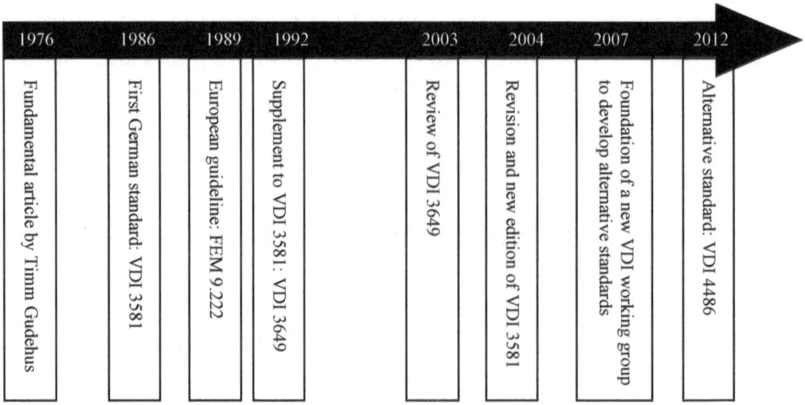

Fig. 7.1 Chronological classification of the development of the standards

Table 7.1 Selected standards for performance and availability measurement

Standards number	Thematic content
FEM 9.221	Proposes a method for the determination of reliability and **availability** of storage and retrieval machines and defines a procedure for carrying out related tests in practice
FEM 9.222	Proposes further methods for determining the **availability**, and regulates the steps for putting into operation, handing over and acceptance of storage and conveying systems
FEM 9.851	Specifies a method for determining the cycle times and thus the handling **performance** of storage and retrieval machines
VDI 3580	Provides instructions on tracking failures for the **availability** calculation
VDI 3581	Contains theoretical basics and generally applicable formulae for the **availability** calculation considering the material flow structure
VDI 3649	Represents a supplement to VDI 3581 and shows the influence of the availability of individual elements on the entire system and possibilities for increasing **availability**
VDI 4486	Introduces the term of **performance availability** and describes a method for availability measurement under consideration of the operator's business process

In guideline VDI 3581 the preparation and realization of availability tests are regulated. Equivalent circuit diagrams are established and calculation schemes in line with the structure of the system, more precisely with or without redundancies, are described. Using Boolean theory, complex structures are divided into simple serial or parallel substructures and thus calculated in several iteration steps. Though, taken into account that the presence of buffers and the coincidence of downtimes of different elements cannot be considered, the Boolean method reaches its limits (Maier et al. 2011). After the development of the first form of VDI 3581 and the FEM 9.222 based on it, VDI 3649 was created as a supplement to evaluate the influence of individual system elements on the overall system availability and discuss how it can be increased by a revised element arrangement.

Regarding systems performance, the existing guidelines and consequently the proposed acceptance tests assume an idealized order structure. However, a deviating structure and internal company strategies for maintenance and monitoring of the system have a significant influence on the performance (Hegmanns et al. 2014). For this reason, an extended performance and availability analysis is required. VDI 4486 introduces the term performance availability and attempts to focus on the business process of the plant operator. Associated with the idea of being able to supply all customers of the logistics service on time and in line with their needs, performance availability indicates the degree of fulfillment of processes under agreed requirements and deadlines. For this purpose, redundancies, performance reserves and buffer capacities are taken into account in the calculation.

The dependencies between the individual components and the subsystems of the plant are critical for determining this parameter, as is the uncertainty about the extent to which various influencing variables affect the overall system. For this reason, and due to the lack of maturity of various approaches, an analytical calculation of performance availability proves to be very complex and not appropriate (Schieweck et al. 2016).

Following the methods presented in the guidelines, the parameters and details of the acceptance tests are determined for each project individually for each customer. This involves a high level of manual effort in the preparation, execution, and processing of the test results. Despite the necessity for manual documentation of the plant malfunctions, examples of practice show that the preparation and execution of the availability tests are personnel- and thus cost-intensive. Further restrictions in the performance measurement lie in standardized measurement scenarios. These are used to generate sufficiently good statements about the system in a reasonable time, yet, they allow only limited conclusions about the performance at full load. The manual execution and the associated effort in determining performance and availability result in only limited performance checks.

Automation regarding the information flow of the internal logistics system can increase the overall efficiency (Granlund and Wiktorsson 2014), hence it is vital to provide readily available information concerning the performance of the plant. Named restrictions can be overcome with the implementation of automated performance measurement. Whereas many research projects deal with concepts for performance measurement in manufacturing processes, a literature review on performance measurement and monitoring in internal logistics systems reveals that there have been relatively few attempts to systematically address the lack of automation and continuity in the measurement of performance and availability of these systems (Table 7.2).

The literature review outlines the shortfall of related work on automated performance measurement in internal logistics systems. Table 7.2 provides some examples of publications on this topic. Mörth et al. (2020) proposes an approach for performance monitoring based on Cyber-Physical Systems (CPS). For testing, a CPS demonstrator was implemented on a real conveyor belt. This offers a small-scale realization of a data process chain, from data generation and processing followed by the estimation of the visualization of appropriate performance monitoring on a dashboard. Pei et al. (2019) develops a method to develop an assessment tool for intralogistics. To analyze and evaluate intralogistics´ current status quo the authors consider Cyber-Physical Production Systems (CPPS) enabling technology. Alves et al. (2015) proposes a framework for mapping the current performance of internal logistics flows. Based on Multicriteria Constructivist methods the approach aims to assist in the identification, organization, measurement, and integration of performance variables.

Table 7.2 Literature on performance measurement in internal logistics systems

Literature	Description
Fabri et al. (2020)	Uses a Discrete-Event-Simulation and a set of KPIs to assess the logistics flows' performance
Mörth et al. (2020)	Introduces a conceptual model for Internet of Things (IoT)-enabled data process chains linked to performance measurement for internal logistics systems
Moons et al. (2020)	Uses a logistics performance measurement framework based on the Analytic Network Process to assess the efficiency of replenishment scenarios
Guerreiro et al. (2019)	For intralogistics process planning the paper presents a Big Data architecture to extract, handle, further process data and apply analytics
Pei et al. (2019)	Develops an assessment tool to analyze and evaluate the intralogistics' performance by considering Cyber-Physical Production Systems enabling technology
Alves et al. (2015)	Based on Multicriteria Constructivist methods, the paper proposes an evaluation framework for performance measurement of the internal logistics for service companies

Synthesizing the findings of the reviewed literature, previous studies are dealing with the evaluation and performance measurement of internal logistics in the scope of manufacturing operations and production systems. Recent works often take Industry 4.0 concepts—Internet of Things, Big Data, Cyber-Physical Systems, Cloud Computing, etc.—into consideration. The main distinction of the present work is the focus on internal logistics consisting of storage and conveying systems. A specific selection of input parameters for the evaluation of the combination of machine, storage and picking systems is investigated and the automation and real-time data availability for performance measurement is discussed.

7.5   Deduction of a Model for Availability and Performance Assessment

The design of a system concerning its performance is one of the most important tasks in the planning of internal logistics systems. For this purpose, there is a multitude of possibilities to make statements about the expected performance and availability. As addressed in the previous section, common practice procedures are linked to high personnel and cost intensity. Nevertheless, the effort required for this purpose should be in appropriate relation to the quality of the result.

Performance, information, and process factors are seen as key success factors for internal logistics systems (Granlund and Wiktorsson 2014). The logistics performance determination is subject to uncertainties. Performance in this context can be very diverse and therefore difficult to measure. Logistics performance can be shown by a specific selection of objectively measurable variables (Weber and Wallenburg 2010).

To be able to monitor and evaluate the two factors of availability and performance over the life cycle of the plant, key figures and corresponding monitoring are necessary (Müller and Lenz 2013). For controlling and management, the key figure system is an important instrument for making changes more transparent and monitoring the effects of decisions through target/actual comparisons. The approach also allows activities in individual areas of responsibility to be reviewed and weaknesses to be identified. This enables effective control (Vollmuth 2007). Due to its adaptability to different application purposes, the KPI system is also suitable as a basis for the sought-after model for performance and availability evaluation. Thus, a dashboard with key performance indicators is designed based on this concept.

To gain the input parameters for the searched conceptual model, a qualitative research approach is applied. The explorative interview aims to help raise awareness of the problem and generate hypotheses. To facilitate this, the interviews are conducted relatively openly and the respondents are given the opportunity for digressions and changes of topic. Nevertheless, a conversation guideline is used to ensure the comparability and completeness of the data (Bogner 2005).

Table 7.3 Selection of interview partners, field of activity, location

Expert	Department	Country
Expert 1	Research area plant management	Austria
Expert 2	Project Realization	Germany
Expert 3	Spare parts management	Germany
Expert 4	Sales	Germany
Expert 5	Project Realization	Austria
Expert 6	Maintenance and Support	France
Expert 7	Maintenance and Support	Switzerland
Expert 8	Sales	Germany
Expert 9	Software development	Austria

Semi-structured expert interviews were conducted to collect data and input parameters. For this purpose, a representative selection of experts from different departments of a leading intralogistics provider with multiple offices in Europe was made in advance, which was complemented by experts in the field of plant management (Table 7.3). All interviews were carried out as one-on-one interviews.

Furthermore, the systematic conduct of the interviews was ensured by the preparation of an interview guideline based on the examples and design recommendations discussed in (Kruse 2009). The following issues have been taken into account in the preparation of the guideline:

- No closed questions
- No alternative or multiple questions
- No direct suggestive questions
- No judgmental questions
- A simple choice of words adapted to the sociolinguistic level of the interviewee.

In total nine interviews were conducted, in which the interlocutors were asked about their experience with the determination of performance and availability in internal logistics systems and its components. The questions focused on the identification of KPIs and information that allow statements about the performance and availability of individual components as well as about the entire plant. Findings and named parameters were aggregated into factors and ranked based on the frequency of their

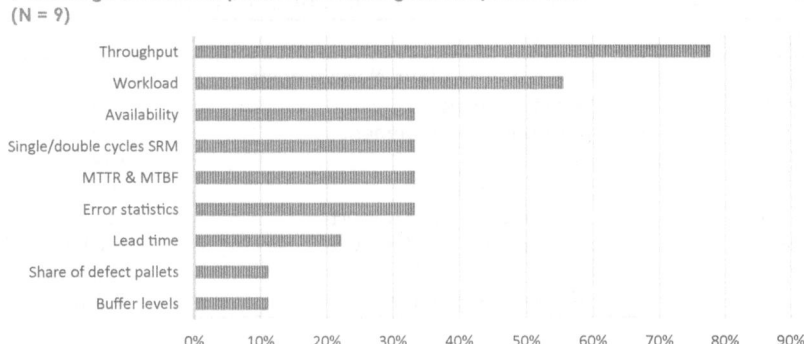

Percentage of interview partners mentioning named parameters
(N = 9)

Fig. 7.2 KPI factors by frequency of mentions

mentions in the interviews (Fig. 7.2). In the process, if one parameter or a synonym was mentioned multiple times in an interview, nevertheless it was counted as one mentioning.

Additionally, four out of nine experts mentioned that for better significance KPIs should be formulated for specific subsystems—storage and retrieval machines, picking/manual working stations and conveying technology. The following five most mentioned factors were selected for describing the conceptual model:

- **Throughput:**
  As the main parameter for performance measurement throughput was mentioned in seven interviews. Further, five experts pointed out the need for historical data-comparison in form of maximum throughput and current throughput.
- **Workload:**
  Knowing the actual order strain of the entire system to evaluate a plant´s performance was considered as an important factor by five of the experts.
- **Single/double cycles SRM:**
  According to 33% of the interlocutors, the throughput and, respectively, the performance of storage and retrieval machines (SRM) are best measured by the amount of single/double cycles.

- **Availability and MTTR & MTBF:**
  As reported by the interview findings, availability should be calculated using MTTR (Mean Time To Repair) and MTBF (Mean Time Between Failure) and composing the ratio. Nevertheless, three interlocutors mentioned that MTTR and MTBF should be displayed separately for the evaluation of error handling.
- **Error statistics:**
  Three of the interlocutors explained the necessity of knowing the ten most frequently occurring errors. This was considered helpful for error handling and advanced planning.

Visualizing the surveyed factors for performance measurement, in Fig. 7.3 the authors propose a dashboard design. The clear depiction of relevant key figures makes it possible to quickly assess the condition of the plant and evaluate the strategies adopted.

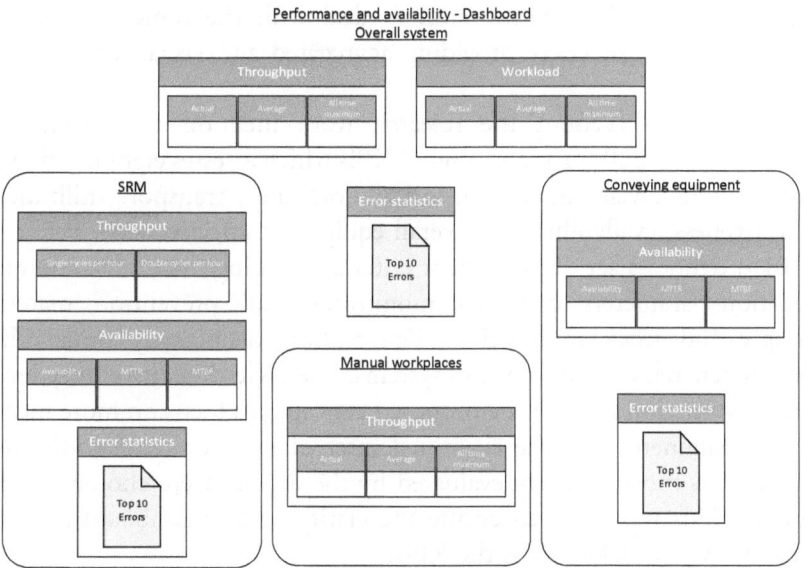

Fig. 7.3 Proposed dashboard for performance measurement

The expert interviews show that 33% of the respondents would consider the availability of the mentioned factors on a real-time basis as helpful for decision making and evaluation of operating strategies. This is in line with findings in other researches (Keivanpour and Ait-Kadi 2019; Dallasega et al. 2019a; Lee et al. 2018). Decision making and thus productivity loss can be improved by providing and processing real-time data (Syafrudin et al. 2018).

7.6   Discussion and Further Research Directions

The concepts of automation can be classified either into mechanization, relating to the automation of physical flows, or into computerization, referring to the automation of the information flow (Granlund and Wiktorsson 2014; Zsifkovits and Woschank 2019; Woschank et al. 2020). In this chapter, the authors deal with the topic of computerization, more precisely providing aggregated and visualized real-time data.

Taking into account the research work mentioned in Sect. 7.4, Mörth et al. (2020) proposes a set of KPIs which assesses eight key figures (throughput, cycle time, transport performance, transport utilization, effectiveness, availability and overall equipment effectiveness). Pei et al. (2019) defines a set of assessment criteria including, inter alia, communication parameters, condition monitoring, fault prevention, material supply and stock monitoring. The authors deliberately omit other parameters relevant for logistics systems, like cycle time, inventory level, and information flow, in the proposed concept. To focus on more meaningful parameters for the examined internal logistics systems, the five parameters most critically evaluated by the experts were chosen for the model. This was done to ensure the clarity of the dashboard and the possibility to quickly assess the KPIs.

An integrated monitoring system assists to prevent performance degradation and failures (Jain and Lad 2019). With the integration of the model into the existing software, control technology and information and communication systems, the collection of KPIs can be automated.

Thus, the manual effort and associated costs are reduced, the accessibility is increased and the continuous improvement in efficiency is facilitated.

Literature (Orellana and Torres 2019; Helo and Shamsuzzoha 2020; Huang et al. 2019; Lee et al. 2018; Hwang et al. 2017) shows that the availability of indicators to assess the impact of the operation on the set objectives at real-time is used to facilitate continuous improvement processes. Fawcett and Cooper (1998) conducted a survey study with 111 firms, in which the higher-performing firms were found to place greater emphasis on performance measurement. The relation between efficient logistics processes and the access to measurement information emerges and the comprehensive performance measurement is revealed as a requirement for improvement of efficiency and operational performance.

The continual condition monitoring of a plant enhances the identification of causes for downtimes. Thus, actions can be developed to eliminate them. This leads to less downtime. Maintenance strategies can also be compared, evaluated and, if necessary, adapted to changing conditions (Jünemann and Beyer 1998). Using automation and digitization enables in-depth reliability analysis and condition-based maintenance. Based on the inspection of operating conditions, optimal maintenance actions are suggested (Wang et al. 2020). Due to the described function and application, the created model not only supports the maintenance strategy but also serves as a tool for preventive maintenance.

The presented work offers a starting point for further research in the direction of extended performance measurement of storage and conveying systems. Identifying the characteristics of automated high-bay warehouses, automated small-parts warehouses and various combinations of machine, storage and picking systems enables a better understanding of internal logistics systems and their complexity. The authors propose a model extending the site acceptance tests according to the valid set of standards. Some limitations in the calculation of performance and availability parameters have to be addressed in future research work. Overcoming restrictions caused by the simplification of structures and processes of internal logistics systems for calculation purposes and the generation of data are topics for further research.

7.7 Conclusions

This chapter proposes a conceptual model for automated performance measurement of internal logistics systems. Based on an intensive literature review, semi-structured expert interviews were conducted to develop this concept, in which the possibility of monitoring the plant condition over its life cycle using selected key figures was discussed and the input for the concept sought was generated.

The model obtained provides based on meaningful key figures information about the condition of the plant and enables early detection of performance and availability losses. This allows preventive maintenance measures to be introduced, taking specific and resource-saving measures, and avoiding long downtimes.

Integrating the model into the existing software automates the collection of KPIs and hence reduces the manual effort and personnel resources required for the preparation and execution of the tests. The software integration also contributes to the standardization of the process for availability and performance assessment. A natural next step to produce industrially relevant solutions would be the formulation of a methodology to integrate the conceptual model in a software solution and to evaluate the model application in industrial environments under full load conditions. The automated data acquisition, management of data complexity and technology capabilities in this context offer a broad field for further research work.

Acknowledgments This project has received funding from the European Union's Horizon 2020 research and innovation program under the Marie Skłodowska-Curie grant agreement No 734713.

References

Alves, V.T., J.C. Mairesse-Siluk, A.L. Neuenfeldt-Júnior, M. Soliman, and L.D. Dalla-Nora. 2015. Performance assessment of internal logistics for service companies. *Revista Facultad de Ingenieria* 1 (74): 188–199.

Arnold, D., H. Isermann, A. Kuhn, H. Tempelmeier, and K. Furmans. 2008. *Handbuch Logistik*, 3rd ed., revised. VDI-Buch. Berlin: Springer.

Bogner, A. 2005. *Das Experteninterview: Theorie, Methode, Anwendung*, 2nd ed. Wiesbaden: VS Verl. für Sozialwiss.

Bundesvereinigung Logistik. 2018. Expertenbefragung zur künstlichen Intelligenz in der Logistikbranche. Statist.com. https://de.statista.com/progno sen/943372/expertenbefragung-zur-kuenstlichen-intelligenz-in-der-logistikb ranche. Accessed on 7 April 2020.

Dallasega, P., R.A. Rojas, G. Bruno, and E. Rauch. 2019a. An agile scheduling and control approach in ETO construction supply chains. *Computers in Industry* 112: 103–122. https://doi.org/10.1016/j.compind.2019.08.003.

Dallasega, P., M. Woschank, S. Ramingwong, K. Tippayawong, and N. Chonsawat. 2019b. Field study to identify requirements for smart logistics of European, US and Asian SMEs. Proceedings of the International Conference on Industrial Engineering and Operations Management Bangkok, Thailand, March 5–7, 844–855.

Dallasega, P., M. Woschank, H. Zsifkovits, K. Tippayawong, and C.A. Brown. 2020. Requirement analysis for the design of smart logistics in SMEs. In *Industry 4.0 for SMEs*, ed. D.T. Matt, V. Modrák, and H. Zsifkovits, 147–162. Cham: Springer. https://doi.org/10.1007/978-3-030-25425-4_5.

Eberlin, S., and B. Hock. 2014. *Zuverlässigkeit und Verfügbarkeit technischer Systeme: Eine Einführung in die Praxis*. Wiesbaden: Springer Vieweg.

Fabri, M., H. Ramalhinho, M. Oliver, and J.C. Muñoz. 2020. Internal logistics flow simulation: A case study in automotive industry. *Journal of Simulation*: 1–13. https://doi.org/10.1080/17477778.2020.1781554.

Fawcett, S.E., and M.B. Cooper. 1998. Logistics performance measurement and customer success. *Industrial Marketing Management* 27 (4): 341–357. https://doi.org/10.1016/S0019-8501(97)00078-3.

FEM 9.221. 10.1981. Leistungsnachweis für Regalbediengeräte.

FEM 9.222. 06.1989. Regeln über die Abnahme und Verfügbarkeit von Anlagen mit Regalbediengeräten und anderen Gewerken.

FEM 9.851. 06.2003. Leistungsnachweis für Regalbediengeräte.

Follert, G., and L. Nagel. 2006. Verfügbarkeit intralogistischer Systeme. *Logistics Journal*.

Gottmann, J. 2016. *Produktionscontrolling: Wertströme und Kosten optimieren*, 1st ed. Wiesbaden: Springer Gabler.

Granlund, A., and M. Wiktorsson. 2014. Automation in internal logistics: Strategic and operational challenges. *International Journal of Logistics Systems and Management (IJLSM)* 18 (4): 538. https://doi.org/10.1504/IJLSM.2014.063984.

Gudehus, T. 1976. Zuverlässigkeit und Verfügbarkeit von Transportsystemen: Teil1: Kenngrößen der Systemelemente. *f+h Fördern und Heben* 26 (13): 1029–1033.

Guerreiro, G., P. Figueiras, R. Costa, M. Marques, D. Graça, G. Garcia, and R. Jardim-Gonçalves (eds.). 2019. A digital twin for intra-logistics process planning for the automotive sector supported by big data analytics. ASME International Mechanical Engineering Congress and Exposition, Volume 2B: Advanced Manufacturing. https://doi.org/10.1115/IMECE2019-11362.

Hegmanns, T., A. Kuhn, M. Roidl, S. Schieweck, M. ten Hompel, M. Güller, et al. 2014. Planung und Berechnung der systemischen Leistungsverfügbarkeit komplexer Logistiksysteme. *Logistics Journal: Proceedings*. https://www.logistics-journal.de/proceedings/2014/4062/hegmanns_2014wgtl.pdf.

Helo, P., and A.H.M. Shamsuzzoha. 2020. Real-time supply chain—A blockchain architecture for project deliveries. *Robotics and Computer-Integrated Manufacturing* 63: 101909. https://doi.org/10.1016/j.rcim.2019.101909.

Huang, S., Y. Guo, S. Zha, and Y. Wang. 2019. An internet-of-things-based production logistics optimisation method for discrete manufacturing. *International Journal of Computer Integrated Manufacturing* 32 (1): 13–26. https://doi.org/10.1080/0951192X.2018.1550671.

Hwang, G., J. Lee, J. Park, and T. Chang. 2017. Developing performance measurement system for Internet of Things and smart factory environment. *International Journal of Production Research* 55 (9): 2590–2602. https://doi.org/10.1080/00207543.2016.1245883.

Inboundlogistics (ed.). 2018. 2018 top 100 logistics IT market research survey. https://www.inboundlogistics.com/cms/article/2018-top-100-logistics-it-market-research-survey/. Accessed on 20 May 2020.

Jain, A.K., and B.K. Lad. 2019. A novel integrated tool condition monitoring system. *Journal of Intelligent Manufacturing* 30 (3): 1423–1436. https://doi.org/10.1007/s10845-017-1334-2.

Jünemann, R., and A. Beyer. 1998. *Steuerung von Materialfluß- und Logistiksystemen: Informations- und Steuerungssysteme, Automatisierungstechnik.* Logistik in Industrie, Handel und Dienstleistungen. Berlin and Heidelberg: Springer.

Keivanpour, S., and D. Ait-Kadi. 2019. Internet of Things enabled real-time sustainable end-of-life product recovery. *IFAC-PapersOnLine* 52 (13): 796–801. https://doi.org/10.1016/j.ifacol.2019.11.213.

Klaus, P., W. Krieger, and M. Krupp. 2012. *Gabler Lexikon Logistik.* Wiesbaden: Gabler Verlag.

Koch, S. 2012. *Logistik: Eine Einführung in Ökonomie und Nachhaltigkeit.* Berlin and Heidelberg: Springer.

Kruse, J. 2009. *Reader "Einführung in die Qualitative Interviewforschung".* Freiburg: Institut für Soziologie.

Lee, C.K.M., Y. Lv, K.K.H. Ng, W. Ho, and K.L. Choy. 2018. Design and application of Internet of Things-based warehouse management system for smart logistics. *International Journal of Production Research* 56 (8): 2753–2768. https://doi.org/10.1080/00207543.2017.1394592.

Maier, M.M., W.M. Scheid, B. Bertsche, and W.A. Neuhaus. 2011. Praxisgerechte Abnahmeprozeduren für intralogistische Systeme unter Berücksichtigung der Zuverlässigkeits- und Verfügbarkeitstheorie. Ilmenau, Techn. Univ., Diss., Univ.-Bibliothek, Ilmenau.

Moons, K., G. Waeyenbergh, P. Timmermans, D. de Ridder, and L. Pintelon. 2020. Evaluating replenishment systems for disposable supplies at the operating theater: A simulation case study. In *Health care systems engineering*, ed. V. Bélanger, N. Lahrichi, E. Lanzarone, and S. Yalçındağ, vol. 316, 147–162. Springer Proceedings in Mathematics & Statistics. Cham: Springer. https://doi.org/10.1007/978-3-030-39694-7_12.

Mörth, O., C. Emmanouilidis, N. Hafner, and M. Schadler. 2020. Cyberphysical systems for performance monitoring in production intralogistics. *Computers & Industrial Engineering* 142: 106333. https://doi.org/10.1016/j.cie.2020.106333.

Müller, Roland M., and Hans-Joachim Lenz. 2013. Methoden der Unternehmenssteuerung. In *Business intelligence*, ed. R.M. Müller and H.J. Lenz, 121–235. Berlin and Heidelberg: Springer. https://doi.org/10.1007/978-3-642-35560-8_4.

Orellana, F., and R. Torres. 2019. From legacy-based factories to smart factories level 2 according to the industry 4.0. *International Journal of Computer Integrated Manufacturing* 32 (4–5): 441–451. https://doi.org/10.1080/095 1192X.2019.1609702.

Pei, S., J. Zhao, N. Zhang, and M. Guo. 2019. Methodology on developing an assessment tool for intralogistics by considering cyber-physical production systems enabling technologies. *International Journal of Computer Integrated Manufacturing* 32 (4–5): 406–412. https://doi.org/10.1080/095 1192X.2019.1605200.

Schieweck, S., E.N. Jung, and M. ten Hompel. 2016. Leistungsverfügbarkeit zwischen Theorie und Praxis. *Logistics Journal: Proceedings 2016*. https://doi.org/10.2195/lj_Proc_schieweck_de_201602_01.

Schulte, C. 2013. *Logistik: Wege zur Optimierung der Supply Chain*, 6th ed., revised. Vahlens Handbücher der Wirtschafts- und Sozialwissenschaften, München: Vahlen.

SCI Verkehr. 2019. Megatrends in der Logistik in Deutschland 2019. https://de.statista.com/statistik/daten/studie/980502/umfrage/megatrends-in-der-logistik-in-deutschland/. Accessed on 15 May 2020.

Staufen AG, Staufen Digital Neonex GmbH. 2018. Deutscher Industrie 4.0 Index 2018. https://www.staufen.ag/fileadmin/HQ/02-Company/05-Media/2-Studies/STAUFEN.-Studie-Industrie-4-0-index-2019-de.pdf. Accessed on 10 May 2020.

Syafrudin, M., G. Alfian, N. L. Fitriyani, and J. Rhee. 2018. Performance analysis of IoT-based sensor, big data processing, and machine learning model for real-time monitoring system in automotive manufacturing. *Sensors* 18 (9). https://doi.org/10.3390/s18092946.

ten Hompel, M., H. Büchter, and U. Franzke. 2008. *Identifikationssysteme und Automatisierung*. VDI-Buch. Berlin and Heidelberg: Springer-Verlag.

VDI 3580. 10.1995. Grundlagen zur Erfassung von Störungen an Hochregalanlagen.

VDI 3581. 12.2004. Verfügbarkeit von Transport- und Lageranlagen sowie deren Teilsysteme und Elemente.

VDI 3649. 01.1992. Anwendung der Verfügbarkeitsrechnung für Förder- und Lagersysteme.

VDI 4001. 07.2006. Terminologie der Zuverlässigkeit.

VDI 4486. 03.2012. Zuverlässigkeit in der Intralogistik: Leistungsverfügbarkeit.

Vollmuth, H.J. 2007. *Kennzahlen*, 4th ed. TaschenGuide, 13. Planegg, München: Haufe.

vom Bovert, E.M., and R. Jünemann (eds.). 2001. Modellerstellung zur Verfügbarkeitsprognose komplexer Förder- und Lagersysteme. Logistik für die Praxis. Zugl.: Dortmund, Univ., Diss., 2000. Dortmund: Verl. Praxiswissen.

Wagner, K.W., and G. Patzak. 2015. *Performance Excellence: Der Praxisleitfaden zum effektiven Prozessmanagement*, 2nd ed., revised. München: Hanser.

Wang, Y., Y. Liu, J. Chen, and X. Li. 2020. Reliability and condition-based maintenance modeling for systems operating under performance-based contracting. *Computers & Industrial Engineering* 142: 106344. https://doi.org/10.1016/j.cie.2020.106344.

Weber, J., and C. M. Wallenburg. 2010. *Logistik- und Supply Chain Controlling*, 6th ed., revised. Stuttgart: Schäffer-Poeschel Verlag.

Werner, H. 2014. *Kompakt Edition: Supply Chain Controlling: Grundlagen, Performance-Messung und Handlungsempfehlungen*. Lehrbuch. Wiesbaden: Springer Gabler.

Woschank, M., E. Rauch, and H. Zsifkovits. 2020. A review of further directions for artificial intelligence, machine learning, and deep learning in smart logistics. *Sustainability* 12 (9): 3760. https://doi.org/10.3390/su12093760.

Zsifkovits, H., and M. Woschank. 2019. Smart Logistics – Technologiekonzepte und Potentiale. *Berg Huettenmaenn Monatsh* 164 (1): 42–45. https://doi.org/10.1007/s00501-018-0806-9.

8

A Case Study: Industry 4.0 and Human Factors in SMEs

Helmut Zsifkovits, Manuel Woschank, and Corina Pacher

8.1 Introduction

In recent years, scientists and practitioners developed a multitude of technologies and technological concepts based on the vision of Industry 4.0 which was conceptualized as a part of an international strategy to increase the productivity and long-term competitiveness of companies by focusing on principles like digitalization, interconnectivity, and autonomization (Zsifkovits and Woschank 2019; Woschank and Zsifkovits 2021). Thereby, a special focus is placed on the continuous development of small- and medium-sized enterprises which are considered as the backbone of the European economy as they are contributing significantly to the local added value (Matt et al. 2020). However, the risks and barriers on the way to a digitalized production and logistics system should not be neglected, especially when it comes to integrating the human being into

H. Zsifkovits · M. Woschank (✉) · C. Pacher
Chair of Industrial Logistics, Montanuniversitaet Leoben, Leoben, Austria
e-mail: manuel.woschank@unileoben.ac.at

© The Author(s) 2021
D. T. Matt et al. (eds.), *Implementing Industry 4.0 in SMEs*,
https://doi.org/10.1007/978-3-030-70516-9_8

233

reshaped work processes of Industry 4.0-aligned organizations (Dallasega et al. 2019, 2020; Woschank et al. 2020a).

In this context, a multitude of studies reveals that there still seems to be a missing 'digital maturity' regarding the design and implementation of production- and logistics-related processes in SMEs. In a recent study, only 33% of SMEs working in an industrial environment reported that they have started to implement Industry 4.0-related initiatives. Moreover, most of the participating SMEs consider themselves as relatively underdeveloped in terms of Industry 4.0-strategies by describing themselves as 'digitally aware' or even as 'digital newcomer' while only 10% of the participating SMEs classify themselves as 'digital orientated'. It should be further noted that in this study, no company would describe itself as a 'digital champion' which means that they have advanced knowledge in the field of Industry 4.0 (Wirtschaftskammer Österreich 2019). In this regard, a study by Fraunhofer IIS confirmed this view by revealing that only 32% of the participating logistics service providers considered themselves as well developed regarding the maturity for their (transport) logistics processes (Fraunhofer IIS 2017). Nevertheless, 85% of the participants within an expert survey rate the relevance of the digitalization of the value chain as important or even as highly important (Statista.de 2020) because it will be able to generate a variety of improvement opportunities, as time savings, decreasing susceptibility to errors and failures, physical relief for employees, better service for the end customer, etc. (bitkom 2020b).

Up to now, the digitalization of production and logistics systems is mainly based on the potential usage of the following technologies: warehouse management systems, smart sensors, the usage of tablets and smartphones in logistics operations, electronic freight documents, driverless forklift systems, big data analytics, augmented reality, etc. (bitkom 2020a). Thereby, the main barriers regarding the implementation of Industry 4.0 strategies can be summarized as high investment costs, data protection challenges, lack of knowledge, the complexity of the subject itself, the vulnerability of systems to failures, etc. (bitkom 2020c; Wirtschaftskammer Österreich 2019).

However, a multitude of studies stresses the importance of the human workforce for the successful implementation of Industry 4.0 technologies and technological concepts (Creditreform 2019). In the regard, the study of Hobscheidt et al. focused on the development of risk-optimized implementation paths for Industry 4.0 based on socio-technical patterns. Thereby, they state that the dimensions of humans, technology, and organization interact interdependently so that the risks and their effects become almost unmanageable. Therefore, structured tools, e.g., risk-optimized socio-technical implementation paths or implementation roadmaps, are absolutely necessary (Hobscheidt et al. 2020). Vuksanović Herceg et al. introduced an exploratory research study where they analyzed the most important driving forces and implementation barriers of companies in Serbia. Surprisingly, the participants did not see human resources as the driving force behind the implementation, but rather as a barrier when they lack the necessary competencies and skills (Vuksanović Herceg et al. 2020). Cresnar et al. focused on the usage of management tools to speed up the implementation of Industry 4.0. The empirical results tendentially revealed a significant correlation between the usage of various management tools (e.g., Balanced Scorecard, Six Sigma, TQM, etc.) and the Industry 4.0 readiness in manufacturing organizations (Črešnar et al. 2020).

In this chapter, the authors investigate a set of requirements for the successful implementation of Industry 4.0 in SMEs that are directly, or at least indirectly, related to human factors. After a structured analysis of the recent literature on human factors, the authors outline current knowledge regarding critical success factors of learning processes and discuss the transformation process toward a learning-orientated culture in manufacturing enterprises and describe a case where the role of human factor within an Industry 4.0 approach is analyzed more in detail.

8.2 Problem Formulation

In this subsection, the authors analyze a data set dealing with possible barriers to Industry 4.0 concepts in smart logistics from the perspective of human factors. This can be seen as the starting point of the subsequent investigation regarding the role of human beings in the digital transformation process in manufacturing enterprises. The primary data is based on theoretical research which has been systematically extended by semi-structured expert interviews in international workshops with scientists and practitioners and finally evaluated by using a large-scale survey.

In a global research study, Dallasega et al. exploratively evaluated the requirements for the implementation of Industry 4.0 in SMEs by focusing on the area of smart logistics. Based on the Grounded Theory, the research team conducted a total of six workshops with 37 SMEs and 67 experts in Italy, Austria, the USA, and Thailand leading to a total of 548 statements as an outcome of the subsequent content analysis. The statements were further aggregated to a total of 16 items within the three clusters of (1) 'smart and lean x-to-order supply chains' (SAL), (2) 'intelligent logistics through ICS and CPS' (ICT), and (3) 'smart and automated logistics vehicles' (AUT) (Dallasega et al. 2019, 2020). In a follow-up survey, the items were ranked by logistics and/or supply chain professionals in Europe regarding the importance by using a Likert scale ranging from $1 =$ not important to $5 =$ very important. Therefore, a total sample of 9,032 logistics and/or supply chain managers was contacted via e-mail by using an online-based survey tool leading to 71 valid answers and a total response rate of 0.78%.

In the next step, the items were evaluated by an expert team, consisting of three independent researchers, regarding their relevance in terms of human factors by using the coding $1 =$ highly pertinent, $2 =$ moderately pertinent, and $3 =$ not pertinent. The results are presented in Table 8.1.

The qualitative content analysis revealed that most requirements for the implementation of Industry 4.0 in SMEs are directly, or at least indirectly, related to human factors. 'The importance of specific work instructions for the collaboration throughout the supply chain by using ICT' was ranked as the most important one, followed by 'the training and further qualification of employees focusing on state-of-the-art software and data analysis tools'.

Furthermore, it will be important to train the human workforce regarding tools and methods which will allow an 'identification and avoidance of material flow breaks throughout the supply chain' as was as in 'the usage of advanced planning and control systems (PPC) that

Table 8.1 Human-factor-related success factors

Code	Item	Mean	STD DEV	Pertinance
SAL6	The implementation of specific work instructions for collaboration throughout the supply chain by using ICT	3.84	0.84	1
SAL5	The training and further qualification of employees focusing on state-of-the-art software and data analysis tools	3.84	0.93	1
SAL1	The identification and avoidance of material flow breaks throughout the supply chain	3.93	0.86	2
SAL4	The usage of advanced planning and control systems (PPC) that allow forecasting rapidly demand changes	3.86	0.93	2
ICT6	The usage of decision support systems for planning and controlling logistics (e.g., for supplier selection decisions)	3.84	0.88	2
ICT4	The limitation of data access to different stakeholders in the supply chain	3.80	0.87	2
SAL2	The on-demand (Just-in-Time) production and delivery of products to the customers	3.91	0.90	3
SAL3	The availability of real-time order information regarding the status of production and shipping throughout the supply chain	3.90	0.80	3
ICT5	The alignment of ERP/database systems throughout the supply chain	3.87	0.87	3
ICT3	The transparency of inventory levels and storage locations throughout the supply chain	3.84	0.95	3
ICT2	The digital tracking of products throughout the supply chain	3.83	0.92	3
ICT7	The ensurance of data security throughout the supply chain	3.83	0.81	3

(continued)

Table 8.1   (continued)

Code	Item	Mean	STD DEV	Pertinance
ICT1	The digital connection of customers and suppliers for improved collaboration throughout the supply chain	3.83	0.95	3
AUT1	The usage of automated ordering systems	3.68	0.94	3
AUT3	The self-control of warehousing processes (autonomous processes)	3.68	0.91	3
AUT2	The self-control of material flow processes (autonomous processes)	3.65	0.99	3

allow forecasting rapidly demand changes'. The increasing complexity of logistics processes generally requires support in cognitive activities, for example by 'the usage of decision support systems for planning and controlling logistics (e.g., for supplier selection decisions)'. Furthermore, employees should receive ongoing training in data security to adequately address this important subject. This also includes 'the limitation of data access to different stakeholders in the supply chain'.

8.3   Related Work

In this subsection, the authors review the recent literature on human factors for the successful implementation of Industry 4.0 strategies in smart logistics. Therefore, the keywords 'human factor', 'human capital', 'human integration', or 'human*' were used in combination with the keywords 'industrial logistics', 'smart logistics', or 'logistics 4.0'. We focused on the research areas of 'engineering' and 'business management and accounting' and only used studies were written in the English language without a restriction regarding the type of study within the last ten years by using the database Scopus as the main source for our literature analysis. An additional analysis in similar databases (e.g., Web of Science, Science Direct, Emerald, etc.) did not lead to significant differences in the resulting research studies (Woschank et al. 2020b).

Based on the overall research strategy, the search string was formulated as follows: TITLE-ABS-KEY ('human factor' OR 'human capital' OR 'human integration' OR 'human*') AND TITLE-ABS-KEY ('industrial logistics' OR 'smart logistics' OR 'logistics 4.0') AND (LIMIT-TO (SUBJAREA, 'ENGI') OR LIMIT-TO (SUBJAREA, 'BUSI')) AND (LIMIT-TO (LANGUAGE, 'English')). The characteristics of our literature analysis are summarized in Table 8.2.

In the first step, the literature review resulted in a total of 31 identified studies for the initial quantitative analyses. Therefore, the descriptive results of the identified will be presented in the next paragraphs. Figure 8.1 shows the development of the research studies in the time frame from 2010 to 2020.

In general, there is a strong upward trend in the number of identified studies in the time frame from 2010 to 2020. In detail, 6.25% were published in 2013, 3.13% were published in 2014, 3.13% were published in 2016, 9.38% were published in 2017, 21.88% were published in 2018, 9.38% were published in 2019 while most the studies (43.75%) were published 2020.

From the type of study, 48.39% of the identified studies were published as conference proceedings, 38.71% are articles, 6.45% are books, and 6.45% are published as reviews. The results are displayed in Fig. 8.2.

Table 8.2 Characteristics of the literature analysis

Keywords (1)	Keywords (2)	Language	Time frame[a]	Type of Study[a]
Human Factor	Industrial Logistics	—	2010–2020	—
Human Capital	Smart Logistics	English	—	—
Human Integration	Logistics 4.0	—	—	—
Human*	—	—	—	—
—	—	—	—	—
—	—	—	—	—

[a]No further restrictions were defined

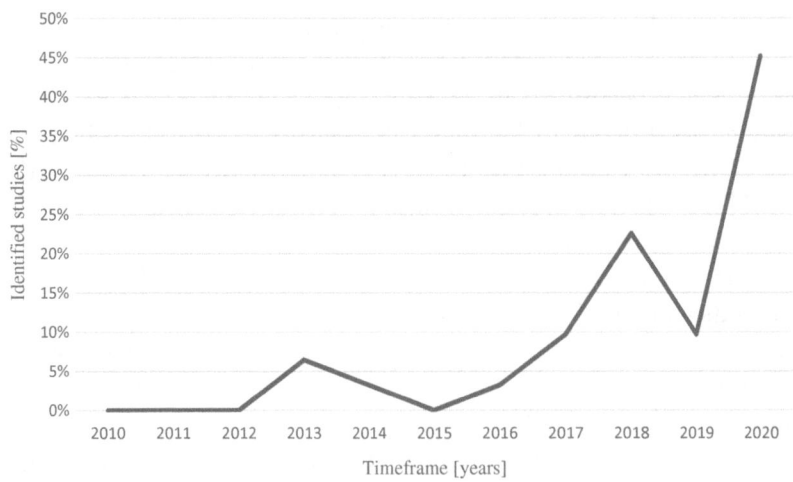

Fig. 8.1 Development of the relevant research studies from 2010 to 2020

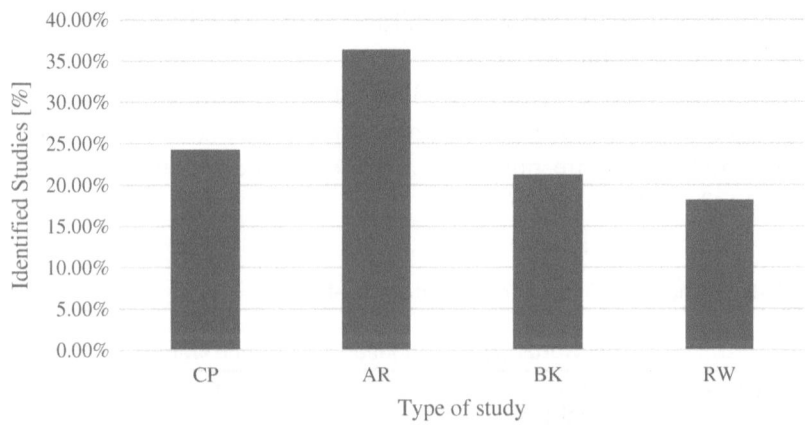

Fig. 8.2 Classification of the relevant research studies per type of study: Conference papers (CP), articles (AR), books (BK), and reviews (RW)

As indicated in Fig. 8.2, most of the publications were either published as a conference paper or as an article. Additional analysis revealed that most of the identified studies (70.97%) were published in a wide variety of media and, therefore, assigned to the category 'others', 19.35% were published in 'Advances in Intelligent Systems and Computing' (19.35%), and 9.68% were published in 'Procedia Manufacturing'. Therefore, Table 8.3 displays the main sources of the identified studies.

Figure 8.3 provides an overview of the identified research collaborations.

From the point of research collaborations, 3.23% of the identified studies were written by one author, 12.90% of the identified studies were written by two authors, 32.26% of the identified studies were written by three authors, 38.71% of the identified studies were written by four authors, 9.68% of the identified studies were written by five authors, and 3.23% of the identified studies were written by seven authors.

Figure 8.4 displays the analysis of research subject-related keywords. Regarding the research of human factors in Industry 4.0, the most important related author keywords are 'Internet of Things', 'Human (integration)', 'Robot (integration)', '(Teaching) and Learning', and

Table 8.3 Distribution of the identified studies

Source	Records (#)	Records (%)
Advances in Intelligent Systems and Computing	6	19.35
Procedia Manufacturing	3	9.68
Others	22	70.97

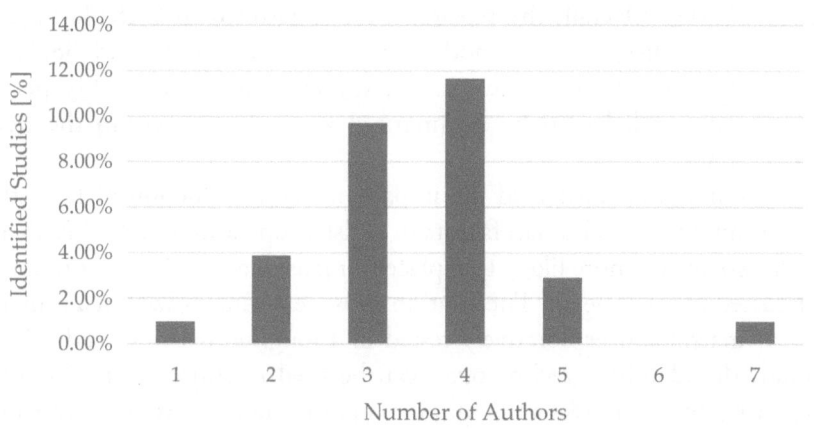

Fig. 8.3 Overview of the identified research collaborations

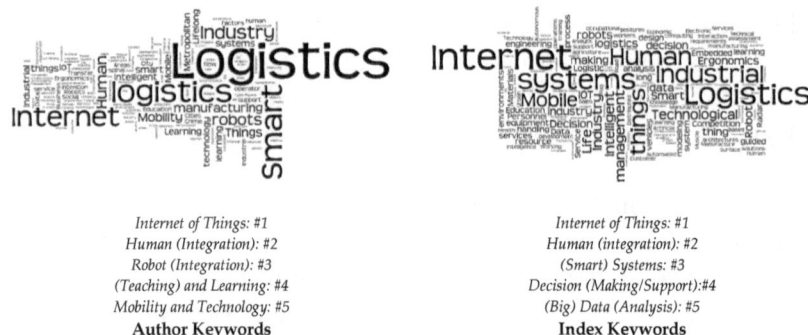

Internet of Things: #1
Human (Integration): #2
Robot (Integration): #3
(Teaching) and Learning: #4
Mobility and Technology: #5
Author Keywords

Internet of Things: #1
Human (integration): #2
(Smart) Systems: #3
Decision (Making/Support):#4
(Big) Data (Analysis): #5
Index Keywords

Fig. 8.4 Analysis of keywords

'Mobility and Technology' and the most important index keywords are 'Internet of Things'. 'Human (Integration)', '(Smart) Systems', 'Decision (Making/Support)', and '(Big) Data (Analysis)'.

In a second step, the abstracts of the identified studies were qualitatively analyzed by the research team and by three additional reviewers and coded with 1 = highly pertinent, 2 = moderately pertinent, and 3 = not pertinent for subsequent full-text analysis. Thereby, studies without significant differences were directly included or excluded in/from the research process. Studies with significant differences were reevaluated by the research team to get unambiguous research results (Woschank et al. 2020b). Overall, the research process resulted in 25.81% highly pertinent studies, 35.48% moderately pertinent studies, and 38.71% not pertinent studies. Consequently, the content of the highly pertinent studies will be briefly summarized and analyzed within the new paragraphs.

Cimini et al. investigated the impact of the introduction of Industry 4.0 technologies on human factors in logistics operations. Whether these technologies are more likely to replace humans or aim to support them is discussed in more detail. The relationships were summarized in a human factor matrix and exploratively tested in a longitudinal case study. As a result, the identified technologies can be used to support stressful and repetitive physical tasks and complex cognitive tasks, as well. In general, the evolution is more toward assistance rather than to a replacement of the human workforce (Cimini et al. 2020).

Winkelhaus and Grosse conducted a systematic literature review for the development of a new logistics system framework. By reviewing 114 articles the authors conceptualized a framework that combines external triggers, main technological innovations, impacts of human interactions, and logistics tasks based on the technologies of the Internet of Things, cyber-physical systems, Big Data, cloud computing, mobile-based systems, social media-based systems, etc. As an outcome, the authors postulate that the human-centric view was not discussed intensively yet. Future research should focus on the design of logistics systems from a human-centric point of view by focusing on topics, e.g., worker shortage, demographic changes, systematic skill development, new ways of learning, and the digital transformation based on the capabilities of the employees (Winkelhaus and Grosse 2020).

Schmidtke et al. evaluated the technical potentials and challenges within internal logistics 4.0 by discussing the future role of human beings in the industrial working environment. They concluded that humans never will be completely replaced within the processes of industrial production. Science and industry must develop working environments that allow the collaboration between humans and assistance systems (Schmidtke et al. 2018).

Delfmann et al. demonstrated why logistics operations will play an even increasingly central role in the future and, thereby, state that logistics must be a crucial element within the Industry 4.0 implementation strategies. Furthermore, eleven central research questions are presented which are of high importance for the entire research community. Thereby, question nine focuses on the interaction of humans and autonomous machines as equal partners in socio-technical systems. Moreover, question eleven is addressed toward the systematic development of qualifications and competencies for Industry 4.0 (Delfmann et al. 2018).

Wrobel-Lachowska et al. stated that because highly qualified workers will be needed in the logistics sector in the future, the fact arises that the educational process must be adapted. Selected logistics companies and universities were studied to conduct and present an analysis of the challenges for logistics education. They concluded that universities should redesign their courses by implementing modern learning methods, change the role from a teacher to a mentor, reinforce cooperation, and shape competencies based on the needs of the industry (Wrobel-Lachowska et al. 2019).

Beham et al. focused on the optimization of slab logistics processes in the steel industry. They further state that automated decision support systems are frequently not accepted by human operators. They developed a cooperative system where human input is required to confirm the status of the material which should be used to reduce data errors, human errors, and the breakdown of machines and vehicles (Beham et al. 2020).

Tran-Dang et al. investigated the role of the Internet of Things for logistics. Thereby, they stated that is important to combine human knowledge with machines that support decision-making. However, besides technical challenges, the coordination and harmonization of control and management policies and regulations are identified as further key success factors (Tran-Dang et al. 2020).

8.4   Learning and Learning Culture

The implementation of Industry 4.0 technologies and concepts in companies entails both a transformation of production processes and the changed work and learning processes for employees. Thereby, it is important to notice that the increasing complexity and the associated changes in technical work processes require operational learning processes for all employees (Ullrich et al. 2018).

The work of the future will be more flexible, more mobile, and more digitally networked. This is a result of the current transformation processes toward digitization affecting all areas of human life. However, humans will not be replaceable by technologies or machines in the future. In this regard, Dengler and Matthes forecast that the current number of 40% of all employees in jobs with low substitutability potential will decrease a maximum of 30% employment level through automation. Conversely, this study makes clear that the human workforce will continue to be regarded as an essential component in global value chains. It seems to be obvious that an ongoing specialization requires adapted operational or organizational learning processes. However, these processes cannot be viewed in isolation, but they must be considered as part of organizational development (Dengler and Matthes 2015; Zehnder 2014).

Consequently, it becomes clear that companies are increasingly required to design, manage, and integrate learning processes and learning environments into their organizations to be able to prepare employees based on the changing fields of work and tasks. Organizations need to transform learning organizations and employees also need to acquire lifelong knowledge. For these reasons, learning processes during the entire life are considered enormously important. The principle of lifelong learning (LLL) affects all dimensions of society and makes a significant contribution to maintaining and increasing workability and, therefore, contributes to the increase of competitiveness. However, this raises the question of to which the extent the respective organization responds to the changing environmental requirements in a learning manner by planning and implementing individual, collective, and organizational learning processes. In this context, Franz states that the process of learning in organizations is essentially dependent on the communication patterns within and between the participants. Therefore, the authors will define the term 'learning processes' and provide implications for organizations regarding the implementation of learning processes and suitable learning environments, as well (Franz 2016).

Learning takes place always and everywhere and includes all areas of daily life. Through the permanent intake of information and everyday experiences, respectively, an expansion of the human scope of action and habitus takes place. Based on this learning process, an ongoing change is generated and serves as a foundation for further learning activities. In general, the learning process by itself can be divided into three phases. In the first phase, which is also referred to as 'acquisition', new knowledge is generated. The incorporation of new knowledge is defined as 'perception'. The knowledge is then subsequently incorporated into the next phase. 'Retention' comprises the second phase. During this process, what has been learned is stored in the memory as a reminder. In the third learning phase, the so-called reproduction, the application of the stored knowledge and the learned competences take place (Geller 1996).

Hereby, the principle of lifelong learning describes the expectation to deal with learning and educational processes in the entire course of life, due to social changes of the knowledge society, such as industrialization 4.0 or the half-life of knowledge and the associated need for action. The European Union defines LLL as 'any purposeful learning activity that serves a continuous improvement of knowledge, skills, and competences'. Accordingly, LLL is both an opportunity and a challenge for all people. The overall process of learning across the lifespan takes place in different following dimensions: (1) Temporal: The focus is on the entire life course, no longer limited to the stages of childhood and adolescence or young adulthood. Learning and educational trajectories must be considered and considered across the lifespan; (2) Content: LLL refers to all learning processes inside and outside educational institutions. Thus, learning takes place not only in educational institutions but also in other places of learning and in all diverse forms; and (3) Spatially: LLL no longer involves the mere acquisition of cultural goods or professional competencies but encompasses all kinds of topics and subjects as well as the development of skills and the change of self-perception and world perception (Hof 2013).

Learning processes need to be investigated from a time- and process-orientated perspective. Learning is a lifelong process, not just the short-term acquisition of new levels of knowledge, but the longer-term confrontation with observations from the world. Learning is integrated into the daily life of the individual learner and, therefore, also into company processes. Therefore, a special emphasis should be placed on human beings as capital for a company as well as for the entire society. Consequently, the human workforce is considered as an 'individual educational subject' including their specific life situations and individual as well as collective goal settings. The LLL approach further leads to a delimitation of learning boundaries and, therefore, to a multitude of massive changes at the institutional level. Organizations should not only focus on the design of appropriate learning environments and learning formats, but also on the constitution of appropriate institutional frameworks. Therefore, in any case, networking, communication, and cooperation are essential at all organizational levels for the integration of individual and collective learning experiences. Human beings acquire knowledge on different levels during their lives. Three specific forms of learning can be divided into (1) informal, (2) formal, and (3) non-formal learning processes. Informal learning takes mainly place in everyday life and mostly by unconscious learning experiences. In educational institutions, formal learning is acquired through predefined curricula and finally confirmed by a formal qualification. Non-formal learning usually takes place within courses, further educational measures, or seminars and is not characterized by a certificate. In the context of learning and educational processes, these forms of learning are interdependent meaning that no precise separation can be made. Learning processes are furthermore dependent on the form of organization which can be divided into the individual, collective, and organizational processes, whereby again the boundaries may overlap (Woschank and Pacher 2020a, b, c, d).

In general, there is no precise definition or theory regarding the concept of individual learning processes. The process of individual learning includes the independent and self-controlled information intake and the subsequent processing of information. The self-controlled learning process is based on two essential prerequisites. On the one hand, the learner must be capable of autonomous learning. This is the case if the learner can decide on the topics as well as on the methods on his responsibility. On the other hand, the learning process requires the necessary willingness to learn on the part of the individual (Eggers 2007). Autonomous learning takes place primarily in private settings and, therefore, mostly outside educational institutions (Haefner 1980). This form of learning is becoming increasingly important today regarding learning over the entire lifespan since we live in a knowledge society and permanent education is essential for maintaining and expanding human working abilities which should lead to increased competitiveness of the society.

Collective learning processes take place in form of interactions between subjects, respectively, in a social group. According to Miller, an individual can only learn something new, if learning processes take place in an integrated manner based on an interaction with a person or group. Thereby, the individual ability to learn of all persons involved is considered as a prerequisite. Collective learning processes can generally be defined as the sum of a wide variety of individual processes allowing 'a universal antigenic sequence of awareness regarding the connection between logic, rational reasoning and a moral view of the word produced by the individual in the course a lifelong process of development and education' (Miller 1986). Asymmetric levels of information within a group can be considered as a starting point for collective learning processing. Thereby, the exchange of information enables collective processes and can subsequently also lead to a change in values regarding the organization and, therefore, affect the entire organizational culture (Miller 1986).

Organizational collective learning processes, often also known as organizational learning processes, describe the learning at the workplace and/or in organizations. In this regard, Probst and Naujoks understand organizational learning as 'the ability of an institution to discover mistakes, to correct them, and to change the organizational values and knowledge base in a way that problem-solving abilities and capabilities for action are generated' (Probst and Naujoks 1993).

Within organizational learning processes, findings from individual learning processes are mostly transferred to organizations (Hodel and Geißler 1998). This process takes place in two stages. First, learning consists of the acquisition of elements to understand a system. In the second stage, collectivization processes take place, which subsequently manifests organizational learning (Hodel and Geißler 1998). All participants are substantially involved in educational success, like superiors, coworkers, or also the training offerors and/or the persons responsible for the training. Organizational learning is thus dependent on a variety of success factors, such as trust, appreciation, various or communication patterns. For learning settings in organizations, Treml distinguishes between (1) functional, (2) extensional, and (3) intentional forms of learning and communication (Treml 2000).

In the functional setting, learning takes place implicitly, e.g., incidentally, in everyday work. This implicit learning and the associated knowledge are usually generated unconsciously through imitation, in the sense of Jean Piaget's process of assimilation. However, this exchange can only take place if there is a certain degree of trust among the members or within the team. Trust correlates with the climate within a group. Accordingly, it becomes clear that learning in organizations is strongly influenced by the working climate. A better climate and appreciation among each other will consequently lead to higher learning success within a team. In contrast to this, an unsatisfactory working climate will most likely lead to both moderate learning processes and deficits in professional competencies (Zehnder 2014). Moreover, extensional learning includes, for example, the intended establishment of inter-disciplinary working groups within business organizations to provide space for exchange and a potential change of individual perspectives. The extensional form of learning and communication aims at a proactive interaction of different (professional) perspectives to promote joint learning and collective development processes. Furthermore, intentional forms of learning must be developed as formal learning processes in organizations, such as quality development processes like strategy meetings (Franz 2016).

In conclusion, it can be stated that the establishment of a learning culture in organizations is indispensable to be able to professionally adapt the human workforce to the permanently changing working environment. Collective learning processes are influenced by a multitude of determinants, e.g., the team climate, feelings, individual motives, and appreciation. It is important to notice that the management is responsible for the shaping of the corporate culture and thus also the associated learning culture. Management must create the necessary conditions for successful learning in and as an organization and involve all participants in this cyclical processes. Within the development process, appreciation and recognition of the people involved and their activities are of essential importance. Recognition creates a sense of belonging to the respective organization and, therefore, contributes significantly to a common commitment.

Within the framework of organizational learning processes, the cooperation of interdisciplinary teams can be promoted through targeted educational events that aim to generate joint learning and, subsequently, joint knowledge. Therefore, potential measures could include discussions, joint projects, courses, or further training measures. Knowledge transfer and (informal) exchange should be promoted, since, in addition to technical competencies, transversal competencies such as communication skills, problem-solving skills, teamwork skills, and creativity will be of essential importance in the future. The organizational learning processes should therefore aim at a holistic understanding of education and support the cooperation between the organizational development team, all affected departments, and human resource management. In this way will it be possible to prepare the workforce for the changing working environment and equip the human workforce with the necessary competencies, thus generating an interdisciplinary learning and communication culture in business organizations.

8.5   A Case of Human Factors in Implementing New Technology

The challenges of introducing Industry 4.0 in an organization are not purely technical. The effort to change processes through digital technologies in a traditional work environment is often met with strong resistance from the humans affected by the change. When implementing new processes, there are often strong mental barriers from people that have been doing the work in a certain manner over the years. Technology is an enabler for innovation, can potentially facilitate or support manual tasks, but it is also seen as a threat, making a human's work and mindless essential, and machines can even replace human work.

To make Industry 4.0 a success, in addition to technological challenges, organizational adaptations are required, and a change in mindsets. The adoption of operating procedures, technologies, and systems as part of the Industry 4.0 concept relies on the human factor. Human workers often have goals that are quite different from those of the organization. They are aiming to enter and remain in the market, build their career, and obtain equitative wages, stability, intellectual growth, learning and/or professional achievement. Companies rather seek the best possible human performance to increase productivity (Silva et al. 2019). Neglecting the human factor will inevitably fail.

Using the case of one machining services company, the implementation scenario is described, with the challenges and barriers met, and the steps required to ensure an effective implementation of new technology.

8.5.1 The Objective of Investigation: The Company 'Precision Machine Products, Inc.' (PMP)

The company was founded 45 years ago, with its operations in a rural, small-town environment. They are industry leaders in CNC machining, with machines including a wide range of brands and dimensions. Their capabilities include drilling, deburring, grinding, milling, turning, contour milling, broaching, thread cutting, slotting, tapping, band sawing, and tube fabrication. Materials worked with include different qualities of steel, aluminum, brass, copper, and plastics. They are capable of working with extremely low tolerances.

Value-added services include CAD design, CAM programming, assembly, laser welding, laser engraving, inventory management, and outsourced finishing services (grinding, plating, heat treating, and anodizing). Industries served to include industrial automation, medical, aerospace, agricultural, electronic, robotics, oil and gas, hardware, plumbing, optics, among others. PMP is a strong and reliable partner to its customers, also specializing in secondary operations, providing JIT, KANBAN, and emergency services.

They have been facing stronger competitive pressure in recent years, with new entrants on the market. These are mostly young companies, driven by innovative technologies. A high level of automation gives these enterprises much flexibility and cost advantage. A narrow specialization in niche products opens them new markets, making them a competitor to PMP.

The management decided to go for a progressive strategy, meeting the challenges by innovating the entire company. The goals to achieve were defined as follows:

- Foster digital technologies (Industry 4.0)
- Improve flexibility of operations
- Make processes more robust
- Increase visibility within the production processes

Some limitations had to be observed. One of these was the budget. PMP at that point could not invest in machines and technology as they wished.

So, most of the equipment had to stay in place, being slowly replaced over the next years. Also, management did not have intentions to make major changes to the workforce. Most of the workers came from the town and had long been employed by PMP, some of them in the second generation.

PMP hired a Continuous Improvement Manager to drive the process. Serge was a brilliant, ambitious guy, with long experience in the automotive industry. He had been managing several major technology projects, including workplace automation, automated warehouses, and robotics.

Next, the search for a technology partner started. What PMP was expecting to realize over the next years was rather Industry 3.5 than 4.0 in terms of technology maturity. Keeping this in mind they needed a partner that could provide turnkey products that integrated seamlessly with how PMP as a manufacturer worked today. Most of their machines and equipment were to stay in operation for the years to come. So, this would have to be an evolutionary approach, making small improvement steps instead of huge disruptive changes.

DataFusion, Inc. is an innovative startup company, founded by two striving entrepreneurs a few years ago. They bring advanced enterprise data skills, have a track record of commercializing technology, and have held executive leadership positions at several startups.

DataFusion started on the premise that Industry 4.0 would require turnkey solutions that integrate seamlessly with how manufacturers work today. Their software platform is an information system for shop floor performance. It fuses data from all machines into a production scoring system, monitoring overall equipment efficiency (OEE) and shop floor productivity in real-time.

Real-time production scoring automatically tracks specific jobs on the machines and establishes a benchmark of efficient production. Performance vs. the benchmark is monitored, focusing the operators' attention on improving the lower scores.

PMP is a key customer for DataFusion, as a startup, they are still quite new on the market and eager to attract some major accounts for strengthening their market presence and further development of solutions.

8.5.2 The Project

PMP at the time of project start had around 100 machines on the shop floor, of various brands. There are 4–5 different classes of machines. Most of the machines run in one shift. Some of them are in two-shift operation, mainly because of long set-up times. These are assigned to one large customer. In the second shift, a small number of staff are assigned.

Set-up is a problem creating bottlenecks in using the capacity. Small production batches or one-piece flow, therefore, it is not realistic. There are some efforts to work on set-up times to improve and get more flexibility.

Most of the machines could be easily connected to the DataFusion platform, using their data collection stations. These support most major machine tool brands. Data transfer is done by WIFI into the cloud-based platform. Around 20 of the machines were not suitable for integration, due to lacking interfaces.

The data integration was achieved in a very short time. The real-time production scoring was up and running within a few days. Beyond machine status, utilization, and OEE numbers, the system enabled benchmarking machines. Workstations with a low performance can be easily identified and selected for focusing on productivity improvement efforts.

The achievements of the project are quite promising. Integration of machine data into one platform offers opportunities to better control performance, establish benchmarks, and detect bottlenecks and problems. The management of visibility is critical for the adoption and success of the system.

8.5.3 The Human Factor

PMP is a family business in the third generation. This is characteristic of the company culture. Workers tell you why something does not work, they have always done it the same way. It is very difficult to change the mindset.

Serge, the Continuous Improvement Manager, guided us through the facility. He is passionate about his work, sees a large potential for improvement of operations, strengthening PMP's position in the competitive market.

The majority of working processes could be automated, but Industry 4.0 is intended to improve human work and not replace people. The workers are still to be convinced, though. They are more difficult to handle than the machines, Serge said.

When Serge came to the company, he asked how the last day was. One of the operators told him it was good, five parts. If he asked again days later, he got the answer that the number of parts was 300, which was also good. He is trying to get in some transparency visibility, to understand the reasons why a process is working or not. If they cannot deliver the operators tell sometimes, the machine is running at full capacity. It is difficult to assess whether this is 20, 25, or 70%.

Serge makes himself unpopular with some of his actions and regulations. There is a customer to satisfy, he says.

His approach to overcome reservation and distrust is visibility. He installed whiteboards where the machine operators themselves indicated performance, with the ability to compare between machines. For each machine and operator, a target to attain is set. If the target number is surpassed, this is marked with a green pen, if the achievement is less, this is indicated with a red pen. Shortcomings against the targets set must be reported. There was a dominance of red on the board.

One workplace showed an achievement of 199 against a target of 225, the deviation of 26 marked in red. The operator said he did not have the required number of parts. The parts arrived from quality control only at 10 am, due to some faulty process, he did not have any influence on that. There is a queue before quality control. With the shift starting at seven, and the parts arriving at 10, these are three lost hours of production, more than 30% of the shift time. Considering that the deviation in the number of parts was only minus 20%, the shift could have been one with good productivity.

There is a large amount of waste involved in many of the processes, delays, faults, non-value-added activities. Analyzing an order with a throughput time of 20 days, you might find it took 8 days just in administration. This is an opportunity to improve. People are aware of that, they know what is working, and what is not. Inertness is stronger than the willingness to change, though.

Workers spend much time searching for parts or tools, sometimes hours. This greatly reduces productivity. One worker is proud of walking several kilometers in one day.

Machines need to be effectively maintained to reduce standstill. These are a major cause for high costs and a loss of capacity, often resulting in delayed completion of customer orders, therefore, another type of waste.

We are now in the second shift. A worker is grinding bars. At the machine, one signal light is red. The operator is supposed to pay attention and act accordingly. He is too much occupied with what he is doing presently. Some machines are running, are not stopped even if there are no parts to be worked on. Much energy is wasted. All the lights are on, there are no energy-saving lamps.

It is not an easy task to find people with the right qualification. Some of them have grown to be highly proficient over the years. They are experienced, they know from the noise and the smell of whether the machine is working properly or not. Now they are losing this capability because they are wearing ear protection. They do not have that feeling anymore. Young people are coming in, they do not have that experience. There are efforts to attract people from technical schools, from colleges, but it is difficult to find qualified people.

So how can Industry 4.0 contribute? End-to-end data integration and networked automation can make operations more effective. Problems can be identified before they become apparent, before some tool breaks. Causes can be analyzed, and potential action suggested, or even taken autonomously. Industry 4.0 should foresee problems, preventive action.

In conclusion, much productive time is wasted or spent on firefighting. Smart technology and data integration are enablers for better visibility and more productive use of valuable resources. Human factors must be considered, though. Resistance and inertness in people are major barriers to achieving improvements.

8.6   Conclusions and Outlook

Industry 4.0 offers various benefits to the human factor, including a reduction of physical efforts, improved decision-making based on defined criteria, more efficient internal and external communication, and the effective usage of tools and data.

However, the establishment of an adequate teaching and learning culture must be considered a prerequisite in the process of digital transformation on the level of the employees and on the level of top management, as well. Within professional education processes, adapted learning conditions and learning environments must be established to create a foundation for successful learning paths.

Moreover, the communication patterns within the company must be regarded as an essential success factor toward the implementation of a new, respectively, of a realigned learning culture. Without trust and commitment on the individual level, but also in groups of companies, no professional learning processes can be realized. Accordingly, companies must be sensitized in terms of both determinants that promote learning and determinants that inhibit learning by focusing on the realignment toward either functional, extensional, or intentional forms of learning and communication or even a combination of them in business organizations.

Problems on the shop floor are often man-made, or organization induced. So, technology is not a solution for itself, implementation must be planned for in a human-centered manner. Human work will continue to be required in Industry 4.0 environments, both for the development of this concept as the management of advanced production systems, and the application of technologies and tools. Change processes necessitate interventions and actions in cognitive, emotional, and psychic aspects (Silva et al. 2019). Adequate conditions and environments for human work must be ensured. Information, motivation, and empowerment of people are critical factors for the effective and efficient introduction of smart industry solutions.

Acknowledgements This research is part of the project 'SME 4.0 – Industry 4.0 for SMEs', which has received funding from the European Union's Horizon

2020 research and innovation program under the Marie Skłodowska-Curie grant agreement No. 734713.

References

Beham, A., S. Raggl, V.A. Hauder, J. Karder, S. Wagner, and M. Affenzeller. 2020. Performance, quality, and control in steel logistics 4.0. *Procedia Manufacturing* 42: 429–433. https://doi.org/10.1016/j.promfg.2020.02.053.

bitkom. 2020a. Digitalisierung der Logistik. https://www.bitkom.org/sites/def ault/files/2019-06/bitkom-charts_digitalisierung_der_logistik_03_06_2019. pdf. Accessed on 10 December 2020.

bitkom. 2020b. Digitalisierung macht Logistik schneller, sicherer und einfacher. https://www.bitkom.org/Presse/Presseinformation/Digitalisier ung-macht-Logistik-schneller-sicherer-und-einfacher. Accessed on 10 December 2020.

bitkom. 2020c. Industrie 4.0—so digital sind Deutschlands Fabriken. https:// www.bitkom.org/sites/default/files/2020-05/200519_bitkomprasentation_ industrie40_2020_final.pdf. Accessed on 10 December 2020.

Cimini, C., A. Lagorio, F. Pirola, and R. Pinto. 2020. How human factors affect operators' task evolution in logistics 4.0. *Human Factors and Ergonomics in Manufacturing & Service Industries*: 1–20. https://doi.org/10. 1002/hfm.20872.

Creditreform. 2019. Digitalisierung und Wirtschaft 4.0. https://www.creditref orm.at/nc/news/news/news-list/details/news-detail/digitalisierung-und-wir tschaft-407404.html. Accessed on 10 December 2020.

Črešnar, R., V. Potočan, and Z. Nedelko. 2020. Speeding up the implementation of Industry 4.0 with management tools: Empirical investigations in manufacturing organizations. *Sensors* 20 (12): 1–25. https://doi.org/10. 3390/s20123469.

Dallasega, P., M. Woschank, S. Ramingwong, K.Y. Tippayawong, and N. Chonsawat. 2019. Field study to identify requirements for smart logistics of European, US and Asian SMEs. Proceedings of the 9th International Conference on Industrial Engineering and Operations Management (IEOM), Bangkok 2019, 844–855.

Dallasega, P., M. Woschank, H.E. Zsifkovits, K.Y. Tippayawong, and C.A. Brown. 2020. Requirement analysis for the design of smart logistics in SMEs. In *Industry 4.0 for SMEs*, ed. D.T. Matt, V. Modrák, H.E. Zsifkovits, 147–162. Cham: Palgrave Macmillan.

Delfmann, W., M. ten Hompel, and W. Kersten. 2018. *Logistics as a science— Central research questions in the era of the fourth industrial revolution*, 9th ed. Berlin: BVL.

Dengler, K., and Britta Matthes. 2015. Folgen der Digitalisierung für die Arbeitswelt. Substituierbarkeitspotenziale. https://doku.iab.de/forschungsbe richt/2015/fb1115.pdf. Accessed on 10 December 2020.

Eggers, H. 2007. *Autonomes Lernen im Englischunterricht an der Grundschule*. München: Grin.

Franz, J. 2016. *Kulturen des Lehrens. Eine Studie zu kollektiven Lehrorientierungen in Organisationen Allgemeiner Erwachsenenbildung*. Bielefeld: Bertelsmann.

Fraunhofer IIS. 2017. Transportlogistik 4.0. https://www.scs.fraunhofer.de/con tent/dam/scs/de/dokumente/studien/Transportlogistik.pdf. Accessed on 10 December 2020.

Geller, B. M. 1996. *Individuelle, institutionelle und metaorganisatorische Lernprozesse als konstituierende Elemente des ganzheitlichen organisatorischen Lernens. Eine modelltheoretische Analyse*. Linz: Trauner.

Haefner, K. 1980. Bildungspolitische und bildungskulturelle Konsequenzen autonomen Lernens. In *Individuelles Lernen und Studieren*, ed. H.-E. Piepho and G. Bauer, 120–149. Alsbach: Zebisch.

Hobscheidt, D., A. Kühn, and R. Dumitrescu. 2020. Development of risk-optimized implementation paths for Industry 4.0 based on socio-technical pattern. *Procedia CIRP* 91: 832–837. https://doi.org/10.1016/j.procir.2020. 02.242.

Hodel, M., and H. Geißler. 1998. *Organisationales Lernen und Qualitätsmanagement. Eine Fallstudie zur Erarbeitung und Implementation eines visualisierten Qualitätsleitbildes*. Frankfurt: Lang.

Hof, C. 2013. Übergänge und Lebenslanges Lernen. In *Handbuch Übergänge*, ed. W. Schröer, B. Stauber, A. Walther, L. Böhnisch, and K. Lenz, 394–415. Weinheim: Beltz Juventa.

Matt, D.T., V. Modrák, and H.E. Zsifkovits, eds. 2020. *Industry 4.0 for SMEs*. Cham: Palgrave Macmillan.

Miller, M. 1986. *Kollektive Lernprozesse. Studien zur Grundlegung einer soziologischen Lerntheorie*. Frankfurt: Suhrkamp.

Probst, G.J.B., and H. Naujoks. 1993. Autonomie und Lernen im entwicklungsorientierten Management. *Zeitschrift für Organisation* 6: 368–374.

Schmidtke, N., F. Behrendt, L. Thater, and S. Meixner. 2018. Technical potentials and challenges within internal logistics 4.0. Proceedings of the 4th IEEE International Conference on Logistics Operations Management, 1–10. https://doi.org/10.1109/GOL.2018.8378072.

Silva, V.L., J.L. Kovaleski, R.N. Pagani, A. Corsi, and M.A.S. Gomes. 2019. Human factor in smart industry: A literature review. *Future Studies Research Journal: Trends and Strategies* 12 (1): 87–111. https://doi.org/10.24023/FutureJournal/2175-5825/2020.v12i1.473.

Statista.de. 2020. Wie bewerten Sie die Relevanz von Digitalisierung der Wertschöpfungskette für Ihr Unternehmen? https://de.statista.com/prognosen/943099/expertenbefragung-zur-digitalisierung-in-der-logistikbranche-in-deutschland. Accessed on 10 December 2020.

Tran-Dang, H., N. Krommenacker, P. Charpentier, and D.-S. Kim. 2020. The Internet of Things for logistics: Perspectives, application review, and challenges. *IETE Technical Review*: 1–29. https://doi.org/10.1080/02564602.2020.1827308.

Treml, A.K. 2000. *Allgemeine Pädagogik. Grundlagen, Handlungsfelder und Perspektiven der Erziehung*. Stuttgart: Kohlhammer.

Ullrich, C., A. Hauser-Ditz, N. Kreggenfeld, C. Prinz, and Igel Christoph. 2018. Assistenz und Wissensvermittlung am Beispiel von Montage und Instandhaltungstätigkeiten. In *Zukunft der Arbeit - eine praxisnahe Betrachtung*, ed. S. Wischmann and E.A. Hartmann, 107–123. Berlin: Springer Vieweg.

Vuksanović Herceg, I., V. Kuč, V.M. Mijušković, and T. Herceg. 2020. Challenges and driving forces for Industry 4.0 implementation. *Sustainability* 12 (10): 1–22. https://doi.org/10.3390/su12104208.

Winkelhaus, S., and E.H. Grosse. 2020. Logistics 4.0: A systematic review towards a new logistics system. *International Journal of Production Research* 58 (1): 18–43. https://doi.org/10.1080/00207543.2019.1612964.

Wirtschaftskammer Österreich. 2019. Digitale Transformation von KMUs in Österreich 2019. Erfassung des Digialisierungsindex 2019. https://www.wko.at/branchen/information-consulting/unternehmensberatung-buc

hhaltung-informationstechnologie/kmu-digitalisierungsstudie-2019.pdf. Accessed on 10 December 2020.

Woschank, M., and C. Pacher. 2020a. A holistic didactical approach for industrial logistics engineering education in the LOGILAB at the Montanuniversitaet Leoben. *Procedia Manufacturing* 51: 1814–1818. https://doi.org/10.1016/j.promfg.2020.10.252.

Woschank, M., and C. Pacher. 2020b. Fostering transformative learning processes in industrial engineering education. Proceedings of the 5th NA International Conference on Industrial Engineering and Operations Management (IEOM), Detroit 2020, 2022–2029.

Woschank, M., and C. Pacher. 2020c. Program planning in the context of industrial logistics engineering education. *Procedia Manufacturing* 51: 1819–1824. https://doi.org/10.1016/j.promfg.2020.10.253.

Woschank, M., and C. Pacher. 2020d. Teaching and learning methods in the context of industrial logistics engineering education. *Procedia Manufacturing* 51: 1709–1716. https://doi.org/10.1016/j.promfg.2020.10.238.

Woschank, M., and H.E. Zsifkovits. 2021. Smart logistics—Conceptualization and empirical evidence. *Chiang Mai University Journal of Natural Sciences* 20 (2): 1–9.

Woschank, M., E. Del Rio, H.E. Zsifkovits, and P. Dallasega. 2020a. Comparison of Industry 4.0 requirements between Central-European and South-East-Asian enterprises. Proceedings of the 5th NA International Conference on Industrial Engineering and Operations Management (IEOM), Detroit 2020. 2013–2021.

Woschank, M., E. Rauch, and H.E. Zsifkovits. 2020b. A review of further directions for artificial intelligence, machine learning, and deep learning in smart logistics. *Sustainability* 12 (9): 1–23. https://doi.org/10.3390/su1209 3760.

Wrobel-Lachowska, M., A. Polak-Sopinska, and Z. Wisniewski. 2019. Challenges for logistics education in Industry 4.0. In *Advances in human factors in training, education, and learning sciences*, ed. S. Nazir, A.-M. Teperi, A. Polak-Sopińska, 329–336. Cham: Springer.

Zehnder, H. 2014. *Betriebliche Bildung. Zwischen Wahrnehmungsverzerrung und Lernresistenz ; was optische Täuschungen über das Lernen verraten*. Berlin: Springer Gabler.

Zsifkovits, H., and M. Woschank. 2019. Smart Logistics – Technologiekonzepte und Potentiale. *Berg- Und Hüttenmännische Monatshefte (BHM)* 164 (1): 42–45. https://doi.org/10.1007/s00501-018-0806-9.

Part III

Organizational and Management Models for Smart SMEs

Part II

9

Transition of SMEs Towards Smart Factories: Business Models and Concepts

Vladimír Modrák and Zuzana Šoltysová

9.1 Introduction

The term Industry 4.0 originates from the final report of the Industrie 4.0 Working Group (Kagermann et al. 2013), and indicates the subset of the fourth industrial revolution (Marr 2016). According to Drath (2014) for the first time an industrial revolution is predicted a-priori, not observed retroactive. This also clarifies whether Industry 4.0 has to be considered as revolution or evolution, respectively. Logically, it depends on which perspective we are looking at. If we concentrate on differences between Industry 3.0 and future trends of Industry 4.0, then Industry 4.0 clearly covers all features of industrial revolution. This standpoint articulates a retrospective view of the future, i.e., operates with projective

V. Modrák (✉) · Z. Šoltysová
Faculty of Manufacturing Technologies, Technical University of Kosice, Presov, Slovakia
e-mail: vladimir.modrak@tuke.sk

Z. Šoltysová
e-mail: zuzana.soltysova@tuke.sk

© The Author(s) 2021
D. T. Matt et al. (eds.), *Implementing Industry 4.0 in SMEs*,
https://doi.org/10.1007/978-3-030-70516-9_9

interpretation of the past. On the other hand, each industrial revolution can be considered a separate milestone that takes some decades from beginnings to a substantial reworking the economy (Stearns 2015). Thus, transformation towards digitization and smart manufacturing is evolutionary in its nature.

Industry 4.0 is frequently discussed from a technological perspective, since advanced technology is indispensable for success of this strategy. In spite of this, only few representative roadmaps for Industry 4.0 technologies adoption are available (Qin et al. 2016) and most of them are not well suitable for SMEs. However, issues that are no less of importance in context of this conception are concerned with *advanced business models* and *human-centred manufacturing* conception.

This chapter aims to analyse implementation success factors of Industry 4.0 especially from business models perspective, and also to address some features of human-centred manufacturing in terms of small- and medium-sized enterprises (SMEs). The motivation of this research is awareness of the importance that just a combination of the selected decisive success factors can significantly help businesses to become more competitive and improving their performance.

When focusing on business models in the context of Industry 4.0 transformation, it is quite obvious that such models will need to adopt new businesses trends, such as mass customization, platform-based businesses, networking manufacturing or creativity-based businesses, respectively. The most related ones are platform-based businesses and mass customization business models. While mass customization practice is relatively well supported by existing methodological frameworks (see, e.g., Pine 1993; Modrak 2017), platform-based business models such as sharing businesses present rather new disruptive approaches, which are not easy to define and categorize. Therefore, the main part of this chapter presented in Sect. 9.3 is devoted to a systematic review on platform-based business models literature using quantitative and qualitative approaches in order not only to map its rapid growth, but also to analyse relation between platform-based business models and traditional business models. The findings of this analysis are summarily presented in Table 9.2. Subsequently, in the same section, existing traditional business models will be analysed how they can be adopted by implementing features of advanced business models. As the result of this analysis for SMEs, the two possible strategies for implementation of platform-based business models are proposed and described also in graphical form, see in Fig. 9.6. Then, in Sect. 9.4, some related aspects of human-centred manufacturing approach will be outlined. Finally, in the Conclusion section, we summarize main ideas from the chapter and provide general findings.

In order to emphasize a comprehensive view of this complex problem, the next section aims to point out the importance of systems approach in implementation of Industry 4.0 concept for SMEs. The necessity to start from high conceptual understanding of the problem is quite clear, but often underestimated. For this reason, the next section can be useful at least for giving an example how systems approach can be used to capture a general conceptual model of the systems thinking for better understanding the relationships between main elements of the smart factory model.

9.2 Importance of Systems Approach in Transforming Organizations

A successful organization transformation, in generally, requires at least an enterprise strategy, executive leadership, a series of decisions and also change in mindset. In addition, an enterprise strategy has to be viewed comprehensively, i.e. as a set of mutually interactive subsystems, components or parts. One of powerful way to see things mutually influential to one another offers system approach, which is also called the structured analysis and design technique. This approach can be also effectively used for understanding of decisive factors influencing transformation of enterprises into smart organizations. The range of the critical success factors (CSFs) depends on specifics of each company, but at least the three of them, identified in previous section, can be considered as crucial elements. These elements are more specifically represented by *advanced manufacturing technologies* including advanced information communication technologies (ICTs), *advanced business models* using online platforms and *human-centred manufacturing* conception.

Nevertheless, the question can arise of why an a priori systems approach in transforming enterprises into smart organizations has considerable potential for success in this effort. The main advantage of this approach lies in the fact that the CSF elements have to be perceived by transforming companies as whole, not merely as a collection of parts, especially at the first stages of the projects. On the contrary, if this approach is not explicitly neither implicitly employed, it usually leads to an atomic way of thinking that is perceived as a syndrome of cognitive immaturity (Maslow 1981). Systems approach can be regularly defined through systems thinking definitions, e.g. as "a framework for seeing interrelationships rather than things, for seeing patterns of change rather than static snapshots" (Bahill and Gissing 1998). According to Halecker and Hartmann (2013) "systems thinking requires a holistic, interdisciplinary and integrated approach". Directly applicable definition for the purpose of our study states that systems thinking consists of the three kinds of conceptual resources, which are: elements, interconnections and a purpose (Fig. 9.1a).

Then, model of systems thinking can be converted for identifying of decisive factors influencing transformation of enterprises into smart factories, as it is shown in Fig. 9.1. The meaning of this model is to emphasize that the three specified basic elements are neither complementary nor alternative to each other, but they are mutually related in a complete causal structure. In order to show their importance, the

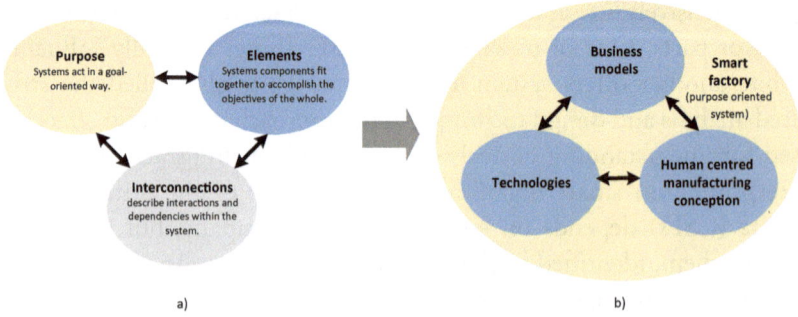

Fig. 9.1 **a** Basic components of systems thinking, **b** Systems model of smart factory with its main elements and interconnections

following two sections are dedicated to describe two of these elements in more detailed way.

9.3 Transition of SMEs Towards Platform-Based Business Models

As this topic of platform-based business models is widely studied in recent literature, we firstly map different approaches to sharing economy practice and structurally analysed them. Subsequently, in Sect. 9.3.2, qualitative analysis of studied literature sources will be provided. The next Sect. 9.3.3 summarizes typical features of platform-based business models.

9.3.1 A Quantitative Analysis of Platform-Based Business Models

Due to the fact that platform-based business models are widely discussed in literature, the quantitative review is an efficient way to analyse research directions and anticipated tendencies. In this order, we started with mapping of number related publications by years. For the purpose, the Web of Science (WOS) database was chosen. Firstly, a research strategy was chosen by finding literature sources related to the term "sharing economy" as part of title, abstract or as keyword on WOS portal searched on July 26, 2020. Then, a total of 2166 potentially relevant papers were found through this database, while 632 publications are open access. Distribution of papers by years of publication is graphically depicted in Fig. 9.2.

The literature sources from Fig. 9.2 consist of journal articles (1571 papers); conference proceedings articles (440 papers); book chapters (76 items); review articles (71 papers); editorial materials (66 papers); book reviews and books (28 papers).

In the next step, the top ten journals, where related papers are published, were selected. Subsequently, they are arranged by number of the papers published in these journals in descending order, namely:

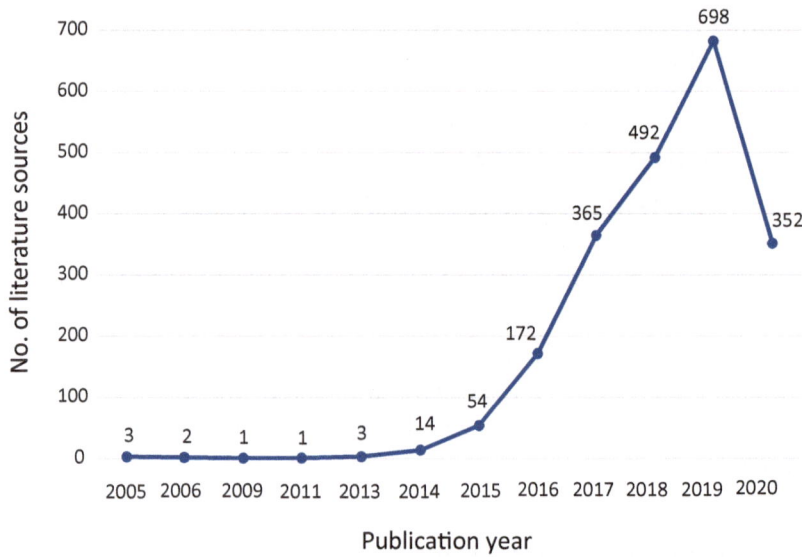

Fig. 9.2 Distribution of papers by years of publication

Sustainability (105 papers); International Journal of Hospitality Management (50 papers); Journal of Cleaner Production (50 papers); International Journal of Contemporary Hospitality Management (32 papers); Current Issues in Tourism (30 papers); Advances in Social Education and Humanities Research (27 papers); Technological Forecasting and Social Change (25 papers); Tourism Management (22 papers); Annals of Tourism Research (20 papers); and Cambridge Handbook of the Law of the Sharing Economy (20 papers). A distribution of the journal papers according to this categorization including journal impact factor (IFs) is shown by graph in Fig. 9.3.

Finally, distribution of literature sources is provided with respect to the top 15 research areas. Based on that, it can be stated that 792 papers are related to Business Economics; 404 papers to Social Sciences; 318 papers to Computer Science; 299 papers to Environmental Sciences Ecology; 284 papers to Engineering; 225 papers to Science Technology; 128 papers to Government Law; 85 papers to Operations Research and Management Science; 79 papers to Transportation; 76 papers to

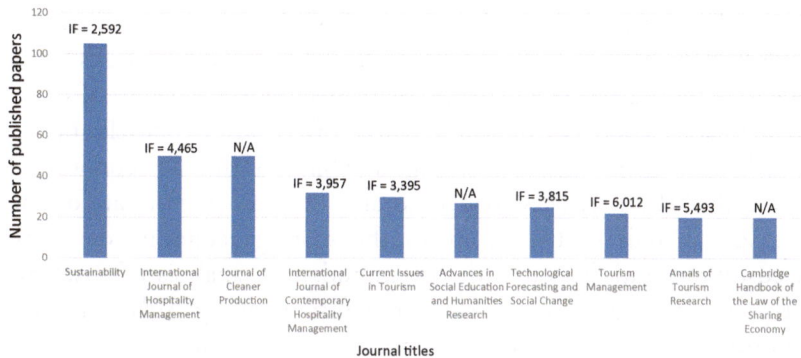

Fig. 9.3 Papers distribution published in the selected journals

Sociology; 73 papers to Information Science and Library Science; 71 papers to Geography; 68 papers to Public Administration; 59 papers to Telecommunications; and 53 papers to Urban Studies. Categorization of the analysed publications from the view of research areas is presented in Fig. 9.4.

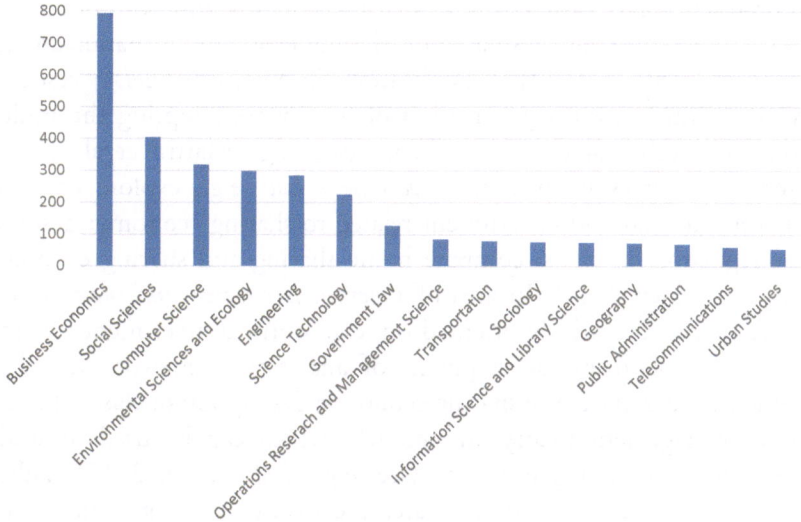

Fig. 9.4 Papers distribution according to research areas

The literature sources shown in previous figure are distributed according to the most frequent research areas. Publications in the research area of *Business economics* are focused on, e.g., analysis of selected marketplaces; analysis of the influences of Airbnb on hotels; business models for the sharing economy; sustainability of sharing economy (see, e.g., Belk 2014a). Related papers to area of *Social sciences* are oriented on, e.g., analysis of the future of the sharing economy; description of customers' satisfaction with accommodation; analysis of customer's perspectives; analysis of review comments; mapping Airbnb in countries (see, e.g., Ert et al. 2016). Publications related to *Computer science* research area are addicted on, e.g., analysis of blockchain technologies for an advanced and cyber-resilient automotive industry; framework for sharing economy based on Internet of things (IoT); designing markets with a focus on exchange platforms (see, e.g., Hawlitschek et al. 2018). Research are of *Environmental sciences and ecology* contains publications focused on, e.g., analysis of motivation for intended sharing economy participation; examination of sustainable business models; sustainability analysis of sharing economy (see, e.g., Lan et al. 2017). In *Engineering* area, we can find papers oriented on, e.g., marketing research on product design; exploration of sharing economy opportunities in the electricity sector; strategies based on sharing economy to manufacturers (see, e.g., Luchs et al. 2016). Publications in *Science technology* research area concentrated on, e.g., creation of framework adopting the multi-level socio-technical transition theory (see, e.g., Martin et al. 2017). Literature sources in area of *Government law*, e.g., explore conflicts between business and government related to sharing economy; describe new regulators in cities; compare home sharing and sharing economy (see, e.g., Posen 2015). In area of *Operations research and management science*, papers are mostly oriented on, e.g., optimal investment strategy for sharing platform; development of analytical framework to select business modes under the sharing economy; description of classical operations management theory and models, which can be used to study applications of sharing economy (see, e.g., Bellos et al. 2017). Publications related to *Transportation* area are focused on, e.g., uncovering motives of business-to-consumer and peer-to-peer car sharing adopters; offering of vehicle-to-vehicle wireless power transfer; description of sharing economy implications in transport sector (see, e.g., Birdsall 2014). *Sociology* area consists of publications related to, e.g., analyse impact of sharing economy on exchange of moral values; analysis of ethnic discrimination in the sharing economy; explore tourists' willingness in providing negative reviews online to express poor experiences (see, e.g., Shuqair et al. 2019). *Information science and library science* contains paper related to, e.g., sharing economy literature reviews; framework for future research, study the role of big data analytics in sharing economy (see, e.g., Sutherland and Jarrahi 2018). Publications in the field of *Geography* are focused on, e.g., digital reputation issues and platform-based tourism; description of sharing economy usage in diverse countries; exploring the regional impact of Airbnb on urban environments (see, e.g., Lima 2019). Publications related to *Public administration* area are oriented on, e.g., examination of consumers' value co-creation in sharing economy; description of civil opportunities in collaborative economy based on sharing economy (see, e.g., Nadeem et al. 2020). In the area of *Telecommunications*, papers are focused on, e.g., description of cloud-based sharing platforms; collaborative consumption through mobile apps; exploring service quality among online sharing economy platforms (see, e.g., Li et al. 2017). In relation to *Urban studies*, there are papers oriented on, e.g., study the understanding the spatial distribution in ride-sharing; exploration of the ride-sharing adaption in urban areas of cities; study the understanding the spatial distribution in ride-sharing; exploration of the ride-sharing adaption in urban areas of cities (see, e.g., Ferreri and Sanyal 2018).

It can be stated that, sharing economy phenomena presented as the exchange relation in the leading world economies is based on the increasing use of innovations and technologies related to Industry 4.0. And thus, brief description of selected publications with identified research domains related to the sharing economy and Industry 4.0 conception is depicted in Table 9.1.

This quantitative overview of the related literature firstly showed that sharing economy (SE) significantly attracted not only practitioners, but also scholar community in recent several years. This is clear evidence that SE phenomena cannot be perceived only as one of possible business approach, but contrariwise, SE-based business approaches vary

Table 9.1 Brief description of selected publications on sharing economy in context of Industry 4.0

Research domains	Selected papers related to the domains	Brief description
Sensors, IoT, big data, artificial intelligence (AI)	First-mover firms in the transition towards the sharing economy in metallic natural resource-intensive industries: Implications for the circular economy and emerging industry 4.0 technologies (Jabbour et al. 2020)	Paper offers novel implications for the theory and practice of operations management for the sharing economy; and provides case study on two companies transitioning to the sharing economy
Mobile apps, digital platforms	Crowd working: Jobs in the age of digitalization and platform economy (Wefersova 2020)	This paper is focused on structural assessment of these digital crowd working platforms and analyse the situation of crowd workers
IoT, cloud computing, cyber-physical systems (CPS), data analysis, AI, smart sensors and wireless communication technology	Industry 4.0 to accelerate the circular economy: A case study of electric scooter sharing (Pham et al. 2019)	Article is oriented on the exploration of fundamental concepts of Industry 4.0 and the influential factors of Industry 4.0. According to the obtained results, Industry 4.0 can provide an enabling framework for the sharing economy in CE implementation
Networking, data collected using computer assisted telephone interviews, business model	Implementation of the sharing economy in the B2B sector (Grondys 2019)	Paper analysis and assessment of the sharing economy between enterprises in order to identify the way how the exchange of material resources between them is implemented using statistical testing
Data analysis using statistical software, business models	Development of a business model by introducing sustainable and tailor-made value proposition for SME clients (Bolesnikov et al. 2019)	Article provides a systematic statistical analysis of the proposed business model's aspects on the selected SMEs

Research domains	Selected papers related to the domains	Brief description
Cloud platform, data analysis	Entropy maximization-based capability allocation of clothing production enterprises in sharing economy (Zheng and Song 2019)	This article is focused on capability allocation of suppliers' production resources in order to maximize their utilization. The proposed model promotes the usage of supplier resources in the production system
Blockchain technology, distributed ledger technology, IoT, CPS, cybersecurity, business model	A review on blockchain technologies for an advanced and cyber-resilient automotive industry (Fraga-Lamas and Fernández-Caramés 2019)	This paper proposed recommendations with the aim of guiding researchers and companies in future cyber-resilient automotive industry developments
Digital economy, AI, technological singularity, IoT, big data	Industry 4.0: socio-economic junctures (Budanov et al. 2017)	The paper is focused on analysis of various social transformation scenarios
Digitalization, robots, automation, data analysis	How big is the gig? Assessing the preliminary evidence on the effects of digitalization on the labour market (Eichhorst et al. 2017)	Article analysed the importance of digitalization in selected countries with a particular attention to the potential or actual impact on the labour market
Digitalization, technological platform, cloud infrastructure, ICT	On the evolution of regional efficiency potentials (Kuch and Westkämper 2017)	The present article addressed the challenge of combining and integrating two quite different dimensions of the business organization: its technological and its managerial perspective
IoT, Big data, cloud computing, AI, e-business, distributed computing and control, multimedia processing	The internet information and technology research directions based on the fourth industrial revolution (Chung and Kim 2016)	This paper categorized topics of articles related to the fourth industrial revolution based on the keyword frequency of main issues

(continued)

Table 9.1 (continued)

Research domains	Selected papers related to the domains	Brief description
IoT, CPS, data analysis, cloud computing	Shared manufacturing in the sharing economy: Concept, definition and service operations (Yu et al. 2020)	This paper presents a dynamic shared manufacturing service scheduling method in support of the technologies of complex network analysis
ICT, data analysis	The information flow between enterprises in the context of sharing economy (Grondys et al. 2020)	This paper examines the enterprises exchange information in the area of sharing economy; existing barriers in the process of information exchange, and the directions and ways to exchange information
Mathematical model	Modelling of sharing networks in the circular economy (Jayakumar et al. 2020)	This paper is focused on development and optimization of a mathematical model based on a framework integrating the key concepts related to a circular economy and SE
Data mining, AI, IoT, big data, CPS	Analysing the major issues of the fourth industrial revolution (Jeon and Suh 2017)	This paper provided an analysis used data mining as modelling method and it is expected that this paper will be helpful for the researcher and policy maker of the fourth industrial revolution
IoT, data analysis, business models	Sharing economy and "industry 4.0" as the business environment of millennial generation—a marketing perspective (Brkljač and Sudarević 2018)	This paper provided comprehension of the relationship between the sharing economy and the "Industry 4.0"

Research domains	Selected papers related to the domains	Brief description
Manufacturing execution system, data analysis, big data, CPS	Comparative study of crossing the chasm in applying smart factory system for SMEs (Choi and Choi 2019)	This article is focused on the practical methodology encouraging manufacturing SMEs in adopting smart factory system utilizing rapidly changing innovative technology
ICT, data analysis, business model	Change through digitization—Value creation in the age of Industry 4.0 (Kagermann 2015)	This paper discusses the impact, challenges and opportunities of digitization
IoT, big data, cyber security, industrial augmented reality, CPS, cloud computing, blockchain	A review on the application of blockchain to the next generation of cybersecure Industry 4.0 smart factories (Fernández-Caramés and Fraga-Lamas 2019)	This paper analysed the benefits and challenges that arise when using blockchain and smart contracts to develop Industry 4.0 applications. Paper describes the most relevant blockchain-based applications for Industry 4.0 technologies
Cloud computing, big data, IoT, CPS, ICT	Italy's Industry 4.0 plan: An analysis from a labour law perspective (Seghezzi and Tiraboschi 2018)	Article identified actions and perspectives to manage current changes, focusing on workers rather than on technologies. Paper describes new functions of labour law in the Industry 4.0 era

Table 9.2 Sharing economy types in three marketplaces

Sharing business model types		Web platform based	Marketplace type		
			C2C	B2C/C2B	B2B
Traditional sharing practice		-	Yes	-	-
Sharing economy	On-demand-based sharing business models	Yes	Yes	Yes	Yes
	Second-hand-based sharing business models	Yes	Yes	-	Yes
	Product-service sharing business models	Yes	Yes	Yes	Yes

depending on specific business conditions. Another interesting finding is that SE penetrated into wide research disciplines. This is quite promising, since it can lead to multidisciplinary exchanges of experiences and bring new stimulus for further development of these phenomena.

9.3.2 A Qualitative Analysis of Platform-Based Business Models

In this subsection it is intended to provide better understanding of the platform-based business models. The term "sharing" has become very popular in recent times, but as known this term is not new. One could see this positive concept in the past where, for instance, overconsumption in households lead to sharing practice to use their resources more efficiently. In this case, we are talking about traditional sharing. But commonly people act in their self-interest solely no matter what consequences arise from this, since Earth's resources are diminishing. On the other hand, when sharing becomes a group effort, then such practice brings positive results for everyone.

SE can be defined, e.g., as "a marketplace that consists of entities that innovatively and sustainably shape how marketing exchanges of valuable products and resources are produced and consumed through sharing, which can occur when entities take part in the actual or life-cycle use of a product or resource and communicate some form of information, and which can be scaled using technology" (Lim 2020). SE phenomena relates to global economic and sustainability problems and from this reason it is getting increasing attention in our daily lives. Moreover, sharing economy practice becomes important driver of local economies, what can be documented, e.g., by the fact that only in Europe SE platform generated revenues of nearly four billion euros and transactions of over 28 billion euros (Agarwal and Steinmetz 2019).

Sharing economy development is adequately supported through scientific and popular literature. It is useful to note that different authors use several synonymous terms describing these phenomena. Some of them can be mentioned here. Botsman and Rogers (2010) describe this as "collaborative consumption"; Lamberton and Rose (2012) as "commercial sharing systems"; Humphreys and Grayson (2008) as "co-production"; Lanier and Schau (2007), Prahalad and Ramaswamy (2004) as "co-creation"; Katz (2015), Lobel (2016) as "platform economy"; Mont (2002) as "product-service systems"; Bardhi and Eckhardt (2012) as "access-based consumption"; Fitzsimmons (1985) as "consumer participation"; Schor (2014), Frenken and Schor (2019) as "stranger sharing"; and Postigo (2003) as "online volunteering". In order to extract specific knowledge from the existing literature, the term sharing economy will be further divided into two main sub-categories: traditional sharing practice (TSP) and sharing economy (SE) (Stanoevska-Slabeva et al. 2017). While in traditional approach products and services are shared based on mutual deal or agreement between both sides of consumers, sharing economy uses payments and feedbacks or complaints through the platforms based on Web 2.0 technologies (Belk 2014a). The concept based on sharing economy opened doors to the rise of numerous for-profit and non-profit businesses. However, there is some confusion or scepticism about this business phenomena among academics and the public due to its novelty and there is no unambiguously view on what exactly the sharing economy will bring for all of us. Belk (2014b) differentiates terms sharing and pseudo-sharing by using epistemological viewpoint. He is explaining that traditional sharing is about helping and building human relations, while pseudo-sharing is a business relationship masquerading as communal sharing. As it was mentioned, traditional sharing is about solving problems related to overconsumption and efficient resources usage. Other characteristics and differences between TSP and SE are discussed by Demary (2015). According to her, SE companies present an important part of business model portfolio and thanks to them competition in most markets they are active insignificantly increased.

Frenken et al. (2015) identified three types of SE business models, which are on-demand-based sharing economy, product-service-based sharing economy and second-hand-based sharing economy. In line with this categorization the following classification of SE business models can be offered (see Fig. 9.5).

The typical features of sharing business models depicted in Fig. 9.5 are as follows.

On-demand-based sharing business model is using web platforms and apps and present the intersection of tendencies towards peer-to-peer (P2P) or consumer-to-consumer (C2C) exchange and access economy. For example, when ordering the taxi through, e.g., Uber company, BlaBlaCar company.

Second-hand-based sharing business model can be characterized as traditional second-hand business extended through web platform and apps. Typical provider of services based on this business model is Momox GmbH company, which is offering an online buying-and-selling service for second-hand garments across some Western Europe

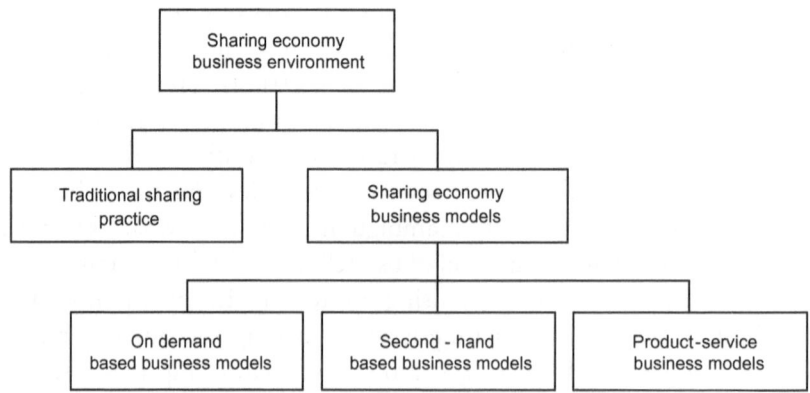

Fig. 9.5 Classification of sharing economy business models

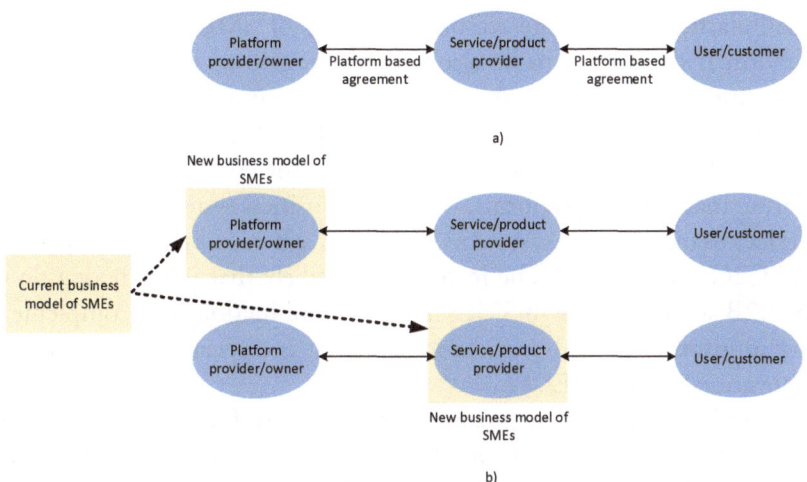

Fig. 9.6 **a** The three actors of platform-based business model, **b** The two possible new strategies for SMEs

countries. Another well-known web-based platforms are for example Ebay or Facebook.

Product-service sharing business models are based on leasing a good from a company on business-to-consumer marketplace (B2C)/consumer-to-business marketplace (C2B) rather than C2C. Consumer utilizing this business model obtains temporary access to a product, while the company retains ownership. An example is car-rental via Hertz or Zipcar.

In order to help SMEs to follow above-described business model it is useful to identify their relations with different online marketplaces. For this purpose, the following comparison of these sharing business models is provided by authors in Table 9.2.

The most relevant sharing business models for transition of SMEs towards sharing businesses are those which operate on B2B and C2C marketplaces. C2C sharing business models are mostly based on virtual networks, through which individual consumers and individual suppliers are connected.

Similarly, for the same reason, transition of SMEs into smart organization is considered to be also suitable in case of B2C/C2B marketplace models. Naturally, it requires the adoption of the E-business solutions allowing many SMEs to respond to these challenging opportunities (Gutowska and Sloane 2009). Nevertheless, the number of B2B sharing marketplace platforms, where one business system sells goods and services to other business systems, is still low when comparing to the B2C and C2C counterparts. The reason why is likely that implementation of the B2B sharing model in SMEs requires combination with complementary innovation-based business models, what is especially challenging for innovation-based SMEs.

In order to formulate practical implications for SMEs, which are acting in B2B and/or B2C markets, it is firstly useful to define main actors of platform-based business model as shown in Fig. 9.6a. Then, in principle, there are the two possible strategies for implementation of platform-based business models (see Fig. 9.6b), which SMEs can choose from.

The first of them is an exploitation at least one of existing online business model platforms and the second one is based on development of own online business model platform.

9.3.3 Typical Features of Platform-Based Business Models

There is no doubt that platform-based business models themselves have a number of inherent advantages over traditional business models. Typical features of platform-based business models can be characterized as follows: they are scalable, networked, intelligent, and with open architecture.

In general, Industry 4.0 prioritizes business models which incorporate the following attributes (Ibarra et al. 2018):

- *A service-oriented approach.* Such business model orientation enables manufacturing companies to provide services through global network to other cyber-physical systems, humans or companies. This approach emphasizes the long-term need for a change from product sales to service-oriented businesses.

- *A user-driven approach.* This direction means for companies to be more responsive to user-driven demands by learning more about their customers. This approach usually helps to identify new innovation areas and comes up with more individualized products. It can also occur that new innovations are adopted by their suppliers.
- *A network-oriented approach.* Such orientation means that business models are based upon the principles of openness, peering, sharing and acting globally. According to Rauch et al. (2017), especially, distributed manufacturing network models are considered as one of the drivers for the design of the "factory of the future". It means that traditional centralized manufacturing systems will be substituted with more and more decentralized and geographically dispersed manufacturing networks.

9.4 New Work Roles in Industry 4.0 Environment

An introduction of Industry 4.0 into manufacturing significantly affects manufacturing processes in a way which lead to disruptive innovations in work patterns. New work roles and personal tasks of manufacturing staff in Industry 4.0 environment directly result from the necessity of intensive human–machine collaboration requiring new knowledge and working skills. According to Romero et al. (2016) smart factory concepts are placing human operators, named as Operators 4.0, as central actors in manufacturing processes. Such workers will be assisted by automated systems allowing them to utilize and develop their creative, innovative and improvisational skills, without compromising production objectives. In this context, one of significant features of smart factories is *human-centred manufacturing* conception.

The term human-centred manufacturing can be characterized as quasi autonomous manufacturing unit, e.g. cell in which a group of multi-skilled operators works as team (Hancke et al. 1990). The team also includes robots, where both actors collaborating on frequently changing operational tasks. Presently, human–robot collaboration is a wide research field, which brings high economic benefits. Due to these reasons, transition of SMEs towards smart organizations cannot succeed only by introduction advanced technologies, but need to be also targeted at designing and developing smart workstations based on human centeredness with incorporation of different types of human Operators 4.0 into autonomous manufacturing units.

Romero et al (2016) proposed the following typology of Operators 4.0:

- *Super-Strength Operator*. This category of operators is represented by intelligent wearable human–robotic exoskeletons for manual for manual handling work. Exoskeletons are helping workers across a variety of industry to prevent workplace injuries and illnesses.
- *Tech-Augmented Operator*. This type of operators is strongly supported by augmented reality (AR) technology. As it is assumed that a number of tasks in manufacturing will be increasingly automated, then augmented reality technology is able to provide additional capabilities to the human operators. It's thanks to that, that AR technology is able to interact with the physical objects in a more intuitive manner where the real objects are accompanied by computer-generated perceptual information.
- *Virtual Operator*. In this case, operators of this type are utilizing virtual reality (VR) technology. VR technology is evenly as AR a vital toll supporting shop-floor operators in the smart factories. For example, it can provide a combination of interactive virtual reality and advanced simulations of realistic scenarios for optimized decision-making for the smart operator.
- *Healthy Operator*. As an example of this type of Operator 4.0 is a human operator using wearable trackers, which are devices dedicated to measure exercise activity, stress, heart rate and other health-related metrics.
- *Smarter Operator*. These operators are supported by Intelligent Personal Assistant (IPA) software, which is dedicated to assist people with basic tasks, usually providing information via online sources. This

software, which is based on artificial intelligence, helps a smart operator in interfacing with machines, computers, databases and other information systems (Myers et al. 2007).

- *Collaborative Operator*. His role lies in co-working with industrial collaborative robots, which provides assistance to the human operator.
- *Social Operator*. In this case, operators are used for communication enterprise social networks, which enable faster cooperation between smart operators and smart machines.
- *Analytical Operator*. The specifics of this operator lie in organizing and analysing large sets of data to identify useful information and predict important events. Usually, Analytical Operator is connected to several other applications using advanced data analytics.

These types of operators 4.0 present ambitious nomenclature of work roles in smart manufacturing environment assuming that physical and software components are deeply intertwined in cyber-physical systems and supported by human–machine interaction technologies, such as dialogue systems, multimedia-multimodal displays, adaptive interfaces and others.

9.5 Conclusions

When coming back to the tractate idea about the three crucial factors for successful transition of SMEs towards smart factory which are namely *advanced manufacturing technologies, advanced business models* and *human-centred manufacturing* conception, now it is more clear that the main dilemmas that SMEs have to face are: to which technologies they need to invest; and which advanced business model is for them suitable. The both of them need to be solved undependably, and this chapter wants to offer some insight into opportunities for exploitation of sharing economy platform-based models.

In this context, when focusing on the second dilemma concerning a business model selection and application, it can be pointed out here that sharing economy is developing promisingly for the better, since it is changing the consumer behaviour towards green practices. Hopefully, this fact can positively motivate SMEs in their transition towards smart factory. However, technological development, as precondition of further development of SE, is not always positively perceived among people. It is due to the fact that advanced technology and related industrialization brought many negative impacts on the environment. On the other hand, further technological development is considered as an important impetus to facilitate transition of SMEs towards the Industry 4.0 conception, which is considered as sustainable growth factor. The root of this contradiction lies in classical dilemma what to prefer—technological development or environmental protection, but optimally both. So thinking optimistically, further successful implementation of Industry 4.0 concept can bring promising benefits for everyone.

Acknowledgements This project has received funding from the European Union's Horizon 2020 research and innovation program under the Marie Skłodowska-Curie grant agreement No 734713.

References

Agarwal, N., and R. Steinmetz. 2019. Sharing economy: A systematic literature review. *International Journal of Innovation and Technology Management (IJITM)* 16 (06): 1–17.

Bahill, A.T., and B. Gissing. 1998. Re-evaluating systems engineering concepts using systems thinking. *IEEE Transactions on Systems, Man, and Cybernetics, Part C (Applications and Reviews)* 28 (4): 516–527.

Bardhi, F., and G. Eckhardt. 2012. Access based consumption: The case of car sharing. *Journal of Consumer Research* 39: 881–898.

Belk, R. 2014. You are what you can access: Sharing and collaborative consumption online. *Journal of Business Research* 67 (8): 1595–1600.

Belk, R. 2014b. Sharing versus pseudo-sharing in Web 2.0. *The Anthropologist* 18 (1): 7–23.

Bellos, I., M. Ferguson, and L.B. Toktay. 2017. The car sharing economy: Interaction of business model choice and product line design. *Manufacturing & Service Operations Management* 19 (2): 185–201.

Birdsall, M. 2014. Carsharing in a sharing economy. Institute of Transportation Engineers. *ITE Journal* 84 (4): 37–40.

Bolesnikov, M., M. Popović Stijačić, M. Radišić, A. Takači, J. Borocki, D. Bolesnikov, and J. Dzieńdziora. 2019. Development of a business model by introducing sustainable and tailor-made value proposition for SME clients. *Sustainability* 11 (4): 1–16.

Botsman, R., and R. Rogers. 2010. Beyond Zipcar: Collaborative consumption. *Harvard Business Review* 88 (10): 1–30.

Brkljač, M., and T. Sudarević. 2018. Sharing economy and "Industry 4.0" as the business environment of millennial generation—A marketing perspective. *Annals of DAAAM & Proceedings* 29: 1092–1101.

Budanov, V., I. Aseeva, and E. Zvonova. 2017. Industry 4.0.: Socio-economic junctures. *Economic Annals-XXI* 168: 33–37.

Choi, Y. H., and S.H. Choi. 2019. Comparative study of crossing the chasm in applying smart factory system for SMEs. *International Journal of Pure and Applied Mathematics* 118: 469–487.

Chung, M., and J. Kim. 2016. The internet information and technology research directions based on the fourth industrial revolution. *KSII Transactions on Internet & Information Systems* 10 (3): 1311–1320.

Demary, V. 2015. Competition in the sharing economy (No. 19/2015). Contributions to the political debate by the Cologne Institute for Economic Research. IW policy paper, 1–27.

Drath, R. 2014. Industrie 4.0–eine Einführung. *Open Automation* 3: 17–21.

Eichhorst, W., H. Hinte, U. Rinne, and V. Tobsch. 2017. How big is the gig? Assessing the preliminary evidence on the effects of digitalization on the labor market. *The International Review of Management Studies* 28 (3): 298–318.

Ert, E., A. Fleischer, and N. Magen. 2016. Trust and reputation in the sharing economy: The role of personal photos in Airbnb. *Tourism Management* 55: 62–73.

Fernández-Caramés, T.M., and P. Fraga-Lamas. 2019. A review on the application of blockchain to the next generation of cybersecure industry 4.0 smart factories. *IEEE Access* 7: 45201–45218.

Ferreri, M., and R. Sanyal. 2018. Platform economies and urban planning: Airbnb and regulated deregulation in London. *Urban Studies* 55 (15): 3353–3368.

Fitzsimmons, J. 1985. Consumer participation and productivity in service operations. *Interfaces* 15: 60–67.

Fraga-Lamas, P., and T.M. Fernández-Caramés. 2019. A review on blockchain technologies for an advanced and cyber-resilient automotive industry. *IEEE Access* 7: 17578–17598.

Frenken, K., and J. Schor. 2019. Putting the sharing economy into perspective. In *A research agenda for sustainable consumption governance*, 121–135. Cheltenham: Edward Elgar.

Frenken, K, T. Meelen, M. Arets, and P. Van de Glind. 2015. Smarter regulation for the sharing economy. *The Guardian*. 20 May. See https://www.theguardian.com/science/political-science/2015/may/20/smarter-regulation-for-the-sharing-economy. Accessed 6 February 2017.

Grondys, K. 2019. Implementation of the sharing economy in the B2B sector. *Sustainability* 11 (14): 1–16.

Grondys, K., P. Bajdor, and M. Starostka-Patyk. 2020. The information flow between enterprises in the context of sharing economy. *European Research Studies* 23 (2): 781–794.

Gutowska, A., and A. Sloane. 2009. Modelling the B2C marketplace: Evaluation of a reputation metric for e-commerce. International Conference on Web Information Systems and Technologies. Springer, Berlin and Heidelberg, 212–226.

Halecker, B., and M. Hartmann. 2013. How can systems thinking add value to business model innovation? XXIV ISPIM Conference—Innovating in Global Markets: Challenges for Sustainable Growth, Helsinki, Finland, 16–19.

Hancke, T., C.B. Besant, M. Ristic, and T.M. Husband. 1990. Human-centred technology. *IFAC Proceedings Volumes* 23 (7): 59–66.

Hawlitschek, F., B. Notheisen, and T. Teubner. 2018. The limits of trust-free systems: A literature review on blockchain technology and trust in the sharing economy. *Electronic Commerce Research and Applications* 29: 50–63.

Humphreys, A., and K. Grayson. 2008. The intersecting roles of consumer and producer: A critical perspective on co-production, co-creation and prosumption. *Sociological Compass* 2: 963–980.

Ibarra, D., J. Ganzarain, and J.I. Igartua. 2018. Business model innovation through Industry 4.0: A review. *Procedia Manufacturing* 22: 4–10.

Jabbour, C.J.C., P.D.C. Fiorini, C.W. Wong, D. Jugend, A.B.L.D.S. Jabbour, B.M.R.P. Seles, and H.M. da Silva. 2020. First-mover firms in the transition towards the sharing economy in metallic natural resource-intensive industries: Implications for the circular economy and emerging industry 4.0 technologies. *Resources Policy* 66: 1–13.

Jayakumar, J., K. Jayakrishna, K.E.K. Vimal, and S. Hasibuan. 2020. Modelling of sharing networks in the circular economy. *Journal of Modelling in Management* 15 (2): 407–440.

Jeon, J., and Y. Suh. 2017. Analyzing the major issues of the 4th industrial revolution. *Asian Journal of Innovation and Policy* 6 (3): 262–273.

Kagermann, H. 2015. Change through digitization—Value creation in the age of Industry 4.0. In *Management of permanent change*, 23–45. Wiesbaden: Springer Gabler.

Kagermann, H., J. Helbig, A. Hellinger, and W. Wahlster. 2013. Recommendations for implementing the strategic initiative Industrie 4.0: Securing the future of German manufacturing industry. Final Report of the Industrie 4.0 Working Group. Forschungsunion, 1–82.

Katz, V. 2015. Regulating the sharing economy. *Economics*: 1–60. https://doi.org/10.15779/Z38HG45.

Kuch, B., and E. Westkämper. 2017. On the evolution of regional efficiency potentials. *Procedia Manufacturing* 11: 1528–1535.

Lamberton, C., and R. Rose. 2012. When is ours better than mine? A framework for understanding and altering participation in commercial sharing systems. *Journal of Marketing* 76: 109–125.

Lan, J., Y. Ma, D. Zhu, D. Mangalagiu, and T.F. Thornton. 2017. Enabling value co-creation in the sharing economy: The case of mobike. *Sustainability* 9 (9): 1–20.

Lanier, C., and H. Schau. 2007. Culture and co-creation: Exploring consumers' inspirations and aspirations for writing and posting on-line fan fiction. In *Consumer culture theory: Research in consumer behavior*, ed. R. Belk and J. Sherry Jr., vol. 11, 321–342. Amsterdam: Elsevier.

Li, Y., L. Zhu, M. Shen, F. Gao, B. Zheng, X. Du, and S. Yin. 2017. Cloud-Share: Towards a cost-efficient and privacy-preserving alliance cloud using permissioned blockchains. International Conference on Mobile Networks and Management. Springer, Cham, 339–352.

Lim, W. M. 2020. Sharing economy: A marketing perspective. *Australasian Marketing Journal* 28: 4–13.

Lima, V. 2019. Towards an understanding of the regional impact of Airbnb in Ireland. *Regional Studies, Regional Science* 6 (1): 78–91.

Lobel, O. 2016. The law of the platform. *Minnesota Law Review* 101 (1): 87–96.

Luchs, M.G., K.S. Swan, and M.E.H. Creusen. 2016. Perspective: A review of marketing research on product design with directions for future research. *Journal of Product Innovation Management* 33 (3): 320–341.

Marr, B. 2016. Why everyone must get ready for the 4th industrial revolution. *Forbes Tech.* Available online: https://www.forbes.com/sites/bernardmarr/2016/04/05/why-everyone-must-get-ready-for-4th-industrial-revolution/#7f6daba03f90. Accessed on 5 April 2016.

Martin, C.J., P. Upham, and R. Klapper. 2017. Democratising platform governance in the sharing economy: An analytical framework and initial empirical insights. *Journal of Cleaner Production* 166: 1395–1406.

Maslow, A.H. 1981. *Motivation and personality*, 1–354. Prabhat Prakashan.

Modrak, V. (ed.). 2017. *Mass customized manufacturing: Theoretical concepts and practical approaches*, 1–412. Boca Raton: CRC Press.

Mont, O. 2002. Clarifying the concept of product-service system. *Journal of Cleaner Production* 10: 237–245.

Myers, K., P. Berry, J. Blythe, K. Conley, M. Gervasio, D.L. McGuinness, et al. 2007. An intelligent personal assistant for task and time management. *AI Magazine* 28 (2): 47–61.

Nadeem, W., M. Juntunen, F. Shirazi, and N. Hajli. 2020. Consumers' value co-creation in sharing economy: The role of social support, consumers' ethical perceptions and relationship quality. *Technological Forecasting and Social Change* 151: 1–13.

Pham, T.T., T.C. Kuo, M.L. Tseng, R.R. Tan, K. Tan, D.S. Ika, and C.J. Lin. 2019. Industry 4.0 to accelerate the circular economy: A case study of electric scooter sharing. *Sustainability* 11 (23): 1–16.

Pine, B.J. 1993. *Mass customization*, vol. 17. Boston: Harvard Business School Press.

Posen, H.A. 2015. Ridesharing in the sharing economy: Should regulators impose Uber regulations on Uber. *Iowa Law Review* 47: 777–780.

Postigo, H. 2003. Emerging sources of labor on the Internet: The case of America Online volunteers. *International Review of Social History* 48: 205–223.

Prahalad, C.K., and V. Ramaswamy. 2004. Co-creation experiences the net practice in value creation. *Journal of Interactive Marketing* 18: 5–14.

Qin, J., Y. Liu, and R. Grosvenor. 2016. Changeable, agile, reconfigurable and virtual production. A categorical framework of manufacturing for Industry 4.0 and beyond. *Procedia CIRP* 52: 173–178.

Rauch, E., P. Dallasega, and D.T. Matt. 2017. Distributed manufacturing network models of smart and agile mini-factories. *International Journal of Agile Systems and Management* 10 (3–4): 185–205.

Romero, D., J. Stahre, T. Wuest, O. Noran, P. Bernus, P., A. Fast-Berglund, and D. Gorecky. 2016. Towards an operator 4.0 typology: A human-centric perspective on the fourth industrial revolution technologies. Proceedings of the International Conference on Computers and Industrial Engineering, 1–11.

Seghezzi, F., and M. Tiraboschi. 2018. Italy's industry 4.0 plan: An analysis from a labour law perspective. *Journal of International and Comparative Labour Studies* 7 (1): 1–29.

Schor, J. 2014. Debating the sharing economy. *Journal of Self-Governance and Management Economics* 4 (3): 7–22 (October, www.greattransition.org, Retrieved October 23, 2016).

Shuqair, S., D.C. Pinto, and A.S. Mattila. 2019. Benefits of authenticity: Post-failure loyalty in the sharing economy. *Annals of Tourism Research* 78: 1–15.

Stanoevska-Slabeva, K., V. Lenz-Kesekamp, and V. Suter. 2017. Platforms and the sharing economy: An analysis EU H2020 Research Project Ps2Share: Participation, privacy, and power in the sharing economy, 2017. *SSRN Electronic Journal* 2007: 1–92. https://doi.org/10.2139/ssrn.3102184.

Stearns, P.N. 2015. *Debating the industrial revolution*, 1–176. London: Bloomsbury Publishing.

Sutherland, W., and M.H. Jarrahi. 2018. The sharing economy and digital platforms: A review and research agenda. *International Journal of Information Management* 43: 328–341.

Wefersova, J. 2020. Crowdworking: Jobs in the age of digitalization and platform economy. Vplyv Industry 4.0 na tvorbu pracovnych miest 2019, 448–453.

Yu, C., X. Xu, S. Yu, Z. Sang, C. Yang, and X. Jiang. 2020. Shared manufacturing in the sharing economy: Concept, definition and service operations. *Computers and Industrial Engineering* 146: 1–13.

Zheng, F., and Q. Song. 2019. Entropy maximization based capability allocation of clothing production enterprises in sharing economy. *IFAC-PapersOnLine* 52 (13): 211–216.

10

Toward SME 4.0: The Impact of Industry 4.0 Technologies on SMEs' Business Models

Philipp C. Sauer, Guido Orzes, and Laura Davi

10.1 Introduction

Industry 4.0 (I4.0) represents a radical change in firms' operations that is particularly challenging for small- and medium-sized enterprises (SMEs) due to the substantial need for capital and knowledge (Müller and Däschle 2018; Orzes et al. 2020). Nevertheless, I4.0 and the technologies subsumed below this concept promise to enhance the productivity, flexibility, and competitiveness of SMEs (Kagermann et al. 2013; Weking et al. 2019). To realize such a far-reaching change in an economically sustainable way I4.0 implementation should ideally be complemented

P. C. Sauer (✉) · G. Orzes · L. Davi
Industrial Engineering and Automation (IEA), Faculty of Science and
Technology, Free University of Bozen-Bolzano, Bolzano, Italy
e-mail: philippchristopher.sauer@unibz.it

G. Orzes
e-mail: guido.orzes@unibz.it

L. Davi
e-mail: laura.davi@natec.unibz.it

© The Author(s) 2021   **293**
D. T. Matt et al. (eds.), *Implementing Industry 4.0 in SMEs*,
https://doi.org/10.1007/978-3-030-70516-9_10

by a modification in the Business Model (BM) of the firm, i.e., the firm's logic of creating and capturing value (Zott and Amit 2010). Moreover, the I4.0 concept and the underlying technologies are constantly innovated (Kagermann et al. 2013; Culot et al. 2020) thus requiring a corresponding continuous evaluation of BMs that enable its implementation and sustained viability (Müller and Däschle 2018).

While this need is widely acknowledged in literature, it is hindered by the substantially different nature of the concepts of I4.0, originating from the engineering domain, and BMs, originating from the business strategy domain. Extant definitions of I4.0 technologies (see Culot et al. 2020 for an extensive review) and of BMs reveal indeed a wide range of underlying elements that need to be considered. However, there is still no uniform "recipe" of technology to BM element interrelations. Due to the limited availability of resources of SMEs, the majority of them struggles to build and maintain sufficient expertise in both fields that are required to simultaneously modify both the operations and BM (Müller and Däschle 2018). Nevertheless, SMEs build the backbone of an economy and represent even in the highly industrialized European Union 99% of all businesses (European Commission 2020).

This chapter aims therefore to investigate the interrelation between I4.0 technologies and BMs with a specific focus on SMEs. In detail, the addressed research question is: *How can the implementation of Industry 4.0 technologies drive the modification of the business models of small- and medium-sized enterprises?* To answer this question, a combined approach relying on literature and on web-based secondary data is applied. This enables to take stock of the current literature to identify hotspots and gaps in the academic investigation of the interrelation of I4.0 technologies and BMs in SMEs. Subsequently, this will be complemented by an empirical investigation of 30 SMEs websites to validate the literature-based findings.

The chapter is organized as follows: Sect. 10.2 introduces the theoretical background for the study consisting of the concepts of I4.0, BM and SMEs. Section 10.3 presents the applied methodology and Sect. 10.4 summarizes the results. Finally, Sect. 10.5 presents the discussion and conclusion of the study including the contributions to research and practice, its limitations, and the derived research directions.

10.2   Background

In this section, we present three concepts that are of particular relevance for our study: I4.0 technologies (Sect. 10.2.1), BMs (Sect. 10.2.2), and SMEs (Sect. 10.2.3).

10.2.1   Industry 4.0

The term "Industrie 4.0" was first coined in 2011 to strengthen the competitiveness of the German industry and since then diffused (with some adaptations) all over the world (Tirabeni et al. 2019). In its core, I4.0 represents a synonym for the Fourth Industrial Revolution that has been preceded by other fundamental steps such as taking advantage of mechanical innovations like the steam power, cotton spinning, and railroads (First Industrial Revolution), allowing mass production through assembly lines and electricity (Second Industrial Revolution), and automating production lines by using electronic systems and computer technologies (Third Industrial Revolution) (Ingaldi and Ulewicz 2020).

Starting from this basis, I4.0 *"describes the increasing digitization and automation of the manufacturing environment, as well as the creation of digital value chains to enable communication between products, their environment and business partners"* (Lasi et al. 2014, p. 240). Similarly, I4.0 can be defined as *"a collective term for technologies and concepts of value chain organization"* (Ślusarczyk 2018, p. 234). Moreover, the distinction between industry and services becomes less relevant as digital technologies are connected with industrial products and services forming hybrid products which are neither exclusively goods nor services (Strandhagen et al. 2017).

The main features of I4.0 are related to integration, real-time operability, flexibility, servitization, customer orientation, and expertization that facilitate the connection and communication between humans and machines through the Internet of Things and Cyber-Physical Systems (Hermann et al. 2016). As a result, products become customized, processes are networked, and knowledge barriers are reduced among users. The "context-aware" Smart Factory takes into consideration the position as well as status of a product within the process and assists machines as well as people performing their tasks (Rejikumar et al. 2019; Weking et al. 2019).

Consequently, I4.0 arises from its elementary technologies, which lay the foundation for the integration of intelligent machines, humans, physical objects, production lines, and processes to form a new kind of value chain across organizational boundaries, featuring intelligent, networked, and agile production (Alcácer and Cruz-Machado 2019). Resulting from the described breadth of the concept and its rapid development, there is an increasing and steadily changing range of technologies that form the entirety of I4.0.

In our study, we consider a set of technologies that form the core of I4.0 and build the basis for the majority of today's I4.0 technology development and application. Adopting such an established and focused set of core technologies enables a high validity and generalizability of the analysis. The considered technologies are drawn from a widely adopted and highly cited Boston Consulting Group report on I4.0 technology (Rüssmann et al. 2015) and a very recent and more academic operationalization of it proposed by Culot et al. (2020) based on a systematic review on I4.0 definitions: Internet of Things, Cyber-Physical Systems, Visualization technologies (among which Augmented Reality), Cloud Computing, Cybersecurity, Blockchain Technology, Simulation and Modelling, Machine Learning and Artificial Intelligence, Big Data Analytics, Additive Manufacturing systems (among which 3D-Printing), and Advanced Robotics.

These core technologies of I4.0 are defined in the following, also to establish enhanced traceability and reliability for the subsequent analysis. Moreover, all I4.0 technologies are written in capital letters in the remainder of the chapter to facilitate their recognition in the text.

Internet of Things (IoT) refers to physical objects connected through the Internet. These (smart) objects have their own intelligence, can collect information, interact with the surrounding environment, connect to one another, exchange data in real time, and trigger actions through the Internet. Therefore, IoT connects people and things anytime, anyplace, with anything and anyone, based on any network and any service (e.g., Weking et al. 2019).

Cyber-Physical Systems (CPS) express the interconnection between physical and virtual environments. CPS integrate, control, and coordinate operations and processes while simultaneously providing and using data-accessing and data-processing services. Integrating CPS within production, logistics, and services enables a connection across all levels of production between autonomous and cooperative elements across the entire supply chain (e.g., Rejikumar et al. 2019).

Visualization Technologies, such as Augmented Reality (AR) and Virtual Reality (VR), enhance the perception of the physical world through visual elements. Visual tools provide a virtual representation of entire (production) systems and their interactions within the supply chain. This enables a transformation of how enterprises serve customers, train employees, design, and create products (e.g., Brenner 2018; Dallasega et al. 2020).

Cloud Computing (CC) includes data storage, servers, databases, networking, and software that enable remote information access on a virtual space. This cloud can connect different entities, which is reflected in its four main types of access: public; private within the same (meta-)organization; hybrid, if public and private clouds are combined; and community, shared by multiple organizations and supported by common interests and concerns (e.g., Armbrust et al. 2010).

Cybersecurity aims to protect private information applied to hardware and software to avoid the misuse of data and devices. Cybersecurity relies on protecting, detecting, and responding to attacks. It has become essential since virtual environments, remote access, and stored data on cloud systems represent increasing vulnerabilities (e.g., Kotarba 2018).

Blockchain Technology is based on decentralizing the storage of data to prevent such data to be owned, controlled, or manipulated by a central actor, thus enabling the immutability and integrity of data across several distributed nodes that are linked in a peer-to-peer network. It is expected to change the way in which ownership, privacy, uncertainty, and collaboration are conceived in the digital world (e.g., Ahram et al. 2017).

Simulation and Modelling facilitate the validation of products, processes, or system design and configuration. Furthermore, they enable cost reduction and increased product quality, while reducing development time, designing and engineering errors, and wastes (e.g., Alcácer and Cruz-Machado 2019).

Machine Learning (ML) and Artificial Intelligence (AI) refer to the simulation of human intelligence in machines that mimic human thinking and actions. ML is in fact a sub-field of AI expressing the idea that a computer program can learn and adapt based on data without human interference (e.g., Matthyssens 2019).

Big Data Analytics describes the acquisition of large and complex data sets from different sources and at different times. It includes the collection, storage, and sharing of data and their analysis and evaluation. Big Data can be defined as *"large volumes of high velocity, complex and variable data that require advanced techniques and technologies to enable the capture, storage, distribution, management, and analysis of the information"* (Mills et al. 2012, p.10). This enables, among other advantages, predictive maintenance and real-time decision-making (e.g., Pisano et al. 2015).

Additive Manufacturing (AM), often simplified as 3D-Printing, is a technology creating three-dimensional components and products directly from raw materials, layer upon layer. It accelerates prototyping in manufacturing and ensures design and product testing, improves creativity in shapes as well as geometry, and allows improved personalization (e.g., Kamble et al. 2018).

Advanced Robotics such as autonomous and collaborative robots interact with one another and are able to work safely with humans in the same workspace (e.g., Kumar 2018).

Besides these technologies, **New Materials** and **Energy Management Solutions** are becoming more and more entwined with the I4.0 concept. A wide range of miniaturized devices that are essential to I4.0 rely indeed on minerals that are criticized for sustainability problems in their production and supply chains (Hiete et al. 2019). These "old" materials will eventually be replaced by substitutes or "new" material currently being developed. In the meantime, such materials are just like the use of energy subject to standardization and management frameworks (Jacob et al.

2019) that aim to control the negative sustainability impact of I4.0 implementation and could drive its success by gradually lowering this downside of the Fourth Industrial Revolution.

10.2.2 Business Model

A BM is *"the basic logic of a company that describes what benefits are provided to customers and partners [...] and how the provided benefits flow back into the company in the form of revenue"* (Schallmo et al. 2017, p. 5). The BM consists of a set of interdependent organizational activities through which human, physical, and/or capital resources are brought together in order to achieve the enterprise's goals (Zott and Amit 2010).

Despite the absence of a unanimous definition, it is widely accepted that a BM reveals how a firm creates, delivers, and captures value. It provides a framework of costs, payments, and revenues together with the company's strategies, ranging from the products or services it offers to how it differentiates from competitors and how it integrates with its partners in the value chain. As a result, *"a good business model yields value propositions that are compelling to customers, achieves advantageous cost and risk structures, and enables significant value capture by the business that generates and delivers products and services"* (Teece 2010, p. 174).

Osterwalder et al. (2005, p. 17) formalized a BM as a *"conceptual tool that contains a set of elements and their relationships and allows expressing the business logic of a specific firm. It is a description of the value a company offers to one or several segments of customers and of the architecture of the firm and its network of partners for creating, marketing, and delivering this value and relationship capital, to generate profitable and sustainable revenue streams."*

Despite the advantages of a well-designed BM, the development and innovation of a BM require substantial expertise and resources causing especially SMEs to struggle with these tasks (Müller et al. 2018). This is also caused by the interrelatedness of the various BM elements, for which previous research found that a change in one BM block likely impacts the other blocks as well (Spieth and Schneider 2016). Moreover, it has been found that changes in the BM can cause drifts in a firm's mission that ultimately results in inconsistent action of the firm with its originally stated mission (Klein et al. 2020). In effect, it has been found that the BM needs to be closely aligned with the firm's orientation to economic, social, and environmental sustainability (e.g., Hahn et al. 2018; Klein et al. 2020). Moreover, changes in a BM can severely affect the customers' brand perceptions such as brand trust, brand loyalty, and brand equity (Spieth et al. 2019).

At the intersection of BMs and I4.0, a recent study by Weking et al. (2019) found a substantial lack of research. Based on 32 cases described in literature, the authors identify three super-patterns of I4.0 business models that are (a) integration, (b) servitization, and (c) expertization. While Weking et al. (2019) analyzed the I4.0 BM in general, they also find that the relationship of I4.0 BMs and I4.0 technologies is under-researched (see also Rayna and Striukova 2016). This particularly applies to SMEs, for which an investigation regarding the implementation of BM adoptions over time should be considered (Müller 2019).

The Canvas model proposed by Osterwalder and Pigneur (2010) highlights the following BM building blocks and related definitions. By using the Canvas model, we again adopt a widely used on well-established framework for the analysis to match the high generalizability and validity of the I4.0 technologies. Moreover, all BM building blocks are written in capital letters in the remainder of the chapter to facilitate their recognition in the text.

Customer Segments are the different groups of individuals or parties that a firm wants to reach and satisfy. The question that the company should ask itself for this block is *"Who are we creating value for? Who are our most important customers?"* (ibid., p. 21). The target market can be a mass, a niche, and/or a segmented market.

Value Propositions define which products and services create value for a given Customer Segment. The relevant questions in this case are: *"What value do we deliver to our customers? To which customers' needs are we going to respond?"* (ibid., p. 23). Each Value Proposition is made up of a specific bundle of goods and/or services peculiar to a Customer Segment. Related values may be quantitative, such as efficiency, or qualitative, such as design (ibid.).

Channels illustrate the way in which the Customer Segments are reached to deliver the appropriate Value Proposition. The right balance of Channels used to satisfy customers' expectations is essential in bringing a Value Proposition to the market. The questions for this building block are: "*Through which Channels do our Customer Segments want to be reached? How are we reaching them now? Which ones are most cost-efficient?*" (ibid., p. 27).

Customer Relationships express the relationships between a firm and a specific Customer Segment. These relationships range from personal to automated ones with the most typical Customer Relationships ranging from personal assistance to self-service. The relevant question is in this case "*What type of relationship does each of our Customer Segments expect us to establish and maintain?*" (ibid., p. 29).

Revenue Streams indicate the income generated from each Customer Segment. There are two main types of Revenue Streams: transaction revenues occurring when a product or service is sold and recurring revenues resulting from ongoing payments or fees. The fundamental questions are: "*For what value are our customers really willing to pay? For what do they currently pay? How much does each Revenue Stream contribute to [the company's] overall revenues?*" (ibid., p. 31).

Key Resources let a firm create and offer a Value Proposition, reach its customers, and earn revenues. Different Key Resources are needed for different BMs. In fact, they can be physical, financial, intellectual, or human. At the same time, they can be owned or leased by the company or acquired from the company's partners. The question that the company should ask itself is "*What Key Resources do our Value Propositions require?*" (ibid., p. 35). That is, what Key Resources are needed by the other building blocks?

Key Activities aim at creating and offering Value Propositions, reaching the different markets, and maintaining Customer Relationships. They range from production to problem-solving, or networking. The relevant question is: What key activities are fundamental for our BM? That is, what key activities are requested by the other building blocks? (ibid.).

Key Partnerships illustrate the network of suppliers and partners that enable the BM. Companies create partnerships to reduce risks or acquire resources. These Key Partnerships can be strategic alliances between non-competitors or competitors and buyer-supplier relationships. The questions are: *"Who are our key partners [and] key suppliers? Which Key Activities do partners perform?"* (ibid., p. 39).

Cost Structure indicates the most important costs after defining Key Resources, Key Activities, and Key Partnerships including costs related to creating and delivering value, maintaining Customer Relationships, and generating Revenue Streams. BM can be cost- or value-driven, though most represent a balance of the two aspects. The relevant questions are: *"What are the most important costs inherent in our BM? Which Key Resources [and] Key Activities are most expensive?"* (ibid., p. 41).

In line with Kotarba (2018), the BM Canvas is often adopted due to its relative simplicity which provides support for quick and efficient content documentation in the process of identifying crucial components of an organization internal structure and relationship with the ecosystem it belongs to.

10.2.3 Small- and Medium-Sized Enterprises

According to the European Commission (2020), SMEs are defined as firms with a maximum number of 250 employees and a maximum annual revenue of 50 million euros.

SMEs play an important role in the economic scenario since they represent 99% of the total enterprises in the European Union (European Commission 2020). According to Müller et al. (2018), SMEs contribute more than 50% of gross value added throughout Europe, but tend to struggle with both the implementation of I4.0 and BM innovations, resulting in insufficient access to external knowledge and unclear innovation strategies, which limit SMEs' efforts in making incremental improvements.

Most SMEs are family-owned and often the owner is also the manager. This may represent an advantage in terms of flexibility and readiness to react to changes due to a flat and clear organization. In fact, being settled around more informal working relationships, communications between managers and employees are quicker and more direct. This allows to share new concepts or innovative ideas more efficiently across the firm and to achieve a deeper engagement of employees. Moreover, SMEs' strength is to create value to the firm, which results in investing in research and development, employees' training, and life-long learning. However, SMEs' limited financial possibilities, if compared to large enterprises, may prevent them from having skilled workers and the necessary economic resources to profitably invest in new technologies, at least at an initial stage (Müller and Däschle 2018; Orzes et al. 2020). In fact, despite training a large percentage of apprentices, they find it difficult to rely on skilled personnel when it comes to specific I4.0 technologies. It seems that the sophisticated technologies require further resources as well as supporting initiatives and may need to be adopted by SMEs at later stages. Organizational changes, together with the involvement of external professionals, must also be considered by SMEs' managers to achieve new goals.

10.3 Methodology

To answer the research question (i.e., *"How can the implementation of Industry 4.0 technologies drive the modification of the business models of small- and medium-sized enterprises?"*), a multi-method approach is applied that takes stock of the currently available literature at the intersection of I4.0 and BMs as well as the information available on companies applying I4.0. While the literature review does not focus a particular firm size, the companies' analysis is exclusively studying SMEs due to their relevance outlined in Sect. 10.2.3. Both approaches are presented in this section, starting with the literature review, followed by the contingency analysis applied to the literature review results, and the secondary data analysis.

10.3.1 Literature Review Methodology

In line with current best practices in the field of operations management, a content analysis-based systematic literature review is conducted as proposed by Seuring and Gold (2012). This contains the sequential steps of (1) material collection, (2) category selection, and (3) material evaluation. Content analysis-based reviews can be applied to academic publications as is regularly done, but also to any kind of documents and written communication (Mayring 2015), such as websites (e.g., Carbone et al. 2017), industry standards (e.g., Sauer and Hiete 2020), or newspaper articles (e.g., Ancarani et al. 2015).

Step 1) material collection encompasses designing the study including specifying the research question(s), defining the search parameters as well as database(s), and obtaining the literature to be analyzed, i.e., the material. This needs to be well documented to satisfy the quality criteria of replicability (Fink 2019; Seuring and Gold 2012). For doing so, a search has been conducted on Elsevier's Scopus, one of the most acknowledged scientific databases. The search aims at finding specific keywords on I4.0 and BMs in a paper's title, abstract, and keywords. In order to select a complete and current list of keywords, it has been decided to adopt the I4.0 search terms provided by Culot et al. (2020), which we find to be exhaustive of the aspects related to I4.0, even if it does not include keywords related to single specific technologies.

Similarly, as regards BM, the term itself is used together with the related terms "business plan" and "revenue model." We acknowledge that other keywords could be relevant, but the choice made satisfied the core of this work. This was determined by adding the keywords sequentially to the search terms while monitoring the increase in papers found. Adding more synonyms to the current search string did not yield additional papers. The resulting keywords and search string are presented in Fig. 10.1 along the other details of the material collection.

Beyond the keywords, the material collection is restricted to publications in English, with a few exceptions made if an English abstract was available while the full paper was available only in German. Furthermore, following state of the art literature reviews (Sauer and Seuring 2017; Seuring and Gold 2012) the type of publication was limited to articles, reviews, editorials, and short surveys to include only peer-reviewed publications that have undergone academic quality checks. Moreover, only publications between 2011, the year of the definition of the term I4.0,

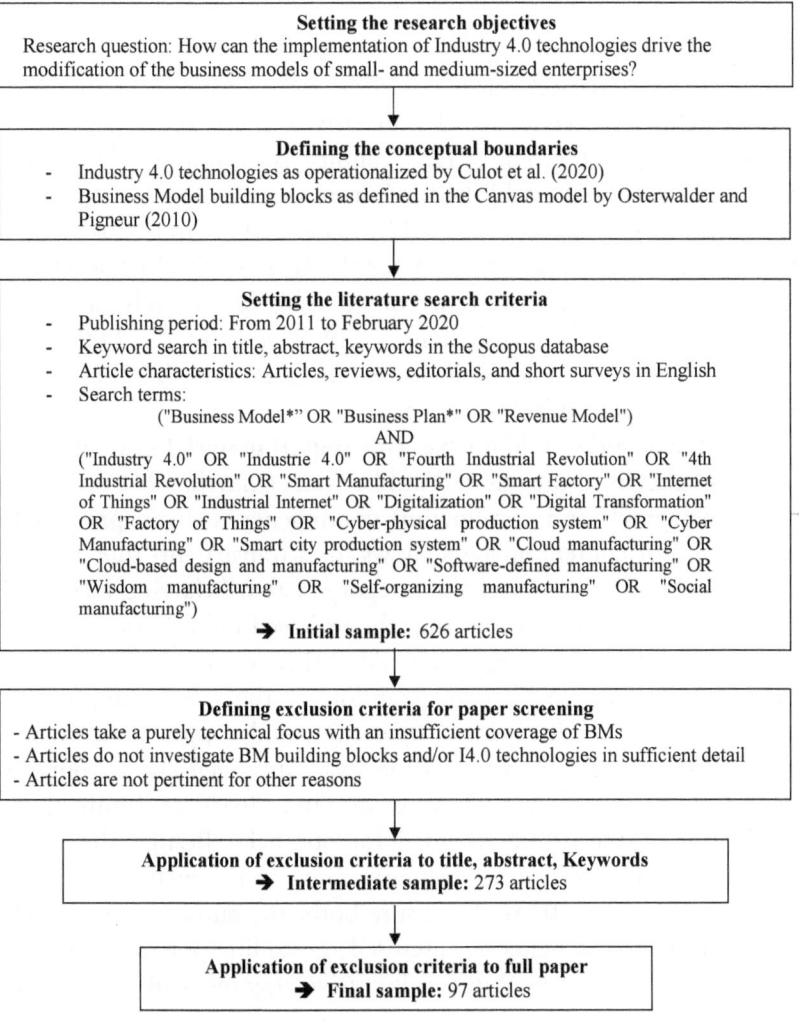

Setting the research objectives
Research question: How can the implementation of Industry 4.0 technologies drive the modification of the business models of small- and medium-sized enterprises?

Defining the conceptual boundaries
- Industry 4.0 technologies as operationalized by Culot et al. (2020)
- Business Model building blocks as defined in the Canvas model by Osterwalder and Pigneur (2010)

Setting the literature search criteria
- Publishing period: From 2011 to February 2020
- Keyword search in title, abstract, keywords in the Scopus database
- Article characteristics: Articles, reviews, editorials, and short surveys in English
- Search terms:
    ("Business Model*" OR "Business Plan*" OR "Revenue Model")
    AND
    ("Industry 4.0" OR "Industrie 4.0" OR "Fourth Industrial Revolution" OR "4th Industrial Revolution" OR "Smart Manufacturing" OR "Smart Factory" OR "Internet of Things" OR "Industrial Internet" OR "Digitalization" OR "Digital Transformation" OR "Factory of Things" OR "Cyber-physical production system" OR "Cyber Manufacturing" OR "Smart city production system" OR "Cloud manufacturing" OR "Cloud-based design and manufacturing" OR "Software-defined manufacturing" OR "Wisdom manufacturing" OR "Self-organizing manufacturing" OR "Social manufacturing")
    ➜ **Initial sample:** 626 articles

Defining exclusion criteria for paper screening
- Articles take a purely technical focus with an insufficient coverage of BMs
- Articles do not investigate BM building blocks and/or I4.0 technologies in sufficient detail
- Articles are not pertinent for other reasons

Application of exclusion criteria to title, abstract, Keywords
    ➜ **Intermediate sample:** 273 articles

Application of exclusion criteria to full paper
    ➜ **Final sample:** 97 articles

Fig. 10.1 Systematic material collection process

until February 2020 were included. This yielded an initial sample of 626 documents fitting the presented criteria.

Subsequently, the first filtering was based on the content of title and abstract as well as on its keywords (when available). This yielded an intermediate sample of 273 papers that were then checked in their full text against the same inclusion criteria displayed in Fig. 10.1, resulting in 97 publications fully relevant.

Step 2) category selection is critical to the validity of the review, since it defines the codes for the analysis of the selected papers (Fink 2019; Seuring and Gold 2012). Validity is established by selecting framework from well-established literature as well as highly ranked and peer-reviewed journals as suggested by Sauer and Seuring (2017). In the extant study, one framework each for the BM and I4.0 perspective are chosen that fulfill this criterion. For the BM side, we rely on the BM Canvas by Osterwalder and Pigneur (2010) and for the I4.0 side we adopted the I4.0 technologies identified by Culot et al. (2020), that have both been introduced in Sect. 10.2.

These frameworks build the basis for **step 3) material evaluation**, i.e., a content analysis of the selected publications (Mayring 2015) described in this paragraph and a follow-up contingency analysis presented in Sect. 10.3.2. A content analysis allows for a transparent, rule governed, and replicable application of the category system (Mayring 2015). This system is the basis for synthesizing the reviewed publication against the research questions (Seuring and Gold 2012; Mayring 2015). Seuring and Gold (2012) provided an adaption of the generic approach by Mayring (2015) to the field of operations and supply chain management. This approach is preferred against other well-known but more generic approaches such as the ones by Fink (2019) or Tranfield et al. (2003) due to the detail provided for our field. Besides the validity, content analysis is also inherently associated to the quality criterion of reliability (Mayring 2015). To ensure both, the author team defined a coding protocol and regularly discussed the coding outcomes and especially unclear cases to establish a discursive alignment of interpretation (Seuring and Gold 2012).

10.3.2 Contingency Analysis of the Literature Review Findings

The content analysis allows for a qualitative and quantitative analysis of the reviewed literature (Mayring 2015), which are presented separately in Sect. 10.4. For the quantitative investigation a contingency analysis is performed that allows to detect *"association patterns between categories, i.e. [...] pairs of categories which occur relatively more frequently together in one paper than the product of their single probabilities would suggest"* (Gold et al. 2010, p. 235). This detection of positive association patterns has later been extended to negative ones by Sauer and Seuring (2017). Such an analysis is based on the coding frequencies from the content analysis. The strength of the association patterns between two codes is evaluated based on the phi-coefficient that is calculated using a chi-square test. To enable this, two quality requirements need to be fulfilled to recognize valid and significant relations. First, the contingency table may not contain any expected counts below five and phi needs to be above 0.3 (Sauer and Seuring 2017).

Still, if a pattern is identified, it does not reveal the underlying causality and the use of both codes in a single paper could even be unintentional. Nevertheless, the significant associations among the codes reveal a connection that needs a literature or theory-based justification. In the end, the contingencies offer insights into the association of codes within the individual papers complementing the content analysis, that focuses on similar contents across different articles (Sauer and Seuring 2017).

Combining content and contingency analysis enables a second level of analysis and interpretation that is particularly interesting as we review a heterogeneous field. The contingencies can reveal statistically significant gaps and links within the reviewed sample. They are therefore essential for answering the research question as is done in the findings and discussion sections.

10.3.3 Secondary Data Analysis Methodology

In order to move beyond the literature-based evidence that has been produced from an academic perspective, the extant study furthermore includes a set of empirical data from practice. Such an approach enables a validation of the literature review results and enhances the generalizability of the findings.

In effect, the applied method is similar to the three-step process outlined for the literature analysis in Sect. 10.3.1. The material collection was realized by analyzing nearly 150 websites of Italian industrial SMEs belonging to the Cluster "Fabbrica Intelligente" (CFI) (https://www.fabbricaintelligente.it/). The CFI is an association including large and small- and medium-sized enterprises, as well as universities, research centers, entrepreneurs' associations, and other stakeholders, dealing with advanced manufacturing and I4.0. The association is recognized by the Italian government as a facilitator of sustainable economic growth since it develops the innovation and specialization of the Italian manufacturing sector.

Among the over 150 SMEs websites visited, 30 have been chosen since they seemed particularly significant for the goal of the research. They describe the digital transformation process of the company with reference to new technologies, together with the enhancement of specific competences and innovative BMs. Other inclusion criteria have considered the pertinence of the information present on the websites as far as the research goal was concerned. The most interesting websites resulted to be those where videos and articles contributed to the completeness of the presentation of the firm itself.

To summarize step 1) material collection, Appendix I lists and characterizes the 30 sample SMEs, indicating them with a progressive number following the alphabetical order and regardless the company's relevance. This number is used to reference the samples' SMEs in the findings section.

Step 2) category selection and the procedure of step 3) material analysis are identical to the literature review part. This is also consistent with the aim to validate the literature-based findings that are presented in the subsequent section.

10.4 Results

10.4.1 Content Analysis of the Reviewed Papers

10.4.1.1 Overarching Trends in the Reviewed Papers

The review reveals four overarching trends that transcend the single BM blocks and I4.0 technologies:

First, I4.0 is still a developing concept, in particular for SMEs, that affects the strategies and operations of businesses, as well as the relationships between enterprises, customers, and suppliers (see also Ingaldi and Ulewicz 2020).

Second, I4.0 does not concern only manufacturing industries; it is rather about the ways in which digital technologies are brought together and, specifically, how organizations can harness them to drive competitive BMs, market, and sustainable growth. This is underlined by the fact that the papers analyzed in the literature review often regard both manufacturing industries and consulting services companies. The distinction between manufactured goods and services is becoming more and more subtle, since products and their functionalities are offered as services, and products themselves are associated to the services they deliver (see also Porter and Heppelmann 2014). This creates hybrid products which are neither exclusively goods nor services.

Third, Big Data gathering and analysis is a critical issue debated in almost 60% of the reviewed literature. Making adequate sense of Big Data means interpreting it, getting insights that lead to better decisions and strategic moves. However, apart from questions related to the expertise required to understand and correctly use this data, there are other concerns about its security. Even if the technologies Big Data and Cybersecurity should be treated as going on hand-in-hand, the latter is much less frequently found in literature, as just less than 20% of the sample papers mention it.

Fourth, the exploitation of I4.0 technologies requires, besides significant investments, a transformation in corporate structure and culture, which have to become open-minded and flexible. Collaborative environments and systematic discussions to innovate established routines are also/equally necessary (Tirabeni et al. 2019). Companies require resources and knowledge from different fields that do not necessarily belong to a single industry. Therefore, relationships must be built within and across industries (Ghanbari et al. 2017). This is particularly true for SMEs, for which an investigation regarding the implementation of BM adoptions over time should be considered (Müller 2019).

10.4.1.2 Business Model Building Blocks Modified by Industry 4.0 Implementation

How do new technologies support this change? Table 10.1 provides an overview of the frequency with which the nine BM building blocks appear in the literature review and some literature-based examples of how I4.0 technologies may influence them.

Among the different building blocks, which are however strictly connected and linked to one another, Customer Relationship and Key Activities are the most frequently cited (both over 70%) by the reviewed papers. This is not surprising since collaborative engagement of customers, namely in the process of co-design, co-engineering, and co-development of products and services is an essential feature for smart factories. As a result of this co-creation process, individual products can be realized thanks for example to Simulation and Modelling. This allows to understand and satisfy customers' needs (Kagermann et al. 2013). In the case of consulting services companies, analyzing customers' data becomes a Key Activity, while manufacturing firms offer product tracking and predictive maintenance thanks to real-time monitoring and automated data analysis (Weking et al. 2019), through IoT, CPS, and Blockchain Technology.

At the same time, many I4.0 technologies, such as Additive Manufacturing, Advanced Robotics, and Visualization Technologies, just to mention a few, enhance flexibility and mass customization capabilities, contributing to a firm's Value Proposition. Together with flexibility, modularity enables a company's adaptation to the changing conditions and prerequisites of different customers' goals/requests/demands.

Table 10.1 Exemplary business model changes through Industry 4.0

Business model building blocks	Appearance in sample papers [%]	Exemplary changes through Industry 4.0 implementation
Key Partners	62.89	• Higher inter-company connectivity and delivery reliability • Increased transparency among multiple actors provided by Blockchain Technology
Key Activities	71.13	• Data-driven systems for quality control enabling real-time decision-making • Production trends reported and summarized by means of data analytics application
Key Resources	67.01	• Servitization with remote monitoring and predictive maintenance enabled by IoT • Equipment purchased according to the specifications of I4.0 • General workforce re-organization with information technology (IT) competencies/qualifications
Value Proposition	68.04	• Interactions between operators and machines and their coexistence • Modular and customer-oriented products and services • Manufacturing of customized complex parts by means of 3D-printing • CPS for higher flexibility in responding to customer wishes and fulfilling orders

(continued)

Table 10.1 (continued)

Business model building blocks	Appearance in sample papers [%]	Exemplary changes through Industry 4.0 implementation
Customer Relationship	70.10	• Measured service, on-demand self-service, and broad network access • Customer adaptation of product features before purchase through virtual product development (Co-design and Co-creation)
Channels	62.89	• Cloud-based platforms for innovating and trading goods and services among users • Single customer followed from evaluation to project and after-sale services • Web sales channels
Customer Segments	44.33	• Larger product spectrum to satisfy different markets • Products tailored on customer demands to reach niche markets
Revenue Streams	50.52	• Different payments methods like pay-per-use, pay-per-feature • IT services generally attractive to create turnover • Increased payment reliability thanks to digital transactions

Business model building blocks	Appearance in sample papers [%]	Exemplary changes through Industry 4.0 implementation
Cost Structure	50.52	• High initial costs to adopt or implement I4.0 technologies • Cost savings through a more flexible and efficient value chain • Cost reduction through the adoption of Cloud Computing

Less than 45% of the articles talk about the importance of I4.0 technologies in order to satisfy different Customer Segments. Furthermore, technological innovations can seldom be used without redesigning the BMs of established companies, requiring them to incorporate new external knowledge into internal activities, which explains why the Key Partners building block results to be often mentioned in literature (over 60%). SMEs in particular are encouraged to cooperate with partners that can help them to transform and exploit external knowledge related to I4.0 (Müller et al. 2020). Dynamic capabilities include a firm's ability to integrate, build, and reconfigure external and internal resources to address and shape rapidly changing environments (Brenner 2018). Key Resources are cited by over 65% of the articles, in which the intellectual resources play a fundamental role. Workers, in fact, possess higher autonomy and levels of participation in decision-making processes, modifying their place in the firm's ecosystem from bare performers to active cooperators.

10.4.1.3 Contingency Analysis of Industry 4.0 Technologies and Business Model Building Blocks

The contingency analysis complements the qualitative analysis as a second analytical step. It discloses connections among the codes and enables the identification of hot topics and gaps in the reviewed literature (Sauer and Seuring 2017).

The analysis focuses on association pattern among the two code sets that are central to this study, i.e., the nine BM blocks and the eleven I4.0 technologies. Both were complemented by an overarching code "Business Model in general" or "Industry 4.0 in general" used if the level of detail of investigation did not justify the more detailed coding into a BM block or I4.0 technology. The contingency analysis investigates the resulting three possible association patterns of (1) I4.0 technology to I4.0 technology, (2) BM block to I4.0 technology, and (3) BM block to BM block. Since the associations do not have a direction, the code sequences do not make a difference. The significant associations are displayed in the following Tables 10.2, 10.3, and 10.4.

Table 10.2 Results of the contingency analysis of I4.0 technologies

	Code 1	Code 2	Count	Expected count	Phi	Approximate significance
Industry 4.0 technology to Industry 4.0 technology	Advanced Robotics	Cloud Computing	14	6.2	0.453	0.000
	Additive Manufacturing	Cloud Computing	14	7.8	0.325	0.001
	Big Data Analytics	Machine Learning and AI	22	14.4	0.373	0.000
	Big Data Analytics	Cloud Computing	32	23.9	0.345	0.001
	Big Data Analytics	Visualization Technologies	17	10.8	0.337	0.001
	Cloud Computing	Visualization Technologies	15	7.4	0.408	0.000
	Cloud Computing	Internet of Things	39	33.4	0.316	0.002

Within the I4.0 technologies, Cloud Computing is found to be contingent to five other technologies. Being positively/negatively contingent means that the two related codes have been/have not been coded together in a significant number of papers, meaning that the appearance of both codes is statistically significantly concentrated/spread across shared/divided parts of the total sample of papers. This is evaluated in a Chi-square test assuming a normal distribution of the two codes across the total sample of papers. This finding underlines the enabling character of Cloud Computing for Advanced Robotics, Additive Manufacturing, Internet of Things, and Visualization Technologies. The final contingency is found with Big Data Analytics, that also exhibits the second highest number of associations. In effect, it can be contended that these two technologies are the core enablers of I4.0 implementation.

Turning back to the central role of Cloud Computing and Big Data Analytics, this becomes also evident in Fig. 10.2, that visualizes the results of the contingency analysis for the I4.0 technologies.

Moreover, the grouping of the technologies by Culot et al. (2020) has been added by means of gray rectangles. This underlines the centrality of the network technologies and adds to the characterization of I4.0 technologies and their interrelations. In effect, all four technology groups by Culot et al. (2020) are represented with at least one technology. Still, the fact that the contingency analysis has not yielded significant results

Table 10.3 Results of the contingency analysis of I4.0 technologies and BM blocks

	Code 1	Code 2	Count	Expected count	Phi	Approximate significance
Industry 4.0 technology to business model block	Advanced Robotics	Channels	4	9.4	−0.321	0.002
	Advanced Robotics	Key Partners	4	9.3	−0.310	0.002
	Cloud Computing	Value Proposition	21	27.6	−0.300	0.003
	Cybersecurity	Channels	5	11.3	−0.347	0.001
	Cybersecurity	Key Resources	6	11.9	−0.329	0.001
	Cybersecurity	Key Partners	5	11.1	−0.335	0.001
	Cyber-physical Systems	Cost Structure	14	21.2	−0.300	0.003
	Visualization Technologies	Customer Relationship	7	12.6	−0.325	0.001
	Visualization Technologies	Value Proposition	7	12.4	−0.312	0.002
	Visualization Technologies	Business Model in general	12	5.4	0.383	0.000
	Cloud Computing	Business Model in general	19	12.0	0.322	0.002
	Cybersecurity	Business Model in general	11	5.4	0.325	0.001
	Big Data Analytics	Business Model in general	24	17.3	0.306	0.003
	Additive Manufacturing	Business Model in general	11	5.7	0.302	0.003
	Digitalization	Business Model in general	9	15.8	−0.310	0.002

Table 10.4 Results of the contingency analysis of BM blocks

	Code 1	Code 2	Count	Expected count	Phi	Approximate significance
Business model block to business model block	Cost Structure	Revenue Streams	45	24.8	0.835	0.000
	Cost Structure	Customer Segments	38	21.7	0.676	0.000
	Cost Structure	Channels	47	30.8	0.691	0.000
	Cost Structure	Customer Relationship	49	34.4	0.660	0.000
	Cost Structure	Value Proposition	46	33.8	0.542	0.000
	Cost Structure	Key Resources	47	32.3	0.638	0.000
	Cost Structure	Key Activities	48	34.9	0.598	0.000
	Cost Structure	Key Partners	46	30.3	0.666	0.000
	Revenue Streams	Customer Segments	38	21.7	0.676	0.000
	Revenue Streams	Channels	47	30.8	0.691	0.000
	Revenue Streams	Customer Relationship	48	34.4	0.615	0.000
	Revenue Streams	Value Proposition	46	33.8	0.542	0.000
	Revenue Streams	Key Resources	46	32.3	0.595	0.000
	Revenue Streams	Key Activities	48	34.9	0.598	0.000
	Revenue Streams	Key Partners	46	30.3	0.666	0.000
	Customer Segments	Channels	39	27.0	0.514	0.000
	Customer Segments	Customer Relationship	43	30.1	0.583	0.000

(continued)

Table 10.4 (continued)

Code 1	Code 2	Count	Expected count	Phi	Approximate significance
Customer Segments	Value Proposition	43	29.7	0.597	0.000
Customer Segments	Key Resources	39	28.4	0.466	0.000
Customer Segments	Key Activities	41	30.6	0.477	0.000
Customer Segments	Key Partners	39	26.6	0.530	0.000
Channels	Customer Relationship	60	42.8	0.803	0.000
Channels	Value Proposition	54	42.1	0.548	0.000
Channels	Key Resources	58	40.2	0.800	0.000
Channels	Key Activities	61	43.4	0.829	0.000
Channels	Key Partners	59	37.7	0.934	0.000
Customer Relationship	Value Proposition	61	47.0	0.684	0.000
Customer Relationship	Key Resources	60	44.9	0.719	0.000
Customer Relationship	Key Activities	65	48.4	0.826	0.000
Customer Relationship	Key Partners	58	42.1	0.739	0.000
Value Proposition	Key Resources	56	44.2	0.555	0.000

Code 1	Code 2	Count	Expected count	Phi	Approximate significance
Value Proposition	Key Activities	61	47.7	0.657	0.000
Value Proposition	Key Partners	54	41.4	0.577	0.000
Key Resources	Key Activities	63	45.5	0.839	0.000
Key Resources	Key Partners	58	39.6	0.825	0.000
Key Activities	Key Partners	60	42.7	0.811	0.000

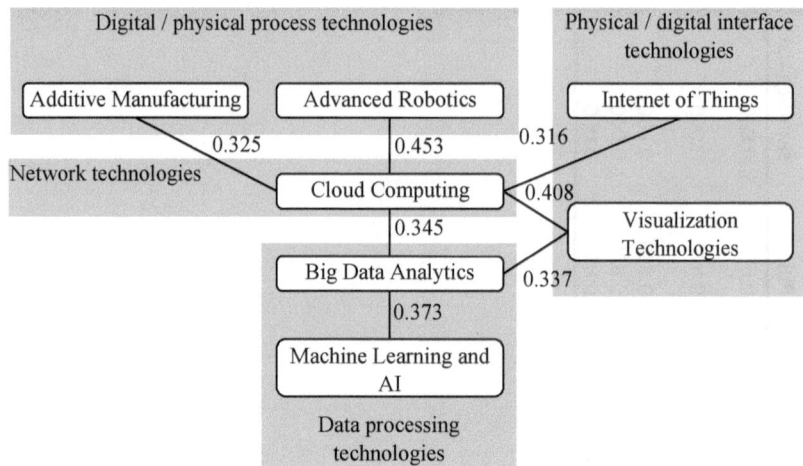

Fig. 10.2 Visualization of the results of the contingency analysis of I4.0 technologies including phi values

for the remaining code combinations underlines that there is no concentration of these combinations throughout the sample. The contingency reveals only outliers in code combinations that are nevertheless displaying the hotspots and gaps in literature. The former is revealed by concentrations of code combinations that signal a relatively intensive investigation of the coded concepts together. Contrastingly, the gaps are revealed if the appearance of the codes is split into separate sub-samples, i.e., the individual concepts are investigated separately, which is indicated by a negative phi value.

The mentioned identification of hotspots and gaps proposed by Sauer and Seuring (2017) can best be seen for this study in the following Table 10.3. Quite distinctively and surprisingly all significant associations among I4.0 technologies and BM blocks exhibit a negative phi value. Contrastingly, there are five significant associations of I4.0 technologies and the overarching code "Business Model in general." This double-sided finding is a strong signal for a lack of detail in the analysis of the intersection of I4.0 technologies and BM blocks or elements. It shows that if the literature is making references to the I4.0 technologies, the respective papers fall short in addressing individual BM blocks but reference a general impact on BMs. However, Table 10.3 also underlines that this is not the case for all the eleven technologies, but only five technologies exhibit the negative associations that indicate an under-researched intersection between these constructs.

Among those five technologies, Cybersecurity is certainly the most interesting one. It has been found only half as often as statistically expected in conjunction with the BM blocks of Channels, Key Partners, and Key Resources. This is surprising since Cybersecurity has been found to be key to the intra- and inter-organizational application of I4.0 (Kagermann et al. 2013; Culot et al. 2020). It should thus take a more central role in future research that could investigate the role of Cybersecurity to enable the implementation of I4.0 and the modification of a firm's BM.

As displayed in Table 10.4 all nine BM blocks are found to be contingent to each other, meaning that relative to the pool of all 22 codes the associations between the BM block codes appear statistically significantly more often than expected against a normal distribution. At the same time, the results are by far more mixed for the I4.0 technologies. This indicates the relative maturity of the BM domain and the high applicability of the BM canvas framework for the literature at the researched intersection. Moreover, the BM literature underlines that a change in one BM block likely impacts the other blocks as well (Spieth and Schneider 2016).

As a maximal used association, 65 papers investigate both Customer Relationships and Key Activities. This is however unsurprising given that one of the core purposes of a firm is the satisfaction of customer needs and thus naturally linking all activities to the relation to the customer. This association is, however, not the strongest with a phi value of 0.826. The strongest one is between Channels and Key Partners (phi = 0.934), which is again due to the closely related content of the codes. Turning to the weakest links, these are found among Customer Segments and Key Resources (phi = 0.466) and Customer Segments and Key Activities (phi = 0.477). Still, all phi values of the associations among the BM blocks are higher than the ones among the I4.0 technologies in Table 10.2 and the intersection of both displayed in Table 10.3 underlining the interconnectedness of the BM blocks in the reviewed literature.

As outlined before, literature reviews and literature-based contingency analysis are limited in their representation of practice. To address this shortcoming, the subsequent section presents the results of the analysis of the company data on the interrelation of I4.0 technologies and BMs.

10.4.2 Secondary Data Analysis

To provide an empirical contribution to the research question, the 30 analyzed SMEs have been categorized in a matrix of I4.0 technologies and BM blocks in Table 10.5. These results are presented in the following and based on them propositions have been built to underline how I4.0 technologies can help to modify an SMEs BM.

Throughout the sample of 30 SMEs, IoT has been found six times, making it one of the most widespread I4.0 technologies in the sample. The companies relate it to a total of four BM blocks. Firstly, IoT is found to enable communication across a wide array of systems and services and allows to develop vertical solutions for connected products, people, and value chains [1]. This makes it a valuable enabler of Key Activities in basically any industry. Moreover, the implementation of IoT is considered useful in the mechanical industry [16] for a better production re-organization to move the entire information flow into the digital sphere. Secondly, IoT also has an impact on the Value Proposition improving business efficiency and performance and supporting innovation in the case of a consulting services company by means of the collection and real-time data analytics through sensors [6]. Thirdly, Customer Relationships are found to benefit from end-to-end solutions including the installation of IoT devices through the enforcement of connectivity, up to data acquisition and analysis [5]. In the mechanical manufacturing industry, IoT has been implemented to establish real-time monitoring solutions thanks to the interconnection between systems, in order to facilitate customers personal assistance [18]. Finally, IoT has been found to enhance production with a series of sensors that enable automatic progressive tracking system from raw materials to the finished product [24].

Table 10.5 Results of the secondary data analysis (n = 30 SMEs, numbers in squared brackets refer to the samples SMEs reported in Appendix I)

	SUM	Key Partners	Key Activities	Key Resources	Value Proposition	Customer Relationship	Channels	Customer Segments	Revenue Streams	Cost Structure
SUM	–	1	14	6	21	6	4	6	0	0
Internet of Things	6		[1] [16]		[6]	[5] [18]	[24]			
Cyber-Physical Systems	4		[29]	[3] [7]	[9]					
Visualization Technologies	5		[8] [27]		[18]	[15]		[27]		
Cloud Computing	4		[17]		[8] [17]		[5]			
Cybersecurity	6	[4]	[17]		[4] [10] [27]	[6]	[22]			
Blockchain Technology	4		[11] [16]					[23]		
Simulation and Modelling	4			[30]	[25]	[11]		[9]		
Machine Learning and AI	4		[26]	[28]	[5] [12]					
Big Data Analytics	5		[1] [14]		[5] [24]	[3]				
Additive Manufacturing	4				[2] [20]		[30]	[2]		

(continued)

Table 10.5 (continued)

	SUM	Key Partners	Key Activities	Key Resources	Value Proposition	Customer Relationship	Channels	Customer Segments	Revenue Streams	Cost Structure
Advanced Robotics	5			[8] [12]	[19] [28]			[5]		
New Materials	3		[13]		[20]			[24]		
Energy Management Solutions	4		[1]		[15] [21] [26]					

In effect, the SME examples underline the potential modifications of BM blocks of SMEs by implementing IoT. This can be summarized in *Proposition 1: Internet of Things can improve Key Activities and Value Proposition, facilitate Customer Relationship, and simplify Channels.*

The application of CPS in an industrial environment has been found to facilitate physical processes by complementing them with digital ones. The four examples found relate to Key Resources and Value Propositions. The former is underlined by the claim that by creating a direct connection between the physical world of machines and products and the virtual world of services and applications, people, processes, and objects can constantly communicate with each other in real time [3]. Similarly, CPS solutions can be used to allow communication and interconnection between different Key Resources (human, physical, and intellectual) in a flexible way [7]. The latter is mainly connected to improved performance. For example, collaborative (man-machine) applications enable the control of robots during complex assembly phases [9]. In a similar way, human-machine collaboration has been found to be essential for the Key Activities of smart testing and optimization of logistics and manufacturing that are again benefitting from the accelerated communication and data exchange as well as the substitution of for example repetitive or dangerous manual labor [29].

These findings from the SME examples can be summarized in *Proposition 2: Cyber-Physical Systems can support Key Resources, Key Activities, and Value Proposition.*

Visualization Technologies like AR facilitate to have the right data at the right time for better efficacy and efficiency [8] (see also Dallasega et al. 2020). In the case of another firm [27], augmented reality visors are programmed to support operators during machine interventions and provide real-time remote support to customers. Moreover, AR allows to be quick, efficient, and performant while interacting with customers, thus enhancing the firm's Value Proposition [18]. As an interesting example, a packaging company [15] guarantees personal guidance through a remote video assistance service. It works through an app and results extremely useful when real-time instructions on maintenance operations have to be shared with customers, supporting a sustained Customer Relationship. For a firm whose products are totally customized

[27], sophisticated vision systems respond to the different customers' needs related to quality control, managing dimensional control, defect detection, chromatic variance, and surface finishing defects allowing to satisfy any Customer Segment.

These findings from the SME examples can be summarized in *Proposition 3: Visualization Technologies can boost Key Activities and Value Proposition, strengthen Customer Relationship, and satisfy different Customers Segments.*

For an integrated systems company [17], Cloud Computing and virtual data storage represents an essential Key Activity to allow information to be accessed remotely anywhere by means of any devices. This activity also improves the Value Proposition offered to customers whose data are collected in the cloud in order to have a unified communication and accessibility [17]. Another example [8] underlines that data storing in Cloud Computing results in major savings since data collection, storage, and analysis, as well as operations and maintenance costs are reduced. Finally, Cloud Computing enables to offer customers a personal assistance in all phases of a project from evaluation to after-sales. It is interesting to see that the edge between Customer Relationship and Channels is often blurred when it comes to consulting services companies, due to the nature of the services involved [5].

These findings from the SME examples can be summarized in *Proposition 4: Cloud Computing can improve Key Activities, Value Proposition, and Channels.*

To secure Key Partners' information and protect their data, Cybersecurity offers innovative solutions for intrusion detection, identity and access management, and antispyware as key activities [17]. This is complemented by the collaboration with professional partners in the field of data protection to offer customers a privacy impact assessment [4]. Cybersecurity related Value Propositions aim at simplifying, optimizing, and accelerating a firms' managing and operations processes, by ensuring disaster recovery and Cybersecurity [4]. This is found to be a continuous challenge for which offers are available that help customers to update their Cybersecurity system keeping it aligned to the evolution of cyber-risks [10].

Similarly, Customer Relationship management can be improved through solutions for backup, disaster recovery, network, and data protection [6]. Finally, customers can be protected and guaranteed a secure connection and communication by means of encryption. In this way, data is protected and intrusion or external interference is avoided [22].

These findings from the SME examples can be summarized in *Proposition 5: Cybersecurity can benefit Key Partners, Key Activities, Value Proposition, Customer Relationships, and Channels.*

Blockchain Technology allows full product traceability through advanced inventory management systems [11]. This can be seen as a fundamental Key Activity as it also improves internal communication thanks to the evolution of machines' data acquisition [16]. Customers are supported in Big Data management, since complex data are easier to understand for both managers and factory workers thanks to personalized and interactive dashboards [27]. In this way, Blockchain Technology enhances Value Proposition since firms can offer their clients fully customized services supported by the identification and tracking of what is going on through the supply chain. Finally, Blockchains enable that the collected data are processed and integrated to generate valuable information for new intelligent maintenance activities and to preempt production delays affecting the availability of goods to the customers [23], thus satisfying different Customer Segments.

These findings from the SME examples can be summarized in *Proposition 6: Blockchain Technology can enhance Key Activities as well as Value Propositions and satisfying different Customer Segments.*

Expertise and experience are among the Key Resources of production firms; however, when dealing with quality and excellence, these can be complemented with Simulation and Modelling especially in material testing and failure investigation [30]. The manufacturing industry applies Simulation and Modelling technologies to create better design and manufacturing solutions. Their use limits design mistakes and empirical technical choices and guarantees a high repeatability of the process over time, thus creating the Value Proposition of reducing internal production costs [25]. Furthermore, Simulation and Modelling are useful tools, together with co-creation, to continuously improve reliability and performance for customers enhancing the Customer Relationships [11]. Finally, this technology is mainly used by engineering industries, since advanced simulation, modelling, and design processes enable to reach both niche markets, such as the aerospace one, and mass markets, such as the automotive one [9].

These findings from the SME examples can be summarized in *Proposition 7: Simulation and Modelling can support the implementation of Key Resources, amelioration of Value Propositions and Customer Relationship, and reaching different Customer Segments.*

Artificial Intelligence plays an important role to predict the maintenance of products and facilities or to control anomalies in the industrial environment, i.e., a firm's Key Activities [26]. It is moreover a Key Resource since the cognitive capabilities of the system provide new paradigms of support to the human operator either in purely manual activities or in hybrid human-robot collaborative stations [28]. In an I4.0 view, firms implement machines that are more and more able to learn while they interact with humans. Therefore, they communicate by means of a more natural language which is accessible and usable by anyone [5]. Artificial Intelligence concentrates on mimicking human decision-making processes and carrying out tasks in ever more human ways, helping to optimize the production process as a Key Resource, and driving an enhanced Value Proposition [12].

These findings from the SME examples can be summarized in *Proposition 8: Machine Learning and Artificial Intelligence can enhance Key Activities, Key Resources, and Value Proposition.*

Big Data can be collected, stored, and analyzed on a platform, thus establishing secure communication between data center sensors [1]. Besides that, real-time data analytics allows to perform several key activities such as predictive maintenance, downtime reduction, analysis performance, and optimization of business processes that are Key Activities in manufacturing [14]. Thanks to data acquisition and analysis, firms can offer more efficient solutions, flexibility, and servitization, as well as constant and continuous improvement of production processes, i.e., the Value Proposition [5, 24]. Finally, Customers Relationships can be enhanced by giving personal assistance throughout the whole activity process from data capture and processing, to planning, execution, real-time monitoring, and quality controls [3].

These findings from the SME examples can be summarized in *Proposition 9: Big Data Analytics can boost Key Activities, Value Propositions, and Customer Relationships.*

Among the Value Propositions, Additive Manufacturing technology can reduce production times and costs, thus optimizing the whole manufacturing process, which is a Key Activity [2]. This is achieved by facilitated creation of complex shapes, while 3D-printing reduces inaccuracies in projects, cuts development costs, decreases human errors, avoids waste of materials, and speeds up product marketing [20]. By evaluating all aspects of a product, from the choice of raw materials to the analysis of the process, it is possible to appreciate how Additive Manufacturing can connect a variety of industrial fields [30]. Finally, a simulation driven engineering company uses additive manufacturing technology to satisfy different Customer Segments' needs for both prototyping and designing of metal components [2].

These findings from the SME examples can be summarized in *Proposition 10: Additive Manufacturing can modify Value Proposition, improve Key Activities as well as Channels, and satisfy different Customer Segments.*

Robots are seen as Key Resources, since they represent "virtual workers" taking on tedious and repetitive tasks [8]. Advanced robotics has to be smartly integrated in the industrial scenario, in order to let workstations become ergonomic and flexible, avoiding stressful jobs for human workers [12]. The introduction of smart robots enhances operating speed and product quality, thus offering changes Value Propositions based on saving time and money, reducing human errors, decreasing waste and rework, and allowing a higher job rotation flexibility [19, 28]. Finally, collaborative Robots, which can interact actively and recognize human voice command, can be used in different sectors being suitable for plenty of businesses, thus satisfying a variety of Customer Segments [5].

These findings from the SME examples can be summarized in *Proposition 11: Advanced Robotics can build Key Resources, change Value Propositions and Customer Segments.*

Even if New Materials can hardly be defined as an I4.0 technology in general, "smart materials" possessing smart properties are an enabler of I4.0 (Culot et al. 2020). New Materials are used in a variety of Key Activities, for example, to enhance surface finishing or perform non-destructive inspection of parts by using magnetic particles, water washable, and post-emulsifiable liquid penetrants and contact and immersion ultrasonic [13]. Most New Materials are developed from existing materials by means of new combinations of elements. The design and production of new polymeric advanced materials for additive manufacturing results in high performance in terms of thermo-mechanical resistance and advanced functional properties, representing new Value Propositions [20]. Finally, New Materials help to reach new Customer Segments and offer infinite mixing possibilities to satisfy any kind of requests [24].

These findings from the SME examples can be summarized in *Proposition 12: New Materials can enrich Key Activities, improve Value Proposition, and reach more Customer Segments.*

Monitor energy consumption and finding energy saving solutions have become Key Activities. This may be obtained by platforms able to collect and remotely control data, as well as calculate the amount of energy consumption and losses in several working conditions [1]. The continuous improvement of environment impact by means of energy produced from different renewable sources, re-used wastewater and photovoltaic systems plays an essential role in the Value Proposition of many firms. For instance, to reduce the consumption of energy and water, a photovoltaic plant can be installed or to minimize the consumption of waste, recycling can be implemented [15, 21]. There are also several technologies that use natural, renewable sources to recharge devices' batteries, thus decreasing the need for maintenance and the negative impact on the environment [26].

These findings from the SME examples can be summarized in *Proposition 13: Energy Management Solutions can improve Key Activities and Value Propositions.*

10.5 Discussion and Conclusion

This study sets out to answer the research question of how can the implementation of Industry 4.0 technologies drive the modification of the business models of small- and medium-sized enterprises?

The findings reveal the double-sided nature of the intersection of I4.0 technologies and BMs including their building blocks. This encompasses on the one hand the concept of BM that is found in this study to be well established in its conception as well as application. Both literature and practice on BMs investigate multiple blocks in conjunction, acknowledging the impact of one on another. This supports previous literature that found the relative high interconnectedness of BM blocks (Spieth and Schneider 2016; Kotarba 2018), no matter if the BM canvas or other frameworks are used. On the other hand, this interconnectedness is particularly striking in comparison to the results obtained in the evaluation of the I4.0 technologies that exhibit a much higher heterogeneity. In line with previous literature, I4.0 is found to be in a constant flux. This finding has been formulated from the beginning of the concept (Kagermann et al. 2013) until some of the most recent publications on it (Culot et al. 2020). The found heterogeneity moreover supports that there are no fixed rules for I4.0 implementation (Botha 2019) and that the revolutions in production require a reflection in the BM (Porter and Heppelmann 2014). Despite this flux, the study at hand supports Müller and Däschle (2018) that underline the value of technical innovations that are associated to I4.0 innovations for strengthening existing BMs or developing new ones. The propositions developed in Sect. 10.4.2 provide detailed guidance based on practice examples of this interrelation.

A particular interesting finding is the mixed results on the relation of Cybersecurity and the BM blocks. In line with literature, the empirical results find the high relevance of this intersection to reduce the vulnerabilities of a digitalized firm (Culot et al. 2020; Götz and Jankowska 2017; Kotarba 2018). Contrastingly, the literature-based contingencies in Sect. 10.4.2 identify an underrepresentation of investigations of Cybersecurity and Key Partners, Channels, and Key Resources. Additionally, in literature there is an overrepresentation of associations of Cybersecurity to BMs in general. Turning to the details of the SME specific empirical findings in Sect. 10.4.2, these identify examples for two of the three combinations. This indicates a need for further investigations and clarifications that gain relevance considering the rapidly rising digitalization of firms and public organization in the course of the Corona crisis (Karabag 2020).

The findings moreover support the notion that I4.0 enabling technologies have the potential to substantially change how organizations and complex systems are managed (Leminen et al. 2018).

Turning from the contributions to the limitations of the study, we contend that any study (in particular literature reviews and secondary data analyses) exhibits limitations and aims at the generation of research direction to enable further work. While the measures to ensure reliability, replicability, and validity have already been elaborated in Sect. 10.3, the extant study still entails three main limitations. Nevertheless, these limitations can guide the way to future research and both are presented in more detail below:

First, the data collection is limited, since it exclusively investigated literature and website contents. Although this represents written communication from relevant practitioner and scientific sources, primary data from interviews for example would be timelier and most likely richer in the description of the relation of the I4.0 technologies and the BM blocks. Based on this, follow-up research could be based on primary data collection such as semi-structured interviews or focus groups. These could be structured into the single I4.0 technologies and BM blocks to enable richer descriptions of the interrelations. Alternatively, the relatively well-established concepts used in this study could build the basis for quantitative investigations. Starting from less formalized approaches like an analytical hierarchy process, the importance of the individual I4.0 technologies for the modification of the BM blocks could be investigated. Moreover, such follow-up empirical research could provide a validation for the propositions given in Sect. 4.2, whose validity and generalizability are limited by the data source.

Second, the data analysis is limited by the adoption of a single framework for I4.0 and BM each and their deductive application in content analysis. This limits the findings to the concepts captured in the frameworks and hinders the investigation of potentially relevant issues beyond.

Starting from here, a replication of the study with an enlarged set of frameworks could be worthwhile. This could also serve to investigate the overlaps of the chosen frameworks and establish a more appropriate one for the researched intersection of I4.0 technologies and BMs and their elements.

Third, the extant study is limited by the restricted granularity of the data collection and analysis, since it analyzed the literature and SMEs without a distinction of industries, countries, or continents from which they originated. Therefore, follow-up research could take an industry as well as country focus to investigate the heterogeneity of implemented I4.0 technologies, BMs, and their intersection. Although the frontier between products and services as well as industries and countries are gradually disappearing, it is still relevant to identify frontrunners and best practices, since their investigation and dissemination have high practical relevance in supporting the innovativeness of SMEs that represent the backbone of our economies.

Finally, the contingency analysis found an underrepresentation of publications on Cybersecurity and the BM blocks of Channels, Key Partners, and Key Resources, that however have been found to be relevant and require further investigations.

In effect, the findings of this study can guide practitioners and in particular managers of manufacturing firms and supply chains. The literature review findings provide an orientation of the state of the art in research on the impact of I4.0 technologies on BM blocks. Moreover, the empirical findings provide a map of 30 innovative SMEs that can be seen at the forefront of I4.0 implementation in one of Europe's main economies, i.e., Italy. This map is complemented by the abstraction of it into a set of propositions providing guidance on how I4.0 technologies have modified the BM blocks after successful implemented in manufacturing companies.

Acknowledgements This project has received funding from the European Union's Horizon 2020 research and innovation program under the Marie Skłodowska-Curie grant agreement No 734713.

Appendix I: 30 Sample SMEs

Firm	Sector
[1] ABO DATA http://www.abodata.com/	Consulting and technology projects to support customers' digital transformation through the implementation of IT solutions
[2] ADDITIVE ITALIA http://www.add-it.tech/	Additive manufacturing firm specialized in simulation driven engineering and design for metal addictive manufacturing
[3] AEC SOLUZIONI http://www.aecsoluzioni.it/wp/	Engineering and development of software solutions to help businesses to manage and to improve their manufacturing processes and efficiency with a view to I4.0 technologies
[4] AGOMIR https://www.agomir.com/	Software solutions: from software application to infrastructure projects, from technical assistance to training, in order to better managing processes and a firm's organization
[5] ALASCOM SERVICES https://www.alascom.it/	System integrator and supplier of technical consulting ICT services, with a specific focus on telecommunications networking and IP technologies
[6] BEANTECH https://www.beantech.it/	Consulting services company, supporting clients in their digital transformation facing the challenges of I4.0

(continued)

(continued)

Firm	Sector
[7] BEATREEX https://www.beatreex.it/	Software systems and cyber-physical software for the digital transformation to satisfy a firm's ever-changing production needs
[8] E.MAGINE https://emagine.ai/	Consulting and management services able to offer tailored solutions, to create and realize innovative projects, supporting customers in strategic choices
[9] EGICON http://www.egicon.com/	Engineering firm for advanced electronic systems, committed to offer their customers the best technology for development and production
[10] FASTERNET SOLUZIONI DI NETWORKING https://www.fasternet.it/	Information and communication technology engineering bound to technological improvement and innovation through customized networking services
[11] FLUID-O-TECH https://www.fluidotech.it/	Engineering and manufacturing firm for a variety of demanding applications, ranging from medical to automotive, industrial, and food service
[12] FRE TOR https://www.fretor.com/	Mechanical firm in automation and industrial robotics suitable for different sectors, from optical to medical, automotive, mechanical, and aerospace industry
[13] FUCINE UMBRE TERNI http://www.fucineumbre.com/	Mechanical solutions for the production of highly stressed structural components for the aerospace industry, forged and finished parts ready for being used in the assembly lines
[14] G2 DI GHIOLDI https://g2team.it/	Automation firm active in different fields dealing with machine manufacturers in a variety of sectors from automotive to food and chemical industries

(continued)

(continued)

Firm	Sector
[15] GALDI https://www.galdi.it/	Packaging firm designing and producing filling solutions in cartons for milk, dairy products, fruit juice, dry food, with the utmost care in food safety and process repeatability over time
[16] INTERMEK https://www.intermek.com/it/	Precision mechanical firm present in various industrial sectors, among which textiles, electrical appliances, industrial vehicles, and meteorological equipment
[17] LAN SERVICE https://www.lanservicegroup.it/ITA	Integrated systems and consulting services company, safely operating on infrastructure and data center of customers directly at their headquarters
[18] MANDELLI SISTEMI https://www.mandelli.com/it/	Mechanical manufacturing firm specialized, among others, in the aeronautic, oil, and manufacturing sectors
[19] MASMEC https://www.masmec.com/	Industrial automation firm specialized in precision technology, robotics and mechatronics, applied to the automotive and biomedical sectors
[20] MAT3D https://mat3d.it/	Additive Manufacturing firm designing and manufacturing new advanced materials for 3D printing in different industrial sectors from prototypes to mold and tools
[21] MECCANICA SBARZAGLIA https://www.meccanicasbarzaglia.com/	Precision Mechanical firm specialized in the processing of composite materials and additive manufacturing
[22] MECT https://www.mect.it/	Mechatronic manufacturing solutions devoted to production of mechatronic measurement and control systems, offering customized solutions, hardware, firmware, and software

(continued)

(continued)

Firm	Sector
[23] ORCHESTRA https://www.retuner.eu/	Integrated smart systems providing I4.0 technologies and solutions to manufacturing SMEs interested in real-time monitoring and control of their own production assets
[24] PERSONAL FACTORY https://www.personalfactory.eu/	Powder mixture industry for the building sector, managing end-to-end processes and offering perfectly customized solutions
[25] PROGIND http://www.progind.it/	Manufacturing firm specialized in molds for plastic material and sheet metal, guarantee high quality solutions and products, carefully designed to respond to customers' needs
[26] QWYDDY TECHNOLOGIES OÜ https://www.qwyddy-tech.com/it/home-en/	Consulting services company associating experience and tradition with upcoming technologies, supporting clients in a customized digitalization process
[27] SMART FACTORY https://www.smartfactory.it/	Mechatronics firm possessing a solid mix of competences in mechanics, electronics, informatics, and mechatronics, helping manufacturers to get closer to I4.0 technologies
[28] SMART ROBOTS http://smartrobots.it/	Advanced robotics firm focusing on the development and commercialization of technologies to support human operators in the factory
[29] STAUFEN.ITALIA https://www.staufen.it/it/	Lean management consulting services company working with their clients to establish a sustainable culture of change inside the business

(continued)

(continued)

Firm	Sector
[30] TEC EUROLAB https://www.tec-eurolab.com/eu-en/default.aspx	Materials and products testing services providing technical support for aerospace and defense, automotive and racing, among many other industries

References

Ahram, T., A. Sargolazei, S. Sargolazei, J. Daniels, and B. Amaba. 2017. Blockchain technology innovations. In *2017 Technology & Engineering Management Conference (TEMSCON)*, 137–141. https://doi.org/10.1109/TEMSCON.2017.7998367.

Alcácer, V., and V. Cruz-Machado. 2019. Scanning the industry 4.0: A literature review on technologies for manufacturing systems. *Engineering Science and Technology, an International Journal* 22 (3): 899–919. https://doi.org/10.1016/j.jestch.2019.01.006.

Ancarani, A., C. Di Mauro, L. Fratocchi, G. Orzes, and M. Sartor. 2015. Prior to reshoring: A duration analysis of foreign manufacturing ventures. *International Journal of Production Economics* 169: 141–155. https://doi.org/10.1016/j.ijpe.2015.07.031.

Armbrust, M., A. Fox, R. Griffith, A.D. Joseph, R. Katz, A. Konwinski, G. Lee, D. Patterson, A. Rabkin, I. Stoica, and M. Zaharia. 2010. A view of cloud computing. *Communications of the ACM* 53 (4): 50–58. http://doi.acm.org/10.1145/1721654.1721672.

Botha, A.P. 2019. Innovating for market adoption in the fourth industrial revolution. *South African Journal of Industrial Engineering* 30 (3): 187–198. https://doi.org/10.7166/30-3-2238.

Brenner, B. 2018. Transformative Sustainable business models in the light of the digital imperative—A global business economics perspective. *Sustainability* 10 (12): 4428. https://doi.org/10.3390/su10124428.

Carbone, V., A. Rouquet, and C. Roussat. 2017. The rise of crowd logistics: A new way to co-create logistics value. *Journal of Business Logistics* 38 (4): 238–252. https://doi.org/10.1111/jbl.12164.

Culot, G., G. Nassimbeni, G. Orzes, and M. Sartor. 2020. Behind the definition of industry 4.0: Analysis and open questions. *International Journal of Production Economics.* https://doi.org/10.1016/j.ijpe.2020.107617.

Dallasega, P., A. Revolti, P.C. Sauer, F. Schulze, and E. Rauch. 2020. BIM, augmented and virtual reality empowering lean construction management: A project simulation game. *Procedia Manufacturing* 45: 49–54. https://doi.org/10.1016/j.promfg.2020.04.059.

European Commission. 2020. What is an SME? https://ec.europa.eu/gro wth/smes/business-friendly-environment/sme-definition_en. Accessed on 27 January 2020.

Fink, A. 2019. *Conducting research literature reviews: From the Internet to paper,* 5th ed. Los Angeles: Sage.

Ghanbari, A., A. Laya, J. Alonso-Zarate, and J. Markendahl. 2017. Business development in the Internet of Things: A matter of vertical cooperation. *IEEE Communications Magazine* 55 (2): 135–141. https://doi.org/10.1109/MCOM.2017.1600596CM.

Gold, S., S. Seuring, and P. Beske. 2010. Sustainable supply chain management and inter-organizational resources: a literature review. *Corporate Social Responsibility and Environmental Management* 17 (4): 230–245. https://doi.org/10.1002/csr.207.

Götz, M., and B. Jankowska. 2017. Clusters and Industry 4.0–do they fit together? *European Planning Studies* 25 (9): 1633–1653. https://doi.org/10.1080/09654313.2017.1327037.

Hahn, R., S. Spieth, and I. Ince. 2018. Business model design in sustainable entrepreneurship: Illuminating the commercial logic of hybrid businesses. *Journal of Cleaner Production* 176: 439–451. https://doi.org/10.1016/j.jcl epro.2017.12.167.

Hermann M., T. Pentek, and B. Otto. 2016. Design principles for Industrie 4.0 scenarios. In *49th Hawaii International Conference on System Sciences (HICSS),* 3928–3937. https://doi.org/10.1109/HICSS.2016.488.

Hiete, M., P.C. Sauer, S. Drempetic, and R. Tröster. 2019. The role of voluntary sustainability standards in governing the supply of mineral raw materials. *GAIA - Ecological Perspectives for Science and Society* 28 (S1): 218–225. https://doi.org/10.14512/gaia.28.S1.8.

Ingaldi, M., and R. Ulewicz. 2020. Problems with the Implementation of Industry 4.0 in Enterprises from the SME Sector. *Sustainability* 12 (1): 217. https://doi.org/10.3390/su12010217.

Jacob, K., A.-L. Guske, I. Antoni-Komar, S. Funcke, T. Gruchmann, J. Kny, et al. 2019. Governance for the sustainable economy. Institutional innovation from the bottom up? *GAIA - Ecological Perspectives for Science and Society* 28 (S1): 204–209. https://doi.org/10.14512/gaia.28.S1.6.

Kagermann, H., W. Wahlster, and J. Helbig. 2013. Recommendations for implementing the strategic initiative Industrie 4.0. Final Report of the Industrie 4.0 Working Group, Forschungsunion.

Kamble, S.S., A. Gunasekaran, and S.A. Gawankar. 2018. Sustainable Industry 4.0 framework: A systematic literature review identifying the current trends and future perspectives. *Process Safety and Environmental Protection* 117: 408–425. https://doi.org/10.1016/j.psep.2018.05.009.

Karabag, S.F. 2020. An unprecedented global crisis! The global, regional, national, political, economic and commercial impact of the coronavirus pandemic. *Journal of Applied Economics and Business Research* 10 (1): 1–6.

Klein S., S. Schneider, and P. Spieth. 2020. How to stay on the road? A business model perspective on mission drift insocial purpose organizations. *Journal of Business Research*. https://doi.org/10.1016/j.jbusres.2020.01.053.

Kotarba, M. 2018. Digital transformation of business models. *Foundations of Management* 10 (1): 123–142. https://doi.org/10.2478/fman-2018-0011.

Kumar, A. 2018. Methods and materials for smart manufacturing: Additive manufacturing, internet of things, flexible sensors and soft robotics. *Manufacturing Letters* 15: 122–125. https://doi.org/10.1016/j.mfglet.2017.12.014.

Lasi, H., P. Fettke, H.G. Kemper, T. Feld, and M. Hoffmann. 2014. Industry 4.0. *Business & Information Systems Engineering* 6 (4): 239–242. https://doi.org/10.1007/s12599-014-0334-4.

Leminen, S., M. Rajahonka, M. Westerlund, and R. Wendelin. 2018. The future of the Internet of Things: Toward heterarchical ecosystems and service business models. *Journal of Business & Industrial Marketing* 33 (6): 749–767. https://doi.org/10.1108/JBIM-10-2015-0206.

Matthyssens, P. 2019. Reconceptualizing value innovation for Industry 4.0 and the Industrial Internet of Things. *Journal of Business & Industrial Marketing* 34 (6): 1203–1209. https://doi.org/10.1108/JBIM-11-2018-0348.

Mayring, P. 2015. *Qualitative Inhaltsanalyse: Grundlagen und Techniken*, 12th ed. Weinheim: Beltz.

Mills, S., S. Lucas, L. Irakliotis, M. Rappa, T. Carlson, and B. Perlowitz. 2012. *Demystifying big data: A practical guide to transforming the business of government*. Washington, DC: TechAmerica Foundation.

Müller, J. M. 2019. Business model innovation in small-and medium-sized enterprises. *Journal of Manufacturing Technology Management* 30 (8), 1127–1142. https://doi.org/10.1108/JMTM-01-2018-0008.

Müller, J.M., and S. Däschle. 2018. Business model innovation of industry 4.0 solution providers towards customer process innovation. *Processes* 6 (12), 260. https://doi.org/10.3390/pr6120260.

Müller, J.M., O. Buliga, and K.-I. Voigt. 2018. Fortune favors the prepared: How SMEs approach business model innovations in Industry 4.0. *Technological Forecasting and Social Change* 132: 2–17. https://doi.org/10.1016/j.techfore.2017.12.019.

Müller, J.M., O. Buliga, and K.-I. Voigt. 2020. The role of absorptive capacity and innovation strategy in the design of industry 4.0 business Models-A comparison between SMEs and large enterprises. *European Management Journal*. https://doi.org/10.1016/j.emj.2020.01.002.

Orzes, G., R. Poklemba, and W.T. Towner. 2020. Implementing Industry 4.0 in SMEs: A focus group study on organizational requirements. In *Industry 4.0 for SMEs*, ed. D.T. Matt, V. Modrák, and H. Zsifkovits, 251–277. Cham: Palgrave Macmillan. https://doi.org/10.1007/978-3-030-254 25-4_9.

Osterwalder, A., Y. Pigneur, and C.L. Tucci. 2005. Clarifying business models: Origins, present, and future of the concept. *Communications of the Association for Information Systems* 16 (1), 1–25. https://doi.org/10.17705/1CAIS. 01601.

Osterwalder, A., and Y. Pigneur. 2010. *Business model generation: A handbook for visionaries, game changers, and challengers.* Hoboken: Wiley.

Pisano, P., M. Pironti, and A. Rieple. 2015. Identify innovative business models: can innovative business models enable players to react to ongoing or unpredictable trends? *Entrepreneurship Research Journal* 5 (3): 181–199. https://doi.org/10.1515/erj-2014-0032.

Porter, M.E., and J.E. Heppelmann. 2014. How smart, connected products are transforming competition. *Harvard Business Review* 99 (11): 64–88.

Rayna, T., and L. Striukova. 2016. From rapid prototyping to home fabrication: How 3D printing is changing business model innovation. *Technological Forecasting and Social Change* 102: 214–224. https://doi.org/10.1016/j.tec hfore.2015.07.023.

Rejikumar, G., V.R. Sreedharan, P. Arunprasad, J. Persis, and K.M. Sreeraj. 2019. Industry 4.0: Key findings and analysis from the literature arena. *Benchmarking: An International Journal* 26 (8): 2514–2542. https://doi.org/ 10.1108/BIJ-09-2018-0281.

Rüssmann, M., M. Lorenz, P. Gerbert, M. Waldner, J. Justus, P. Engel, and M. Harnisch. 2015. Industry 4.0: The future of productivity and growth in manufacturing industries. *Boston Consulting Group* 9 (1): 54–89.

Sauer, P.C., and M. Hiete. 2020. Multi-stakeholder initiatives as social innovation for governance and practice: A review of responsible mining initiatives. *Sustainability* 12 (1): 236. https://doi.org/10.3390/su12010236.

Sauer, P.C., and S. Seuring. 2017. Sustainable supply chain management for minerals. *Journal of Cleaner Production* 151: 235–249. https://doi.org/10.1016/j.jclepro.2017.03.049.

Schallmo, D., C.A. Williams, and L. Boardman. 2017. Digital transformation of business models—Best practice, enablers, and roadmap. *International Journal of Innovation Management* 21 (08): 1740014. https://doi.org/10.1142/S136391961740014X.

Seuring, S., and S. Gold. 2012. Conducting content-analysis based literature reviews in supply chain management. *Supply Chain Management: an International Journal* 17 (5): 544–555. https://doi.org/10.1108/13598541211258609.

Ślusarczyk, B. 2018. Industry 4.0: Are we ready? *Polish Journal of Management Studies* 17 (1): 232–248. http://dx.doi.org/10.17512%2Fpjms.2018.17.1.19.

Spieth, P., and S. Schneider. 2016. Business model innovativeness: designing a formative measure for business model innovation. *Journal of Business Economics* 86 (6): 671–696. https://doi.org/10.1007/s11573-015-0794-0.

Spieth, P., T. Röth, and S. Meissner. 2019. Reinventing a business model in industrial networks: Implications for customers' brand perceptions. *Industrial Marketing Management* 83: 275–287. https://doi.org/10.1016/j.indmarman.2019.04.013.

Strandhagen, J.O., L.R. Vallandingham, G. Fragapane, J.W. Strandhagen, A.B.H. Stangeland, and N. Sharma. 2017. Logistics 4.0 and emerging sustainable business models. *Advances in Manufacturing* 5 (4): 359–369. https://doi.org/10.1007/s40436-017-0198-1.

Teece, D.J. 2010. Business models, business strategy and innovation. *Long Range Planning* 43 (2–3): 172–194. https://doi.org/10.1016/j.lrp.2009.07.003.

Tirabeni, L., P. De Bernardi, C. Forliano, and M. Franco. 2019. How can organisations and business models lead to a more sustainable society? A framework from a systematic review of the industry 4.0. *Sustainability* 11 (22): 6363. https://doi.org/10.3390/su11226363.

Tranfield, D., D. Denyer, and P. Smart. 2003. Towards a methodology for developing evidence-informed management knowledge by means of systematic review. *British Journal of Management* 14 (3): 207–222. https://doi.org/10.1111/1467-8551.00375.

Weking, J., M. Stöcker, M. Kowalkiewicz, M. Böhm, and H. Krcmar. 2019. Leveraging industry 4.0–A business model pattern framework. *International Journal of Production Economics* 225: 107588. https://doi.org/10.1016/j.ijpe.2019.107588.

Zott, C., and R. Amit. 2010. Business model design: An activity system perspective. *Long Range Planning* 43 (2–3): 216–226. https://doi.org/10.1016/j.lrp.2009.07.004.

11

General Assessment of Industry 4.0 Awareness in South India—A Precondition for Efficient Organization Models?

Korrakot Tippayawong, Leoš Šafár, Jakub Sopko,
Darya Dancaková, and Manuel Woschank

11.1 Introduction

As a consequence of business and social evolution during recent years, several topics emerged and gained tremendous attention. Megatrends such as climate change, globalization, technological progress, demographics' dynamics, or mass customization are undoubtedly challenging for society. In reaction to the very volatile and complex business environment, various strategic initiatives took place all over the world, for example, Germany's "High Tech Strategy 2020", "Made in China 2025"

K. Tippayawong
Faculty of Engineering, Department of Industrial Engineering, Chiang Mai
University, Chiang Mai, Thailand
e-mail: korrakot@eng.cmu.ac.th

L. Šafár (✉) · J. Sopko · D. Dancaková
Faculty of Economics, Department of Banking and Investment, Technical
University of Košice, Košice, Slovakia
e-mail: leos.safar@tuke.sk

© The Author(s) 2021
D. T. Matt et al. (eds.), *Implementing Industry 4.0 in SMEs*,
https://doi.org/10.1007/978-3-030-70516-9_11

or the USA's "Industrial Internet Consortium" (Ramsauer 2013), to keep pace with exponential technological development and reach sustainable growth. These concepts aim to develop and implement modern strategies (Industry 4.0) to achieve higher effectiveness, competitiveness, sustainability, and to produce higher value-added (Kiel et al. 2016) with emphasis on minimizing the negative impact on the environment. To add on, extensive possibilities covered by Industry 4.0 could improve enterprises' costs management (Lasi et al. 2014; Posada et al. 2015; Calero Valdez et al. 2015). We consider, among many others, Industry 4.0 as crucial from both the social and manufacturing sector's point of view in the foreseeable future. However, as Hofmann and Rüsch (2017) stated, "the concept of Industry 4.0 still lacks a clear understanding". Such unclear interpretations and misunderstandings could be even more pronounced in emerging economies lacking proper educational and informational level, which could lead to even wider discrepancies between developed and emerging economies' business and social environment.

Conducted research presented in this chapter aims to estimate and analyse informational base and general awareness about Industry 4.0 in the area of South India. Authors argue that sufficient knowledge is an important presumption for the successful development of an effective organization and network models in the future environment, especially from a SMEs perspective. We find papers among the literature using questionnaire-based surveys, addressing mostly readiness of industries or SMEs to Industry 4.0. However, there is a lack of literature covering the informational level, attitudes, and expectations of potential employees, in general, the same as within the examined region. This chapter concerns

J. Sopko
e-mail: jakub.sopko@tuke.sk

D. Dancaková
e-mail: darya.dancakova@tuke.sk

M. Woschank
Montanuniversitaet Leoben, Leoben, Austria
e-mail: manuel.woschank@unileoben.ac.at

the important perspective of inhabitants as potential workers regarding Industry 4.0 in less developed regions, which goes hand in hand with the development of effective strategy from an employers' perspective. Main motivation for such research stems from the lack of existing literature examining such perspectives. We argue that potential workers' attitudes represent the key aspects regarding transformation towards innovative future technologies, while recent studies are mainly oriented on SMEs, not their backbones—employees. The objective of presented research is to determine the state of art of general awareness and expectations in mentioned region, which could help employers and policy makers to conduct proper policies in order to prepare students and potential workforce to Industry 4.0 environment. The questionnaire was created using information from previous industry visits and consultations with entrepreneurs, students, and employers in the mentioned region. Issues expressed by respondents are summarized and the most attention-dragging findings are highlighted.

The chapter is further organized as follows: Sect. 11.2 provides a literature background, Sect. 11.3 describes problems concerned, where issues necessary for further research are stated. Section 11.4 provides a methodology description, while in Sect. 11.5 we present obtained results. Section 11.6 concludes.

11.2 Literature Review

Over the past years, we faced a strong advance of technology among almost all sectors. New business propositions and applications within the business systems were enabled given the new technologies. As Thestrup et al. (2006) stated, the collection and management of both physical and virtual data gathered from users, sensors, or devices, emerged. So-called Internet of Things—IoT (Brock 2001, firstly used the term IoT) then means worldwide network of such objects communicating and operating through standardized communication protocols. However, IoT became recognized after the ITU1 report (ITU 2005), describing IoT as the ability to connect everyday objects, meaning that both people will be able to communicate with objects, the same as objects will be able to communicate among themselves. The prerequisite to such communication is advanced wireless technology (identification technologies and sensors). Logically, IoT can be diversified to Industrial IoT and Commercial IoT, while I4.0 expects all those parts to be interconnected and communicating.

To simplify, the goal of IoT infrastructure, as an essential part of Industry 4.0, is to enable participants (people and objects) to be more flexible, to react appropriately and autonomously, thanks to the information sharing network. Harbor Research (2011) suggests, that two major strands of technological development emerged at the beginning of the twenty-first century; first is mentioned IoT and secondly, "Internet of People" (IoP or social networking). These interconnected devices, processes, machines, products, etc., will have a significant impact on the enterprise's life cycle, efficiency, functioning, and consequently to the broader economy (Safar et al. 2018).

To conclude, Sundmaeker et al. (2010) define the IoT as an integrated part of "Future Internet", or a "dynamic global network infrastructure with self-configuring capabilities based on standard and interoperable communication protocols where physical and virtual 'things' have identities, physical attributes, and virtual personalities and use intelligent interfaces, and are seamlessly integrated into the information network". Internet of Things is already partially adopted by households, with the aim of creating a "smart house", even though not every gadget is appropriately connectable yet (Cui 2016). The same problem can be observed among enterprises, especially SMEs. It is assumed that the main obstacle becoming "smart" both for households and industries will be funding, along with insufficient education and knowledge (Safar et al. 2018).

Such interconnected objects and subjects are just prerequisites for the so-called 4th industrial revolution, where cyber and physical levels should merge (Lasi et al. 2014). The term Industry 4.0 points to the 4th industrial revolution and was first presented on Hannover-Messe (one of the biggest international trade fairs oriented on new and smart technologies) in 2011, while it also indicates initiative of German government to improve the environment in manufacturing sector using new technologies (information about the concept were brought up in 2014 at the World Economic Forum in Davos) (Standhagen et al. 2017). According to BITKOM (Germany's digital association, founded in 1999 as a merger of individual industry associations in Berlin, representing more than 2,500 companies in the digital economy, among them 1,000 SMEs all 400 start-ups), the 4th industrial revolution will allow control over the entire life cycle of the product and value stream, therefore redefine organization entirely. concerning efficiency-oriented on cost-savings and complexity reduction, Modrak and Bednar (2015) conclude that the I4.0 environment will initiate mass customization mainly because of the ability of each entity throughout the value stream to communicate and identify itself. All of these visions and concepts are meant to be environmentally, economically, but mainly socially sustainable. Leaving now a technical standpoint, we emphasize non-technical aspects of proposed changes within the industries. Such transformation should bring new organization models, that should reflect both business perspective and state of mind of potential customers and workers.

As Slusarczyk (2018) suggests, the 4th industrial revolution differs from previous revolutions, because it will apply to all aspects of everyday lives, as a consequence of the environment, where information will be exchanged between objects, between people, and between people and objects. In other words, based on real-time data exchange and horizontal and vertical integration of production systems are the main pillars of I4.0 (Thoben et al. 2017), along with cybersecurity, autonomous systems, the capability of analysing large data sets, virtual reality, and cloud computing. Undoubtedly, such changes would require managerial decisions firstly, due to inevitable initial costs linked to such new technological equipment. Schröder et al. (2015) leave the open question, whether it is even worth to implement I4.0, especially for SMEs, despite the consensus we find among authors describing reduced costs and more efficient processes and the environment as a consequence of I4.0. We argue that such dynamics within the industries should be examined deeply, and various elements of sustainable development, not only economic point of view should be evaluated (Kovacs 2018; Eberhard et al. 2017). The opposite of mentioned cost-saving and cost-reducing is initial need for significant financial expenditures, that are on many occasions out of reach for companies, especially SMEs (Soltes and Gavurova 2014). Either way, to move on with such disruptive changes is conditional by the development of adaptable network and organization models. Secondly, we argue that proper informational basis of the knowledge, attitudes and expectations of inhabitants, mainly potential workforce, is inevitable for such managerial decisions, while existing literature addressing mentioned issue is rather scarce.

Another very important social aspect of such a smart environment is how intelligent machines will affect the labour market (Eberhard et al. 2017; Dallasega et al. 2019, 2020; Woschank et al. 2020). This topic can be being examined from two perspectives, firstly by describing requirements towards workers in I4.0 (Eberhard et al. 2017; Dallasega et al. 2020; Safar et al. 2020), secondly by examining the standpoint of workers and their outlook or current state of mind (Eberhard et al. 2017; Wolter et al. 2015). We argue, that unless a reasonable level of awareness and basic knowledge of Industry 4.0 related concepts and inevitable parts is reached, it will be hard to successfully move towards a smart environment, especially in the case of less developed regions. Insufficient information base of eligible workforce represents an obstacle for potential employers oriented towards I4.0. Inadequate information and knowledge could also lead potential employees towards wrong or misjudged conclusions or attitudes. Probably the most crowded thought is that bringing in the intelligent machines would steal jobs, again, especially in less developed regions with the less qualified, manually involved workforce. Consequently, lack of sympathy towards any modernization steps could hold potential progress off—according to Statista (2019), countries without any problems with unemployment (e.g. Germany, USA, Japan) report the highest numbers of installed industrial robots per 10,000 employees. Again, wrong or insufficient knowledge of workers could lead to negative acceptance of incoming transformation towards Industry 4.0, while we still have no sufficient evidence about state of art of this problem, especially in emerging countries. To add on, as Ramingwong and Manopiniwes (2019) put it, investments in R&D go hand in hand with well informed and educated employers and consequently impact the organization models.

11.3 Problem Description

The emerging economies should leverage their advantages, such as huge markets, attractive conditions for manufacturing, fast-growing economies, and a mainly larger labour force with more favourable demography (Iyer 2018). Admitting that Industry 4.0 will primarily affect the manufacturing sector, we face significant discrepancies among countries and regions. Despite the estimate that India will be the world's fastest-growing economy in following years (World Bank 2018) within the manufacturing sector that could hit 1 trillion US$ in 2025 (IMR 2020), we doubt the ability of successful transformation towards Industry 4.0, hence, we find India and its regions important to examine to 4th industrial revolution (Chandran et al. 2019). For example, having Germany—technology and manufacturing leader, however with an ageing population and lack of labour force; and on the other hand, emerging country as India—suffering from technological gaps, which put India to a level of Industry 2.0 as Iyer (2018) concludes, on the contrary, with strong demography.

There is also the political will to spur manufacturing sector, translated into initiatives such as "Digital India" (Goswami 2016), "Skill India" (2020), or "Make in India", with aims to (among others) create sufficient skill sets within the urban poor and rural migrants for inclusive growth, or to increase technological depth in manufacturing to increase domestic value addition. In addition, there is mentioned demographic factor—India has the best demographic dynamics, with approximately 60% of the population age between 15 and 59 (Directorate of Intelligence 2019). Open question remains, are citizens and workers ready for such development in the foreseeable future? Are they ready for emerging organization models within businesses?

We accept, that within such huge country significant disparities among particular regions exist, hence we applied our research only in Southern part of India (authors were physically present in the state of Tamil Nadu during data collection). South Indian region includes several states and union territories (Kerala, Andaman and Nicobar Islands, Tamil Nadu, Pondicherry, Andhra Pradesh, Karnataka, Telangana, Lakshadweep Islands), which in combined counts for 19.31% of the geographical area of the whole India. With over 250 million people, South India represents around 20% of the country's population (Census 2011). As of 2016, the economic growth of South India was around 17%, compared to 8% growth of whole India, while the GDP of South India accounted for 30% of total Indian's GDP. Some specific industries are even more important from overall perspectives, such as cotton production (48% of India's entire cotton production comes from South India) or agricultural production (36% of whole state's production comes from South India).

Same as for other countries and regions, main employers are SMEs. Unfortunately, as Iyer (2018) states for India's industrial policy in general, it is old, and in lack of critical technology. Many enterprises in this area are old and have long-lasting tradition—especially for those the transformation lies upon the success of new network and organization models. Despite the established reputation and customers created, they are equipped with insufficient and old devices or machines. Internet access and computer equipment within industries in this region are also rather poor. Since the majority of the research has been conducted in the field of needed modernization, especially concerning the SMEs to successfully transform towards Industry 4.0, we would rather point at the necessity of having potential labour force ready for such transformation. It is therefore considered that awareness of I4.0 needs to be continuously expanded and promoted, as confirmed by several authors (Safar et al. 2018; Matt and Rauch 2020; Burgess 2002; Kagermann 2015). Even if obtaining new machines and gadgets would be economically viable, will there be enough sufficiently educated workers or customers? Throughout the literature we find papers addressing similar problems within different regions, f.i. concluding that qualified specialists are often not satisfied with the salary, which causes their outflow in favour of richer economic regions, leaving almost no people able to operate such modern machines (Ingaldi and Ulewicz 2020). We argue, that unless some basic level of knowledge regarding addressed issues is reached within the population, the ability to become competitive in an Industry 4.0 environment is rather limited.

11.4   Methodology

The research was conducted in the area of South India, where industries operate in several segments, with a majority representation of SMEs. This survey-based study aims to examine the level of awareness and general consciousness of Industry 4.0 among South Indian students, workers, entrepreneurs, in other words, a broad spectrum of citizens. We expect that proper analysis of gathered responses could provide us with unique and valuable knowledge of the current state of mind of local citizens, along with their current level of internet/connection requiring gadgets/platforms, and further serve as a guide for finding a suitable implementing strategy for new technologies in such areas.

Results presented in this chapter concern opinions and knowledge of inhabitants living, studying, working, or doing business in the previously described area (Table 11.1). For obtaining responses, a questionnaire was used, and data collection took place from December 2019 to

Table 11.1   Profile of respondents

Profile	N	%	Profile	N	%
Gender			*Age*		
Female	147	26.1	25 or below	466	82.6
Male	417	73.9	26–35	57	10.1
			36–45	24	4.3
Status			46 and more	17	3.0
Student	438	77.7			
Employed and Entrepreneur	105	18.6	*Residential place*		
Houseperson and Retired	7	1.2	Andaman and Nicobar Islands[*]	12	2.1
Unemployed	14	2.5	Andhra Pradesh	6	1.1
			Karnataka	10	1.8
Education			Kerala	14	2.5
Higher Secondary and below	181	32.1	Lakshadweep Islands[*]	4	0.7
Bachelor	256	45.4	Pondicherry[*]	9	1.6
Master	67	11.9	Tamil Nadu	507	89.9
Doctorate, Medical, Law degree or higher	60	10.6	Telangana	2	0.4

Adapted from Safar et al. (2020)
Note [*]Union territory

February 2020. As advised by several authors (Schwarz and Hippler 1987; Schuman et al. 1981), we used fixed-choice questions, to maintain time efficiency and difficulty of evaluation. The questionnaire was distributed within several traceable ways during the stay of authors in Tamil Nadu. The sample contains 564 unique responses (after removing incomplete and inappropriately filled responses—respondents' answers were checked to confirm all required questions had been answered in a prescribed manner). Respondents were notified in advance that providing answers to this questionnaire are anonymous. All answers provided will serve only for research purposes, and no personal details will be required or stored.

We divided the questionnaire into four main parts (Fig. 11.1). In the first part, we focused on the social status of the respondent, education, and the place where the respondent currently works, studies, or stays. In the second part, we were interested in respondents' basic internet communication and usage of social communication applications. In the third and main part, we looked at the awareness of Industry 4.0 in general among respondents. We asked about key terms such as cloud solutions, mass customization, Internet of Things, Industry 4.0, smart manufacturing, smart cities, etc. In the fourth part, we intended to examine what the I4.0 could bring to the south Indian region from the responders' perspective.

Several scales were used due to the substance of the question (full text of the questionnaire and scales of answers is provided in Appendix A). Questions addressing previous experience and general awareness about key terms were scaled binomially (*yes/no*). Other questions addressing

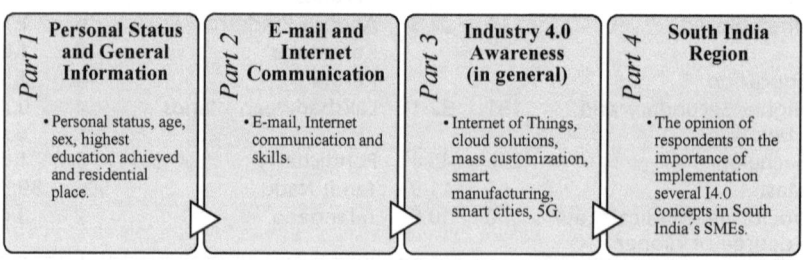

Fig. 11.1 Stages of survey (Adapted from Safar et al. [2020])

south India region were scaled as of 5 levels: "*not at all important/no*"—1; "*slightly important/rather no*"—2; "*no opinion*" (due to lack of information/knowledge, referring also to "*I do not know*")—3; "*fairly important/rather yes*"—4; "*very important/yes*"—5. Supplementary questions regarding usage of social media, email, or e-commerce had specific scales examining the frequency of usage.

To analyse responds, we used tables of counts and percentages for the joint distribution of two (severe combinations) categorical variables. We used custom and contingency tables, statistical testing, and generated bar graphs for easier data presentation. Pearson's chi-square test was performed to test the independence between the row and column variables. Pearson's chi-square test requires a large sample. The main rule regarding the sample size is that not more than 20% of expected cells should be less than 5 and none of the expected cells should be less than 1 (Agresti and Kateri 2011; Armitage et al. 2008). If the relationship was significant, consequently we used z-test to compare the proportion of column pairs to each other (adjusted by Bonferroni correction) according to the social variables and variables reported by Industry 4.0 areas. For 2 × 2 tables, we used Fisher's Exact test. The column proportions test shows whether the ratio in one column is significantly different from the ratio in the other column. The test assigns a letter key (A, B, C) to each category reported in column variables. The definition of each comparison of column proportions is discussed in the following section. All statistical outputs were processed in the IBM SPSS Statistics v25.0.

Further, we will concentrate on presenting the most attention dragging outcomes and dependencies from responses, which were statistically proven as significant.

11.5 Results and Discussion

In order not to confuse respondents and avoid misinterpretations, we provided short descriptions of possibly unknown terms related to our scope (presented in Appendix A). The questionnaire, in its actual form, is composed by thirty-four questions divided into four main areas, mentioned above.

11.5.1 General Awareness, Age and Education

Firstly, we asked our respondents, if they ever heard about key terms related to the 4th industrial revolution. As presented in Fig. 11.2, the term "mass customization" is not known by almost 60% of respondents, while, which is more important, the term "Industry 4.0" is unknown to 49.6% of respondents. Rather than focusing only on simple percentage points-presentations of answers observed, we examined and focused mainly on dependencies between key answers on a statistically significant basis, as presented further in this chapter.

Before examining key aspects of this survey, we took the first step examining dependence between age, education, and such awareness. In all tables below the Chi-square statistic (χ^2) and the p-value is presented for each row question, as an inevitable assumption for further column proportions comparison. χ^2 refers to Pearson's Chi-square statistic value, obtained by the Chi-square test in SPSS, which tests the hypothesis that two variables (row and column) are independent. p-value refers to the significance value, which has the information we are looking for. The lower the p-value, the less likely it is that two variables are unrelated.

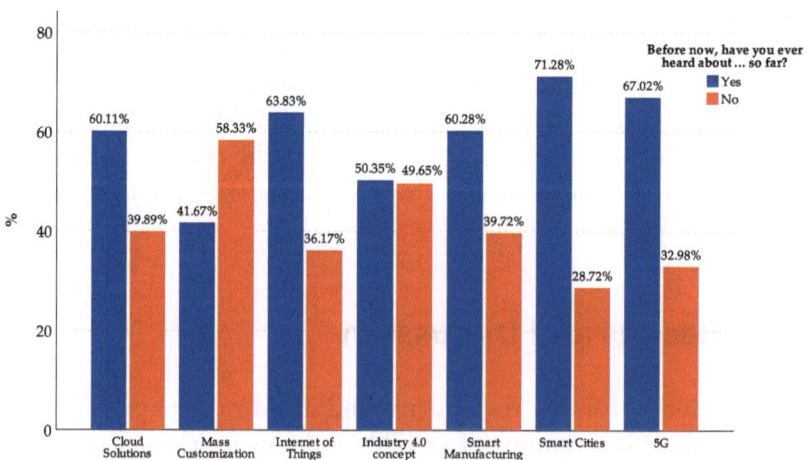

Fig. 11.2 Awareness in general regarding I4.0 related terms (Adapted from Safar et al. [2020])

When the significance value is less than 0.05, we can conclude that there is a relationship between two variables. To understand the relationship between row and column variables we examine the crosstabulation tables with results of the column proportions tests. As we mentioned in the previous section, the column proportions test shows whether the proportion in one column is significantly different from the proportion in the other column. The test assigns a letter key (A, B, C) to each category reported in column variables. We used three significance levels: 0.05^{*}; 0.01^{**}; 0.001^{***}. Column proportion tests are performed by z-test and tests are adjusted for all pairwise comparisons within a row of each innermost sub-table using the Bonferroni correction (see Sedgwick 2012). Below we provide Table 11.2, where the statistically significant relationship between answers "*No*" to above-mentioned general awareness questions and education "*Upper Secondary and lower*" can be observed. We find this in line with basic logic that ongoing and deeper education opens possibilities and provides information about new approaches and cutting-edge trends. Similarly, we find a logical relationship within our answers, that higher education (*Doctorate, Medical or Law degree or higher*) goes with a higher age of the respondent. However, we consider the fact, that 46.4% (45.3%) of the group "*Upper Secondary and lower*" answered, "*No*" when asked about "*Cloud solutions*" ("*Internet of Things*"), as a result of teaching plans that are not updated sufficiently, not the respondents' inability to learn about possibilities linked to I4.0.

In Table 11.2, the column proportions test assigns a letter key, (A) or (B), to each category of question Q10-Q17. (A) refers to the answers "*No*" and (B) to the answers "*Yes*". The row variables are "*Age*" and "*Education*", which have four categories of answers. The two-sided asymptotic significance of Chi-square statistics adjusted by Bonferroni correction is less than 0.05^{*} in all comparisons except of comparison between "*Age*" and "*Mass Customization*" (p-value 0.100). The p-value (0.000^{***}) is less than 0.001, therefore statistically significant. For the column proportions test associated with the age group "*25 or below*" and the answers to question Q10, the B key appears in the column "*No*".

Thus, we can conclude that the proportion of respondents aged "*25 or below*", who answered the question Q10 about cloud solutions negative, is greater than the proportion of respondents answered the question

Table 11.2 Awareness vs age and education

		Before now, have you ever heard about the …							Q17
		Q10	Q11	Q12	Q13	Q14	Q15	Q16	Do you have any previous experience with IoT or I4.0 concept?
		Cloud Solutions so far?	Mass Customization so far?	Internet of Things so far?	Industry 4.0 concept so far? (so-called 4th Industrial revolution)	Smart Manufacturing so far?	Smart Cities so far?	5G so far?	
		No (A) / Yes (B)	No (A) / Yes (B)	No (A) / Yes (B)	No (A) / Yes (B)	No (A) / Yes (B)	No (A) / Yes (B)	No (A) / Yes (B)	No (A) / Yes (B)
X^2 Age	p-value	31.512 0.000***	6.251 0.100	33.499 0.000***	44.270 0.000***	24.811 0.000***	27.827 0.000***	18.893 0.000***	15.197 0.002**
	25 or below	B		B	B	B	B	B	B
	26–35	A		A	A	A	A	A	
	36–45			A	A	A	A	A	A
	46 and more	A				A			A
X^2 Education	p-value	11.945 0.008**	8.559 0.036*	28.072 0.000***	35.004 0.000***	19.037 0.000***	53.918 0.000***	20.402 0.000***	8.560 0.036*
	Upper Secondary and lower	B	B	B	B	B	B	B	B
	Bachelor		A					A	
	Master	A		A	A	A	A	A	A

	Q10	Q11	Q12	Q13	Q14	Q15	Q16	Q17
				Before now, have you ever heard about the ...				
	Cloud Solutions so far?	Mass Customization so far?	Internet of Things so far?	Industry 4.0 concept so far? (so-called 4th Industrial revolution)	Smart Manufacturing so far?	Smart Cities so far?	5G so far?	Do you have any previous experience with IoT or I4.0 concept?
	No Yes (A) (B)	No Yes (A) (B)	No Yes (A) (B)	No Yes (A) (B)	No Yes (A) (B)	No Yes (A) (B)	No Yes (A) (B)	No Yes (A) (B)
Doctorate. Medical or Law degree or higher			B		B	A	B	

Adapted from Safar et al. (2020)

Results are based on two-sided tests. For each significant pair, the key of the category with the smaller column proportion appears in the category with the larger column proportion. X^2 refers to Chi-square statistic. p-value refers to the two-sided asymptotic significance of the chi-square statistic. Significance level for upper case letters (A, B, C): 0.05*; 0.01**; 0.001***. Tests are adjusted for all pairwise comparisons within a row of each innermost subtable using the Bonferroni correction

Q10 positive (aged "*25 or below*"). The same results are listed between the respondents aged "*25 or below*" and other questions except for Q11 regarding mass customization. For the tests associated with "*Education*", the results indicate the same in the case of "*Upper Secondary and lower*" education for all questions Q10-Q17.

We would like to highlight the relationship between the age group "25 or below" and answers "No" to general questions. In absolute terms, 56.0%, and 41.4%, respectively, of the group "25 or below" answered "No" to questions addressing Industry 4.0, and IoT, respectively. We consider this as a very poor informational level especially within the young and flexible group of workers entering labour market. On the contrary, 79.8% (46.6%) of this group is using WhatsApp (Facebook) almost daily, therefore, we cannot explain this level of awareness as a result of insufficient conditions for obtaining information or being digitally isolated. Motyl et al. (2017) surveyed more than 460 students at three different universities in Italy about the Industry 4.0 concept. The authors point out the importance of the digital behaviour of young people, whose relationship with the digital world and services are very important for their further social, but also economic development, ultimately for the development of the region or country. We agree with the authors that in today's environment it is important to empower a broader knowledge of the general I4.0 concepts and bring well-structured action plans into the educational process. These conclusions should emphasis, on the one hand, the role of education, and the SMEs on the other, which are dependent on an educated workforce in the terms of I4.0 and IoT.

11.5.2 Expectations of Importance for SMEs

The second part contains information about the importance of several aspects of doing business from the perspective of respondents, considering SMEs. We present Fig. 11.3 with questions Q18, Q21, Q22, and Q23. We observed a relatively high proportion of responses without any clear opinion regarding each question, while almost one-quarter of respondents consider investing in the training of workers as "*Not at*

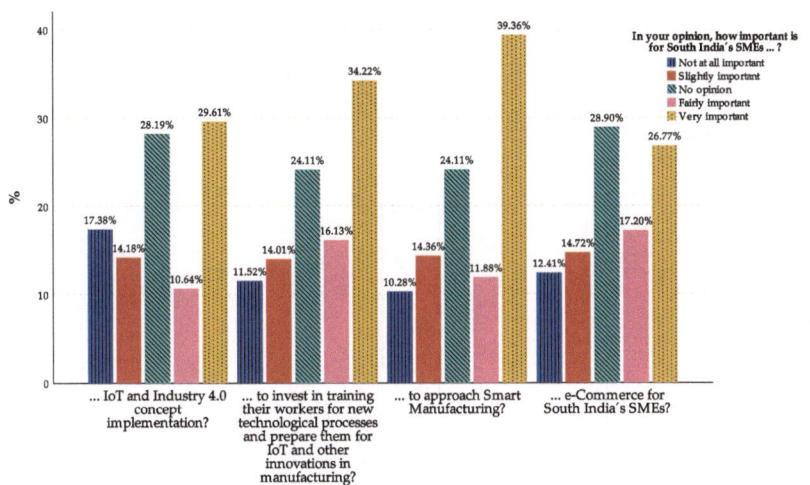

Fig. 11.3 The answers to the questions Q18, Q21–Q23 (Adapted from Safar et al. [2020])

all important", while almost 40% of respondents considered business transformation towards smart manufacturing as very important.

In Table 11.3 below, each column refers to the awareness question mentioned above, and each row refers to questions regarding IoT, I4.0, smart manufacturing, e-commerce, and investing in workers' education. We then expected the row questions (Q18, Q21–Q23) and column variables (Q10–Q17) would suggest some proportional relations. The fact is, that almost in all situations (where the questions Q18, Q21–Q23 were answered "*No opinion*", respectively, "*I do not know*"), the proportion of respondents, who answered the questions Q10–Q17 negatively is greater compared to the proportion of participants who responded positively to these questions. We argue that such statistical evidence of an inability to form an opinion or express expectation stems from an obvious lack of information. On the contrary, the proportion of respondents, who answered the questions Q10–Q17 positively is greater compared to the proportion of respondents with negative answers, if we are considering answers "*Very important*" or "*Fairly important*" regarding questions Q18 and Q21–Q23. A possible and logical explanation could

Table 11.3 Awareness vs importance of modern environment of doing business regarding SMEs (adapted from Safar et al. 2020)

	Q10	Q11	Q12	Q13	Q14	Q15	Q16	Q17
			Before now, have you ever heard about the …					Do you have any previous experience with IoT or I4.0 concept?
	Cloud Solutions so far?	Mass Customization so far?	Internet of Things so far?	Industry 4.0 concept so far? (so-called 4th Industrial revolution)	Smart Manufacturing so far?	Smart Cities so far?	5G so far?	
	No (A) / Yes (B)	No (A) / Yes (B)	No (A) / Yes (B)	No (A) / Yes (B)	No (A) / Yes (B)	No (A) / Yes (B)	No (A) / Yes (B)	No (A) / Yes (B)
X^2 p-value	125.083 0.000***	61.415 0.000***	42.886 0.000***	71.854 0.000***	59.146 0.000***	53.016 0.000***	50.778 0.000***	47.945 0.000***
Q18. In your opinion, is IoT and Industry 4.0 concept implementation important for South India's SMEs?								
Not at all important								
Slightly important								
No opinion	B	B	B	B	B	B	B	
Fairly important	A		A	A		A	A	
Very important	A		A	A	A	A	A	A
X^2 p-value	71.830 0.000***	16.292 0.003**	13.077 0.011*	28.767 0.000***	31.763 0.000***	70.570 0.000***	46.957 0.000***	24.682 0.000***

Before now, have you ever heard about the ...

	Q10	Q11	Q12	Q13	Q14	Q15	Q16	Q17
	Cloud Solutions so far?	Mass Customization so far?	Internet of Things so far?	Industry 4.0 concept so far? (so-called 4th Industrial revolution)	Smart Manufacturing so far?	Smart Cities so far?	5G so far?	Do you have any previous experience with IoT or I4.0 concept?
	No (A) / Yes (B)	No (A) / Yes (B)	No (A) / Yes (B)	No (A) / Yes (B)	No (A) / Yes (B)	No (A) / Yes (B)	No (A) / Yes (B)	No (A) / Yes (B)
Q21. In your opinion, how important is for South India's SMEs to invest in training their workers for new ...?								
Not at all important				B				
Slightly important	A						A	
No opinion	B	B	B	B	B	A	B	B
Fairly important	A				A	A		
Very important	A		A	A	A	A	A	
X^2 / p-value	36.366 0.000***	15.174 0.004**	17.773 0.001***	18.905 0.001***	26.573 0.000***	55.685 0.000***	21.711 0.000***	17.629 0.001***

(continued)

Table 11.3 (continued)

| | Before now, have you ever heard about the … | | | | | | | | | | | | | | Do you have any previous experience with IoT or I4.0 concept? | |
| | Q10 Cloud Solutions so far? | | Q11 Mass Customization so far? | | Q12 Internet of Things so far? | | Q13 Industry 4.0 concept so far? (so-called 4th Industrial revolution) | | Q14 Smart Manufacturing so far? | | Q15 Smart Cities so far? | | Q16 5G so far? | | Q17 | |
	No (A)	Yes (B)	No (A)	Yes (B)	No (A)	Yes (B)	No (A)	Yes (B)	No (A)	Yes (B)	No (A)	Yes (B)	No (A)	Yes (B)	No (A)	Yes (B)
Q22. In your opinion, how important is for SMEs in South India to approach Smart Manufacturing? — Not at all important																
Slightly important		A								A				A		
No opinion	B		B		B		B		B		B		B		B	
Fairly important						A				A		A				
Very important		A		A				A				A		A		A
X^2	104.602		40.564		45.056		61.571		78.993		52.549		63.905		46.675	
p-value	0.000***		0.000***		0.000***		0.000***		0.000***		0.000***		0.000***		0.000***	

	Q10		Q11		Q12		Q13		Q14		Q15		Q16		Q17	
	Before now, have you ever heard about the …														Do you have any previous experience with IoT or I4.0 concept?	
	Cloud Solutions so far?		Mass Customization so far?		Internet of Things so far?		Industry 4.0 concept so far? (so-called 4th Industrial revolution)		Smart Manufacturing so far?		Smart Cities so far?		5G so far?			
	No (A)	Yes (B)	No (A)	Yes (B)	No (A)	Yes (B)	No (A)	Yes (B)	No (A)	Yes (B)	No (A)	Yes (B)	No (A)	Yes (B)	No (A)	Yes (B)
Q23. In your opinion, how important is e-Commerce for South India's SMEs?																
Not at all important											B					
Slightly important				A		A										
No opinion		B		B		B				B		B		B		B
Fairly important		A						A		A		A				
Very important		A		A		A		A		A		A		A		A

Note "No opinion" refers to "I do not know". Results are based on two-sided tests. For each significant pair, the key of the category with the smaller column proportion appears in the category with the larger column proportion. X^2 refers to Chi-square statistic. p-value refers to the two-sided asymptotic significance of the chi-square statistic. Significance level for upper case letters (A, B, C): 0.05^*; 0.01^{**}; 0.001^{***}. Tests are adjusted for all pairwise comparisons within a row of each innermost subtable using the Bonferroni correction

be, that respondents realize the importance of successful transformation of the industries due to previous, at least basic, knowledge about questioned aspects. Special attention was given to the possible relations between answers "*Not at all important*" to questions Q18, Q21–23, and questions Q10–Q17 that were answered as "*No*". The proportion of respondents answering questions Q12, Q14, and Q15 as "*No*" that also answered Q22 (regarding approaching smart manufacturing from SMEs perspective) as "*Not at all important*" was significantly higher than the proportion of respondents answering Q12, Q14, and Q15 as "*Yes*". This brings us to the conclusion, that a better informational level should provide workers and customers with better tolerance towards emerging changes in business and network models throughout SMEs. In total, 10.3% of respondents answered Q22 as "*Not at all important*", 14.4% answered "*Slightly important*" and 24.1% answered, "*No opinion*" (or "*I do not know*"), which makes together 48.8%. We can observe a similar relationship between answers "*No*" to Q12 and Q14 and answers "*Not at all important*" to question Q18 addressing the importance of implementation of IoT and I4.0 from the SMEs perspective. Also, the relationship between respondents answering Q13 regarding I4.0 as "*No*" and Q21 addressing investing in training workers answering as "*Not at all important*" is alarming. This could be seen as a lack of information about inevitable changes in the coming years which translates into unclear visions concerning the crucial role of appropriate education and training for current and potential employees. This is backed up by the evidence in Coşkun et al. (2019), Benesova and Tupa (2017), and Schuster et al. (2016), through which the authors conclude that proper education and requalification is necessary especially regarding current dynamics throughout the industries.

Responds to the questions Q19, Q20, and Q25 are presented further in Fig. 11.4. The respondents were able to choose one of five options: "*No*"; "*Rather no*"; "*No opinion*" (referring also to "*I do not know*"); "*Rather yes*"; "*Yes*". In each of these questions, we can see a high proportion of respondents who replied all questions with the "*No opinion*" ("*I do not know*"). In Q19 it was more than 32% of respondents, in Q20 more than 33% and in Q25 more than 21%. This again points towards a lack of information resulting in the inability to form an opinion

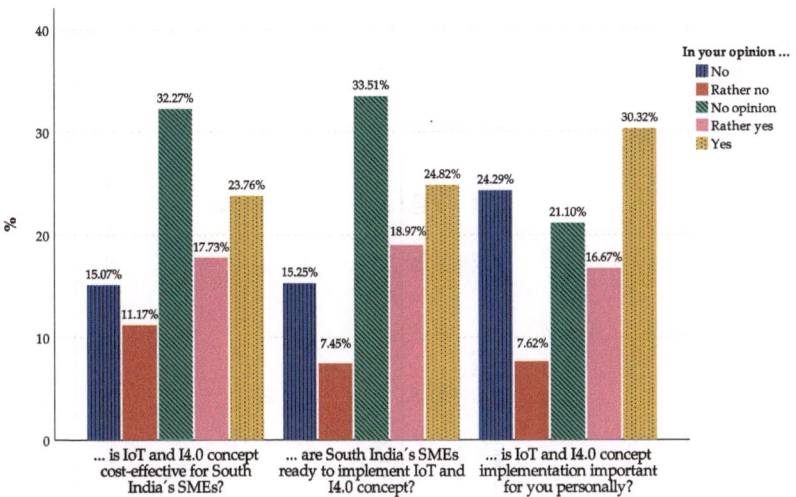

Fig. 11.4 The answers to the questions Q19, Q20–Q25 (Adapted from Safar et al. [2020])

regarding the issue. On the other hand, the answers "*No*" and "*Rather no*" opened further questions that we attempted to examine. Within the age group "*25 or below*", more than 16% of respondents think that IoT concept will be ineffective for South India's SMEs. Almost 34% of the respondents within this group reported "*No opinion*". Examining the performance of this group also on other questions, we observed nearly 17% of the respondents claiming the SMEs in South India are not ready to implement IoT and I4.0 concept, and as many as 36% of the respondents were unable to make a judgement. For more than 27% of the respondents aged "*25 or below*", the I4.0 concept is personally unimportant. More than 24% of respondents from the whole sample do not consider the IoT and I4.0 concept as important from a personal point of view.

These results are further examined against general awareness in Table 11.4. Similarly, we applied the column proportions test. For each combination of testing, we also point to the value of the asymptotic significance statistic (*p*-value), which in all cases is less than 0.05[*] level and thus variables are related. This table includes also a comparison

Table 11.4 Awareness vs questions Q19, Q20, Q25, and Q34

	Q10		Q11		Q12		Q13		Q14		Q15		Q16		Q17	
	Before now, have you ever heard about the …															
	Cloud Solutions so far?		Mass Customization so far?		Internet of Things so far?		Industry 4.0 concept so far? (so-called 4th Industrial revolution)		Smart Manufacturing so far?		Smart Cities so far?		5G so far?		Do you have any previous experience with IoT or I4.0 concept?	
	No (A)	Yes (B)	No (A)	Yes (B)	No (A)	Yes (B)	No (A)	Yes (B)	No (A)	Yes (B)	No (A)	Yes (B)	No (A)	Yes (B)	No (A)	Yes (B)
x^2 p-value	52.087 0.000***		55.046 0.000***		28.232 0.000***		41.303 0.000***		45.401 0.000***		17.383 0.002**		55.819 0.000***		72.579 0.000***	
Q19. In your opinion, is IoT and I4.0 concept cost-effective for South India's SMEs?																
No									B		B		B			
Rather no			A													
No opinion	B		B		B		B		B		B		B			
Rather yes	A		A		A		A		A		A		A		A	
Yes	A		A		A		A		A		A		A		A	
x^2 p-value	52.379 0.000***		16.323 0.003**		48.810 0.000***		37.977 0.000***		46.488 0.000***		35.185 0.000***		59.091 0.000***		31.707 0.000***	

	Q10	Q11	Q12	Q13	Q14	Q15	Q16	Q17
	Cloud Solutions so far?	Mass Customization so far?	Internet of Things so far?	Industry 4.0 concept so far? (so-called 4th Industrial revolution)	Smart Manufacturing so far?	Smart Cities so far?	5G so far?	Do you have any previous experience with IoT or I4.0 concept?
	No (A) / Yes (B)	No (A) / Yes (B)	No (A) / Yes (B)	No (A) / Yes (B)	No (A) / Yes (B)	No (A) / Yes (B)	No (A) / Yes (B)	No (A) / Yes (B)

Before now, have you ever heard about the …

Q20. In your opinion, are South India's SMEs ready to implement IoT and I4.0 concept?

	Q10	Q11	Q12	Q13	Q14	Q15	Q16	Q17
No				B				
Rather no								
No opinion	B	B	B	B	B	B	B	B
Rather yes		A	A		A	A	A	
Yes	A	A	A	A	A	A	A	A
X^2	52.355	26.645	44.466	45.476	44.201	73.733	15.657	26.082
p-value	0.000***	0.000***	0.000***	0.000***	0.000***	.000***	0.004**	0.000***

(continued)

Table 11.4 (continued)

	Q10	Q11	Q12	Q13	Q14	Q15	Q16	Q17
				Before now, have you ever heard about the …				
	Cloud Solutions so far?	Mass Customization so far?	Internet of Things so far?	Industry 4.0 concept so far? (so-called 4th Industrial revolution)	Smart Manufacturing so far?	Smart Cities so far?	5G so far?	Do you have any previous experience with IoT or I4.0 concept?
	No (A) / Yes (B)	No (A) / Yes (B)	No (A) / Yes (B)	No (A) / Yes (B)	No (A) / Yes (B)	No (A) / Yes (B)	No (A) / Yes (B)	No (A) / Yes (B)
Q25. In your opinion, is IoT and I4.0 concept implementation important for you personally? — No								
Rather no	A		B	B		B		
No opinion		B	B	B	B	B	B	B
Rather yes	A	A	A	A	A		A	

Before now, have you ever heard about the ...

	Q10		Q11		Q12		Q13		Q14		Q15		Q16		Q17	
	Cloud Solutions so far?		Mass Customization so far?		Internet of Things so far?		Industry 4.0 concept so far? (so-called 4th Industrial revolution)		Smart Manufacturing so far?		Smart Cities so far?		5G so far?		Do you have any previous experience with IoT or I4.0 concept?	
	No (A)	Yes (B)	No (A)	Yes (B)	No (A)	Yes (B)	No (A)	Yes (B)	No (A)	Yes (B)	No (A)	Yes (B)	No (A)	Yes (B)	No (A)	Yes (B)
Yes		A		A		A		A				A				A
χ^2	17.696		8.219		4.596		8.943		4.793		40.112		15.174		2.905	
p-value	0.000***		0.016*		0.100		0.011*		0.091		0.000***		0.001***		0.234	
Q34. In your opinion, do you see any Smart City in South India in next 10 years? — No							B				B		B			
No opinion				A							B					
Yes		A						A				A				

Adapted from Safar et al. (2020)

Note "No opinion" refers to "I do not know". Results are based on two-sided tests. For each significant pair, the key of the category with the smaller column proportion appears in the category with the larger column proportion. X^2 refers to Chi-square statistic. p-value refers to the two-sided asymptotic significance of the chi-square statistic. Significance level for upper case letters (A, B, C): 0.05*; 0.01**; 0.001*** a. Tests are adjusted for all pairwise comparisons within a row of each innermost subtable using the Bonferroni correction

of the answers to question Q34, which is focused on whether respondents expect any Smart City in South India within the next 10 years. In proportional testing, we found that in three cases (Q12, Q14, and Q17), the p-value is higher than the confidence level 0.05[*] (0.100; 0.091; 0.234). In such cases, we consider these variables as independent.

We highlight a high portion of "*No opinion*" ("*I do not know*") answers observed within the set of questions Q19, Q20 and Q25, related to answers "No" (questions Q10–Q17). A similar pattern was observed and described in Table 11.3. One concern could be potential complexness or difficulty of questions Q19 and Q20, therefore, forming a substantiated opinion could be harder for respondents. On the contrary, the inability to take a personal stance towards I4.0 or IoT we explain as lack of sufficient information, as described previously. Additionally, on a personal level, implementation of I4.0 and IoT (Q25) is not important for respondents answering Q12 and Q13. In total 24.3% (7.6%) of respondents answered "*No*" ("*Rather no*") to a question Q25.

Regarding question Q34, where respondents were asked whether they see any perspective of Smart City transformation within the region in the next ten years, almost 74% of the participants responded positively towards the idea of Smart City transformation. These responses seem to be rather overconfident, in contrast to other studies (Goswami 2016; Iyer 2018) examining the current state of the art in India. Putting this question in the context of questions Q10–Q17 results in similar outcomes as for previous sets of questions, where negative answers to Q10–Q17 are related to negative answers addressing Smart City. On the other hand, the proportion of respondents answering Q34 positively, that answered also Q10–Q17 positively, is higher on the statistically significant basis than the proportion of those who answered Q10–Q17 negatively.

11.5.3 Living Conditions Effects Expectations

Moving towards the next set of questions, Fig. 11.5 summarizes the performance of respondents regarding the questions Q27–Q33, and consequently their opinions on how the IoT and I4.0 will affect several aspects of their lives. The scale of responses used for this set of questions

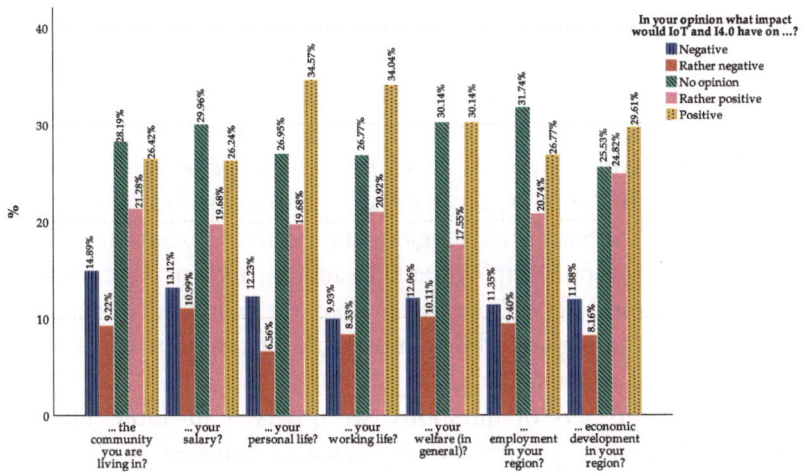

Fig. 11.5 Questions Q27–Q33

consists of five levels: "*Negative*"; "*Rather negative*"; "*No opinion*" ("*I do not know*"); "*Rather positive*", "*Positive*". Same as for previously examined sets of questions, a high frequency of "*No opinion*" answers can be observed. More than 25% of respondents cannot express opinions or expectations of how the 4th industrial revolution will affect the social and economic aspects of their lives in the region. We observed more than 10% of respondents expressing the opinion that IoT and I4.0 could have a negative impact on each questioned aspect of their life. Examining only the age group "*25 or below*", more than 16% of respondents think that IoT and I4.0 will impact their living environment negatively. To add on, almost 30% within the same age group answered: "*No opinion*" ("*I do not know*").

On the other hand, over 25% of the respondents within the age group "*25 or below*" expect a positive impact of IoT and I4.0 on their living environment. Questioning expected impact on salaries, more than 26% of respondents aged "*25 or below*" expect the I4.0 will impact their salary positively. Conversely, nearly 15% of respondents within the same age group express an opinion, that IoT and I4.0 will have a negative impact. Almost 28% of respondents aged "*25 or below*" picked "*No opinion*" ("*I do not know*"). In general, 35% of all respondents believe that IoT and I4.0 will have a positive impact on their personal life, while 27% of the total participants cannot express opinions or expectations of how the 4th industrial revolution will impact their personal life. Similarly, if the possible effect of IoT and I4.0 on the working environment is concerned, 27% of respondents answered: "*No opinion*" ("*I do not know*"). However, 11% of all respondents expect a negative impact of IoT and I4.0 on employment in the South India region, and nearly 32% have no opinion regarding the impact on employment. To add on, 12% of all respondents think that IoT and I4.0 will negatively affect the economic development of the South India region, while almost 26% of respondents answered: "*No opinion*" ("*I do not know*").

As for previous sets of questions, we observed in Table 11.5 and Table 11.6 the same pattern for dependences between answers "*No*" to general awareness questions and "*No opinion*" ("*I do not know*") to questions Q10–Q17. Similarly, for those respondents having prior information about IoT and I4.0 we observed rather positive answers to questions Q10–Q17. Conducting similar research in other regions could provide us with comparable data within the country. However, we find mainly company-oriented questionnaire-based researches also for other emerging countries, which is limiting our space for confrontation of obtained results.

Regarding questions addressing effects on community and salary, a higher proportion of respondents without prior information about IoT and I4.0 expressed "*No opinion*", and a higher proportion of respondents with previous knowledge about IoT and I4.0 expect a positive impact on the community they are living in and the salaries.

We observed that there is a higher proportion of respondents, who answered "*No*" to question related to IoT (Q12), expect a negative impact on their personal lives (Q29) than the proportion of respondents answering Q12 positively. We find it more interesting, that a rather negative impact on personal life (Q29) is expected from a higher proportion of respondents with prior knowledge of smart manufacturing (Q14). Similarly, a higher proportion of respondents already familiar with mass customization (Q11) expect a rather negative impact on their working lives. Rather a negative impact of IoT and I4.0 (Q31) expect a higher

Table 11.5 Awareness vs questions Q27, Q28, Q29, and Q30

	Q10	Q11	Q12	Q13	Q14	Q15	Q16	Q17
	Before now, have you ever heard about the ...							Do you have any previous experience with IoT or I4.0 concept?
	Cloud Solutions so far?	Mass Customization so far?	Internet of Things so far?	Industry 4.0 concept so far? (so-called 4th Industrial revolution)	Smart Manufacturing so far?	Smart Cities so far?	5G so far?	
	No (A) / Yes (B)	No (A) / Yes (B)	No (A) / Yes (B)	No (A) / Yes (B)	No (A) / Yes (B)	No (A) / Yes (B)	No (A) / Yes (B)	No (A) / Yes (B)
X^2	50.862	42.920	18.519	36.924	26.533	37.277	23.614	25.142
p-value	0.000***	0.000***	0.001***	0.000***	0.000***	0.000***	0.000***	0.000***
Q27. In your opinion, what impact would IoT and I4.0 concept have on the community you are living in?	Negative							

(continued)

Table 11.5 (continued)

	Q10		Q11		Q12		Q13		Q14		Q15		Q16		Q17	
	Before now, have you ever heard about the …														Do you have any previous experience with IoT or I4.0 concept?	
	Cloud Solutions so far?		Mass Customization so far?		Internet of Things so far?		Industry 4.0 concept so far? (so-called 4th Industrial revolution)		Smart Manufacturing so far?		Smart Cities so far?		5G so far?			
	No (A)	Yes (B)	No (A)	Yes (B)	No (A)	Yes (B)	No (A)	Yes (B)	No (A)	Yes (B)	No (A)	Yes (B)	No (A)	Yes (B)	No (A)	Yes (B)
Rather negative																
No opinion		B		B		B		B		B		B		B		B
Rather positive													A			
Positive		A		A		A		A		A		A		A		A
X^2	36.200		35.229		10.655		36.390		20.219		23.369		38.126		43.991	
p-value	0.000***		0.000***		0.031*		0.000***		0.000***		0.000***		0.000***		0.000***	

	Q10		Q11		Q12		Q13		Q14		Q15		Q16		Q17	
	Before now, have you ever heard about the ...														Do you have any previous experience with IoT or I4.0 concept?	
	Cloud Solutions so far?		Mass Customization so far?		Internet of Things so far?		Industry 4.0 concept so far? (so-called 4th Industrial revolution)		Smart Manufacturing so far?		Smart Cities so far?		5G so far?			
	No (A)	Yes (B)	No (A)	Yes (B)	No (A)	Yes (B)	No (A)	Yes (B)	No (A)	Yes (B)	No (A)	Yes (B)	No (A)	Yes (B)	No (A)	Yes (B)
Q28. In your opinion, what impact would IoT and I4.0 have on your salary?																
Negative																
Rather negative																
No opinion			B				B		B		B		B		B	
Rather positive			B						B						B	
Positive		A		A		A		A		A		A		A		A
X^2	23.992		35.533		28.588		28.330		23.843		31.829		18.948		26.895	
p-value	0.000***		0.000***		0.000***		0.000***		0.000***		0.000***		0.001***		0.000***	

(continued)

Table 11.5 (continued)

		Q10		Q11		Q12		Q13		Q14		Q15		Q16		Q17	
		Cloud Solutions so far?		Mass Customization so far?		Internet of Things so far?		Industry 4.0 concept so far? (so-called 4th Industrial revolution)		Smart Manufacturing so far?		Smart Cities so far?		5G so far?		Do you have any previous experience with IoT or I4.0 concept?	
		No (A)	Yes (B)	No (A)	Yes (B)	No (A)	Yes (B)	No (A)	Yes (B)	No (A)	Yes (B)	No (A)	Yes (B)	No (A)	Yes (B)	No (A)	Yes (B)
Q29. In your opinion, what impact would IoT and I4.0 have on your personal life?	Negative					B											
	Rather negative										A						
	No opinion			B		B		B		B		B		B		B	
	Rather positive		A				A		A		A		A		A		
	Positive				A				A				A				A

Before now, have you ever heard about the …

	Q10		Q11		Q12		Q13		Q14		Q15		Q16		Q17	
	Before now, have you ever heard about the …														Do you have any previous experience with IoT or I4.0 concept?	
	Cloud Solutions so far?		Mass Customization so far?		Internet of Things so far?		Industry 4.0 concept so far? (so-called 4th Industrial revolution)		Smart Manufacturing so far?		Smart Cities so far?		5G so far?			
	No (A)	Yes (B)	No (A)	Yes (B)	No (A)	Yes (B)	No (A)	Yes (B)	No (A)	Yes (B)	No (A)	Yes (B)	No (A)	Yes (B)	No (A)	Yes (B)
x^2 p-value	37.731 0.000***		23.305 0.000***		10.322 0.035*		20.012 0.000***		23.300 0.000***		46.071 0.000***		25.497 0.000***		13.098 0.011*	
Q30. In your opinion, what impact would IoT and I4.0 have on your working life? Negative																
Rather negative			A													
No opinion	B		B		B		B		B		B		B		B	

(continued)

Table 11.5 (continued)

Before now, have you ever heard about the …

	Q10 Cloud Solutions so far?		Q11 Mass Customization so far?		Q12 Internet of Things so far?		Q13 Industry 4.0 concept so far? (so-called 4th Industrial revolution)		Q14 Smart Manufacturing so far?		Q15 Smart Cities so far?		Q16 5G so far?		Q17 Do you have any previous experience with IoT or I4.0 concept?	
	No (A)	Yes (B)	No (A)	Yes (B)	No (A)	Yes (B)	No (A)	Yes (B)	No (A)	Yes (B)	No (A)	Yes (B)	No (A)	Yes (B)	No (A)	Yes (B)
Rather positive		A								A				A		
Positive		A		A				A		A		A		A		A

Note "No opinion" refers to "I do not know". Results are based on two-sided tests. For each significant pair, the key of the category with the smaller column proportion appears in the category with the larger column proportion. X^2 refers to Chi-square statistic. *p*-value refers to the two-sided asymptotic significance of the chi-square statistic. Significance level for upper case letters (A, B, C): 0.05*; 0.01**; 0.001*** a. Tests are adjusted for all pairwise comparisons within a row of each innermost subtable using the Bonferroni correction

Source Prepared by authors

Table 11.6 Awareness vs questions Q31, Q32, and Q33

	Q10		Q11		Q12		Q13		Q14		Q15		Q16		Q17	
	Before now, have you ever heard about the …														Do you have any previous experience with IoT or I4.0 concept?	
	Cloud Solutions so far?		Mass Customization so far?		Internet of Things so far?		Industry 4.0 concept so far? (so-called 4th Industrial revolution)		Smart Manufacturing so far?		Smart Cities so far?		5G so far?			
	No (A)	Yes (B)	No (A)	Yes (B)	No (A)	Yes (B)	No (A)	Yes (B)	No (A)	Yes (B)	No (A)	Yes (B)	No (A)	Yes (B)	No (A)	Yes (B)
χ^2 p-value	18.762 0.001***		37.818 0.000***		13.151 0.011*		17.364 0.002**		15.664 0.004**		50.336 0.000***		23.580 0.000***		29.795 0.000***	
Q31. In your opinion, what impact would IoT and I4.0 have in general on your welfare?																
Negative																
Rather negative			A					A		A						
No opinion		B	B		B		B				B		B		B	
Rather positive						A										
Positive	A		A				A					A		A		A
χ^2 p-value	25.236 0.000***		20.126 0.000***		20.314 0.000***		41.497 0.000***		24.680 0.000***		28.743 0.000***		29.366 0.000***		27.450 0.000***	

(continued)

Table 11.6 (continued)

Before now, have you ever heard about the ...

	Q10		Q11		Q12		Q13		Q14		Q15		Q16		Q17	
	Cloud Solutions so far?		Mass Customization so far?		Internet of Things so far?		Industry 4.0 concept so far? (so-called 4th Industrial revolution)		Smart Manufacturing so far?		Smart Cities so far?		5G so far?		Do you have any previous experience with IoT or I4.0 concept?	
	No (A)	Yes (B)	No (A)	Yes (B)	No (A)	Yes (B)	No (A)	Yes (B)	No (A)	Yes (B)	No (A)	Yes (B)	No (A)	Yes (B)	No (A)	Yes (B)
Q32. In your opinion, what impact would IoT and I4.0 concept have on employment in your region?																
Negative					B											
Rather negative										A				A		
No opinion			B				B		B		B		B		B	
Rather positive							B					A				
Positive		A		A		A		A		A		A		A		A
X^2	43.742		18.121		27.159		40.205		32.625		62.107		42.842		26.604	
p-value	0.000***		0.001***		0.000***		0.000***		0.000***		0.000***		0.000***		0.000***	

Q33. In your opinion, what impact would IoT and I4.0 concept have on economic development in your region?	Before now, have you ever heard about the ...														5G so far?		Do you have any previous experience with IoT or I4.0 concept?	
	Cloud Solutions so far? (Q10)		Mass Customization so far? (Q11)		Internet of Things so far? (Q12)		Industry 4.0 concept so far? (so-called 4th Industrial revolution) (Q13)		Smart Manufacturing so far? (Q14)		Smart Cities so far? (Q15)		(Q16)		(Q17)	
	No (A)	Yes (B)	No (A)	Yes (B)	No (A)	Yes (B)	No (A)	Yes (B)	No (A)	Yes (B)	No (A)	Yes (B)	No (A)	Yes (B)	No (A)	Yes (B)
Negative					B		B									
Rather negative																
No opinion			B		B		B		B		B		B		B	
Rather positive												A		A		A
Positive		A		A		A		A		A		A		A		A

Note "No opinion" refers to "I do not know". Results are based on two-sided tests. For each significant pair, the key of the category with the smaller column proportion appears in the category with the larger column proportion. X^2 refers to Chi-square statistic. p-value refers to the two-sided asymptotic significance of the chi-square statistic. Significance level for upper case letters (A, B, C): 0.05^*; 0.01^{**}; 0.001^{***} a. Tests are adjusted for all pairwise comparisons within a row of each innermost subtable using the Bonferroni correction
Source Prepared by authors

proportion of respondents with prior information about mass customization (Q11), I4.0 (Q13) and smart manufacturing (Q14). Regarding employment (Q32), the negative impact is expected from the higher proportion of respondents without prior information of IoT (Q12) than from those with such information. On the contrary, we cannot satisfactorily explain the negative expected impact on employment from respondents with prior information of "5G" (Q16). Addressing economic development in the South India region in general, a higher proportion of respondents without prior information about IoT (Q12) and I4.0 (Q13) expect negative impact compared to respondents having such previous information.

Thus, we find the implementation of any I4.0 related features and organization or network models challenging from a non-technical point of view, if respondents' expectations are negative towards key aspects of their lives. On the other hand, throughout each set of questions, we observe a significantly higher proportion of respondents expecting rather positive impacts within questioned aspects, that have previous knowledge or information about key terms addressed in the first part compared to those without such information. To add on, respondents with previous experience with IoT and I4.0 expressed positive expectations with a higher frequency compared to those without such experience. On the contrary, we consider some responses to questions addressing Smart cities in South India (Q34), or readiness of SMEs for implementing IoT and I4.0 (Q20), as rather over-confident, considering current state of art not only in South India (Iyer 2018). Such observations could however stem from possible drawbacks as sampling error. Thus, we recommend further examination of the mentioned region because of its huge demographic potential. Possible improvement of the conducted research should be expanding the sample or expert surveys with representatives of employee associations and other social parties. Because of scarcity in existing literature, we also find contribution in examining other regions and emerging countries from presented perspectives.

11.6 Conclusions

In this chapter, we attempted to examine general awareness, opinions, and attitudes of South India's inhabitants towards the Industry 4.0 and its features. By conducting a survey, we gathered unique answers containing crucial information about the current state of art regarding addressed issues the same as future expectations. Besides simple counts of answers, we provided also testing of interdependencies between general awareness questions and several sets of questions addressing various issues.

The main findings suggest that general awareness is quite low (almost 50% of respondents have no prior information of Industry 4.0), which consequently leads to the inability to form any opinion regarding effects of such new trends on working and personal life, same as on living and business environment. Respondents with insufficient knowledge of IoT and I4.0 then tend to answer negatively regarding questions about possible effects on their lives or salary, or they are unable to form an opinion regarding addressed aspects. On the contrary, respondents possessing prior information or knowledge regarding IoT and I4.0 expressed positive expectations in general.

Based on examined interdependences, we argue that proper education and relevant information dissemination is non-technical, however crucial, to form an applicable organization and network models as a part of the transformation process of the current environment in South India towards Industry 4.0.

Acknowledgements This project has received funding from the European Union's Horizon 2020 research and innovation programme under the Marie Skłodowska-Curie grant agreement No 734713.

The shortened version of this chapter with partial results was published as an article in the journal "Sustainability (MDPI)"—Special Issue: Industry 4.0 for SMEs—Smart Manufacturing and Logistics for SMEs.

The authors express gratitude for tremendous support and help to prof. Naavendra Krishnan, prof. Sudhakara Pandian and all SACS MAVMM staff during the authors' stay in Madurai, Tamil Nadu.

References

Agresti, A., and M. Kateri. 2011. Categorical data analysis. In *International encyclopedia of statistical science*, ed. Miodrag Lovric, 206–208. Springer, Berlin, and Heidelberg: Springer. https://doi.org/10.1007/978-3-642-048 98-2.

Armitage, P., G. Berry, and J.N.S. Matthews. 2008. *Statistical methods in medical research*. Chichester: Wiley. https://doi.org/10.1002/978047077 3666.

Benesova, A., and J. Tupa. 2017. Requirements for education and qualification of people in Industry 4.0. In *27th International Conference on Flexible Automation and Intelligent Manufacturing, Faim2017* 11: 2195–2202. https://doi.org/10.1016/j.promfg.2017.07.366.

Brock, D. 2001. The compact electronic product code—A 64-bit representation of the electronic product code. https://pdfs.semanticscholar.org/952e/23bfd97e803c27e7100e4464720c6917405c.pdf. Accessed on 8 January 2020.

Burgess, S. 2002. Information technology in small business: Issues and challenges. In *Managing information technology in small business: Challenges and solutions*. Hershey, PA: IGI Global. https://doi.org/10.4018/978-1-930708-35-8.ch001.

Calero Valdez, A., P. Brauner, A.K. Schaar, Andreas Holzinger, and M. Ziefle. 2015. Reducing complexity with simplicity—Usability methods for Industry 4.0. In *Proceedings 19th Triennial Congress of the IEA*. https://doi.org/10.13140/RG.2.1.4253.6809.

Census. 2011. (Final Data)—Demographic details, literate population (total, rural & urban). Planning Commission, Government of India. https://niti.gov.in/planningcommission.gov.in/docs/data/datatable/index.php?data=dat atab. Accessed on 12 January 2020.

Chandran, S., R. Poklemba, J. Sopko, and L. Safar. 2019. Organizational innovation and cost Reduction analysis of manufacturing process—Case study.

Management Systems in Production Engineering 27 (3): 183–188. https://doi. org/10.1515/mspe-2019-0029.

Coskun, S., Y. Kayikci, and E. Gencay. 2019. Adapting engineering education to Industry 4.0 Vision. *Technologies* 7 (1). https://doi.org/10.3390/techno logies7010010.

Cui, X. (2016) The Internet of Things. In *Ethical ripples of creativity and innovation*. London: Palgrave Macmillan. https://doi.org/10.1057/978113 7505545_7.

Dallasega, P., M. Woschank, H. Zsifkovits, K. Tippayawong, and CH. A. Brown. 2020. Requirement analysis for the design of smart logistics in SMEs. In *Industry 4.0 for SMEs: Challenges, opportunities and requirements*, ed. Dominik T. Matt, Vladimír Modrák, and Helmut Zsifkovits, 147–162. Cham: Springer. https://doi.org/10.1007/978-3-030-25425-4_5.

Dallasega, P., M. Woschank, S. Ramingwong, K. Tippayawong, and N Chonsawat. 2019. Field study to identify requirements for smart logistics of European, US and Asian SMEs. In *Proceedings of the International Conference on Industrial Engineering and Operations Management*, 844–855. Bangkok, Thailand. http://www.ieomsociety.org/ieom2019/papers/241.pdf. Accessed on January 2020.

Directorate of Intelligence. 2019. CIA—World Factbook. Available at https:// www.cia.gov/library/publications/the-world-factbook/index.html. Accessed 12 January 2020.

Eberhard, B., M. Podio, A.P. Alonso, E. Radovica, L. Avotina, L. Peiseniece, M.C. Sendon, A.G. Lozano, and J. Solé-Pla. 2017. Smart work: The transformation of the labour market due to the fourth industrial revolution (I4.0). *International Journal of Business and Economic Sciences Applied Research (IJBESAR)* 10 (3): 47–66. https://www.econstor. eu/bitstream/10419/185671/1/v10-i3-p47-66-smart-work.pdf. Accessed on December 2019.

Goswami, H. 2016. Opportunities and challenges of digital India programme. *International Education and Research Journal* 2 (11): 78–79. http://ierj.in/ journal/index.php/ierj/article/view/541. Accessed on December 2019.

Harbor Research. 2011. Machine-to-machine (M2M) and smart systems market opportunity 2010–2014. Harbor Research, Inc. https://www.scribd. com/document/35372236/Harbor-Research-Machine-to-Machine-M2M-amp-Smart-Systems-Market-Forecast. Accessed on December 2019.

Hofmann, E., and M. Rusch. 2017. Industry 4.0 and the current status as well as future prospects on logistics. *Computers in Industry* 89: 23–34. https:// doi.org/10.1016/j.compind.2017.04.002.

India Manufacturing Report by Indian Brand Equity Forum. https://www. ibef.org/industry/manufacturing-sector-india.aspx. Accessed on 12 January 2020.

Ingaldi, M., and R. Ulewicz. 2020. Problems with the implementation of Industry 4.0 in enterprises from the SME sector. *Sustainability* 12 (1). https://doi.org/10.3390/su12010217.

ITU Internet Reports. 2005. *The Internet of Things*. Geneva: International Telecommunication Union (ITU). https://www.itu.int/osg/spu/public ations/internetofthings/. Accessed on December 2019.

Iyer, A. 2018. Moving from Industry 2.0 to Industry 4.0: A case study from India on leapfrogging in smart manufacturing. *Procedia Manufacturing* 21: 663–670. https://doi.org/10.1016/j.promfg.2018.02.169.

Kagermann, H. (2015). Change through digitization—Value creation in the age of Industry 4.0. In *Management of permanent change*, ed. H. Albach, H. Meffert, A. Pinkwart, and R. Reichwald. Wiesbaden: Springer Gabler. https://doi.org/10.1007/978-3-658-05014-6_2.

Kiel, D., C. Arnold, and K.I. Voigt. 2017. The influence of the Industrial Internet of Things on business models of established manufacturing companies—A business level perspective. *Technovation* 68: 4–19. https://doi.org/10.1016/j.technovation.2017.09.003.

Kovacs, O. 2018. The dark corners of Industry 4.0-Grounding economic governance 2.0. *Technology in Society* 55: 140–145. https://doi.org/10.1016/j.techsoc.2018.07.009.

Lasi, H., H. G. Kemper, P. Fettke, T. Feld, and M. Hoffmann. 2014. Industry 4.0. *Business & Information Systems Engineering* 6 (4): 239–242. https://doi.org/10.1007/s12599-014-0334-4.

Matt, D.T., and E. Rauch. 2020. SME 4.0: The role of small- and medium-sized enterprises in the digital transformation. In *Industry 4.0 for SMEs: Challenges, opportunities and requirements*, ed. Dominik T. Matt, Vladimír Modrák, and Helmut Zsifkovits, 3–36. Cham: Palgrave Macmillan. https://doi.org/10.1007/978-3-030-25425-4_1.

Modrak, V., and S. Bednar. 2015. Using axiomatic design and entropy to measure Complexity in mass customization. In *9th International Conference on Axiomatic Design (Icad 2015)* 34: 87–92. https://doi.org/10.1016/j.pro cir.2015.07.013.

Motyl, B., G. Baronio, S. Uberti, D. Speranza, and S. Filippi. 2017. How will change the future engineers' skills in the Industry 4.0 framework? A questionnaire survey. In *27th International Conference on Flexible Automation and Intelligent Manufacturing, Faim2017* 11:1501–1509. https://doi.org/10.1016/j.promfg.2017.07.282.

Posada, J., C. Toro, I. Barandiaran, D. Oyarzun, D. Stricker, R. de Amicis, E. B. Pinto, P. Eisert, J. Dollner, and I. Vallarino. 2015. Visual computing as a key enabling technology for Industrie 4.0 and industrial internet. *IEEE Computer Graphics and Applications* 35 (2): 26–40. https://doi.org/10.1109/Mcg.2015.45.

Ramingwong, S., and W. Manopiniwes. 2019. Supportment for organization and management competences of ASEAN community and European Union toward Industry 4.0. *International Journal of Advanced and Applied Sciences* 6 (3): 96–101. https://doi.org/10.21833/ijaas.2019.03.014.

Ramsauer, C. 2013. Industrie 4.0–Die Produktion der Zukunft. *WINGbusiness* 3: 6–12. http://lampx.tugraz.at/~i371/industrie40/7521_0_DieProduktionderZukunft_ChristianRamsauer.pdf. Accessed on December 2019.

Safar, L., J. Sopko, S. Bednar, and R. Poklemba. 2018. Concept of SME business model for Industry 4.0 environment. *Tem Journal-Technology Education Management Informatics* 7 (3): 626–637. https://doi.org/10.18421/Tem 73-20.

Safar, L., J. Sopko, D. Dancakova, and M. Woschank. 2020. Industry 4.0— Awareness in South India. *Sustainability* 12 (8): 1–18. https://doi.org/10.3390/su12083207.

Schröder, C., S. Schlepphorst, and R. Kay. 2015. *Bedeutung der Digitalisierung im Mittelstand* (No. 244). IfM-Materialien, Institut für Mittelstandsforschung (IfM) Bonn. https://en.ifm-bonn.org/uploads/tx_ifmstudies/IfM-Materialien-244_2015.pdf. Accessed on December 2019.

Schuman, H., S. Presser, and J. Ludwig. 1981. Context effects on survey responses to questions about abortion. *Public Opinion Quarterly* 45 (2): 216–223. https://doi.org/10.1086/268652.

Schuster, K., K. Groß, R. Vossen, A. Richert, and S. Jeschke. 2016. Preparing for Industry 4.0—Collaborative virtual learning environments in engineering education. In *Automation, communication and cybernetics in science and engineering 2015/2016*, ed. Sabina Jeschke, Ingrid Isenhardt, Frank Hees, and Klaus Henning, 417–427. Cham: Springer. https://doi.org/10.1007/978-3-319-42620-4_33.

Schwarz, N., and H.-J. Hippler. 1987. What response scales may tell your respondents: Informative functions of response alternatives. In *Social information processing and survey methodology*, ed. Hans-J. Hippler, Norbert

Schwarz, and Seymour Sudman, 163–178. New York, NY: Springer. https://doi.org/10.1007/978-1-4612-4798-2_9.

Sedgwick, P. 2012. Statistical question multiple significance tests: The Bonferroni correction. *British Medical Journal* 344. https://doi.org/10.1136/bmj.e509.

Skill India. 2020. Portal. https://www.skillindia.gov.in. Accessed on 12 January 2020.

Slusarczyk, B. 2018. Industry 4.0—Are we ready? *Polish Journal of Management Studies* 17 (1): 232–248. https://doi.org/10.17512/pjms.2018.17.1.19.

Soltes, V., and B. Gavurova. 2014. Innovation policy as the main accelerator of increasing the competitiveness of small and medium-sized enterprises in Slovakia. *Emerging Markets Queries in Finance and Business (Emq 2013)* 15: 1478–1485. https://doi.org/10.1016/S2212-5671(14)00614-5.

Strandhagen, J. W., E. Alfnes, J. O. Strandhagen, and L. R. Vallandingham. 2017. The fit of Industry 4.0 applications in manufacturing logistics: A multiple case study. *Advances in Manufacturing* 5 (4): 344–358. https://doi.org/10.1007/s40436-017-0200-y.

Sundmaeker, H., P. Guillemin, P. Friess, and S. Woelfflé. 2010. Vision and challenges for realising the Internet of Things. *Cluster of European Research Projects on the Internet of Things, European Commission* 3 (3): 34–36. https://doi.org/10.2759/26127.

The Countries with the Highest Density of Robot Workers. 2019. Available online: https://www.statista.com/chart/13645/the-countries-with-the-highest-density-of-robot-workers/. Accessed on 10 January 2019.

Thestrup, J., T.F. Sorensen, and M. De Bona. 2006. Using conceptual modeling and value analysis to identify sustainable m>business models in industrial services. In *2006 International Conference on Mobile Business*. https://doi.org/10.1109/ICMB.2006.49.

Thoben, K.-D., S. Wiesner, and T. Wuest. 2017. "Industrie 4.0" and smart manufacturing—A review of research issues and application examples. *International Journal of Automation Technology* 11 (1): 4–16. https://doi.org/10.20965/ijat.2017.p0004.

Wolter, M.I., A. Mönnig, M. Hummel, C. Schneemann, E. Weber, G. Zika, R. Helmrich, T. Maier, and C. Neuber-Pohl. 2015. *Industry 4.0 and the consequences for labour market and economy: scenario calculations in line with the BIBB-IAB qualifications and occupational field projections (Industrie 4.0 und die Folgen für Arbeitsmarkt und Wirtschaft: Szenario-Rechnungen im Rahmen der BIBB-IAB-Qualifikations-und Berufsfeldprojektionen)* (No. 201508_en). Institut für Arbeitsmarkt-und Berufsforschung (IAB), Nürnberg [Institute for Employment Research, Nuremberg, Germany]. http://doku.iab.de/for schungsbericht/2015/fb0815_en.pdf. Accessed on December 2019.

World Bank. 2018. Global Economic Prospects, June 2018: The turning of the tide? *Global Economic Prospects*, June. https://doi.org/10.1596/978-1-4648-1257-6.

Woschank, M., E. Rauch, and H. Zsifkovits. 2020. A review of further directions for artificial intelligence, machine learning, and deep learning in smart logistics. *Sustainability* 12 (9): 1–23. https://doi.org/10.3390/su12093760.

12

Implementation Strategies for SME 4.0: Insights on Thailand

Apichat Sopadang, Sakgasem Ramingwong,
Tanyanuparb Anantana, and Krisana Tamvimol

12.1 Introduction

Industry 4.0 or the 4th Industrial Revolution is referred as the advanced manufacturing environment toward the smart technology such as Cyber-Physical Systems (CPS), Internet of Things (IoT), Information and Communications Technology (ICT), Enterprise Architecture (EA), and

A. Sopadang · S. Ramingwong (✉)
Center of Excellence in Logistics and Supply Chain Management, Chiang Mai University, 239 Huay Kaew Road Rd., Muang District, Chiang Mai 50200, Thailand
e-mail: sakgasem.ramingwong@cmu.ac.th

A. Sopadang
e-mail: apichat.s@cmu.ac.th

A. Sopadang · S. Ramingwong · T. Anantana
Department of Industrial Engineering, Faculty of Engineering, Chiang Mai University, 239 Huay Kaew Road Rd., Muang District, Chiang Mai 50200, Thailand
e-mail: tanyanuparb@step.cmu.ac.th

© The Author(s) 2021
D. T. Matt et al. (eds.), *Implementing Industry 4.0 in SMEs*,
https://doi.org/10.1007/978-3-030-70516-9_12

Enterprise Integration (EI) (Lu 2017; Rüßmann et al. 2015). Industry 4.0 employs modern "push" technologies in "pull" applications, i.e., Internet-based and Internet of Services, which is mostly influenced by the computational power, cloud computing, and services. Industry 4.0 allows the company to foresee future products and to appropriately respond to the variety and complexity at low cost and low impact (Ganzarain and Errasti 2016). Production and logistics systems can be decentralized and integrated horizontally and vertically with the use of interconnected sensors, actors, and autonomous systems (Gilchrist 2016; Jazdi 2014). However, the integration of physical and software systems and modeling the intelligence system can be highly expensive and complicated due to the complex, dynamic, and integrated information systems (Lasi et al. 2014; Rauch et al. 2020).

The extension of Industry 4.0 goes from Smart Manufacturing to Smart Logistics, including organization and management (SME4.0 2020). The constraints comprise of SME focus, mass customization, and X-to-order environment (Mihiotis 2014), economic, ecological, and social sustainability (Brozzi et al. 2020; Gabriel and Pessl 2016; Prause 2015), lean philosophy, changeability, and flexibility. The enablers include IoT, Big Data, CPS (Lee et al. 2015), smart sensors, digitalization, and automation.

Industry 4.0 has become the new normal for large enterprises where organization and business models can be redesigned and investment can be made viably (Safar et al. 2018). However, implementing Industry 4.0 to SMEs is yet challenging due to their limited resources, knowledge, and investment (Bär et al. 2018; Ganzarain and Errasti 2016; Ramingwong et al. 2019; Ramingwong and Manopiniwes 2019). In such a quest, SME

T. Anantana
Science and Technology Park, Chiang Mai University, 155 Moo 2 Mae Hia, Muang District, Chiang Mai 50100, Thailand

K. Tamvimol
Wangree Agriculture Technology Institute, Wangree Health Factory Co., Ltd., 143 Moo 12 Khaopra, Muang District, Nakornnayok 26000, Thailand
e-mail: krisana@wangreefresh.com

4.0 has been simply defined as the implementation of Industry 4.0 to SME (Matt and Rauch 2020; Sopadang et al. 2020).

Therefore, it is the aim of this study to investigate how SMEs can become SME 4.0. The development and implementation strategies for SME 4.0 are of interest. The study explores and discusses the success of a Thai start-up SME as case study by aligning with the developed meta-model of implementation strategies for SME 4.0.

12.2 Implementation Strategies for SMEs

It is important that SMEs must develop and implement Industry 4.0 strategies to become SME 4.0 according to their strength, resources, and investment. To date, there are extensive works regarding the SME 4.0 implementation including Industry 4.0 maturity models for SMEs (Chonsawat and Sopadang 2019; Ganzarain and Errasti 2016; Rauch 2020), the procedure of manufacturing resources migration toward Industry 4.0 (Pérez et al. 2018), smart SME 4.0 implementation toolkits (Sopadang et al. 2020) as well as requirement mapping and roadmaps for SME 4.0 (Modrak et al. 2019). With different views of the cause, the proposed model or methodology in this literature is diversified. However, most of the focus is on the organization itself. This chapter further investigates the external bodies, by which in this case the collaboration of universities and tech-development agencies are enveloped. The collaboration is assumed as the triple-helix model of innovation (Galvao et al. 2019; Leydesdorff 2010; Nakwa and Zawdie 2016).

Figure 12.1 illustrates the meta-model of implementation strategies for SME 4.0 developed and used in this case study. The model is triple helix where the organization works with universities and tech-development agencies on the inside-out and outside-in approaches.

To develop suitable strategies for SME implementation, the development plan and analysis phases are required. The development plan is to develop technology blueprint, which is the result of industrial research and capacity development. Whereas the analysis phase refers to gap analysis from business trend analysis and business foresight. The following sections discuss the three phases of the model, i.e., Analysis, Development Plan, and Implementation Strategies.

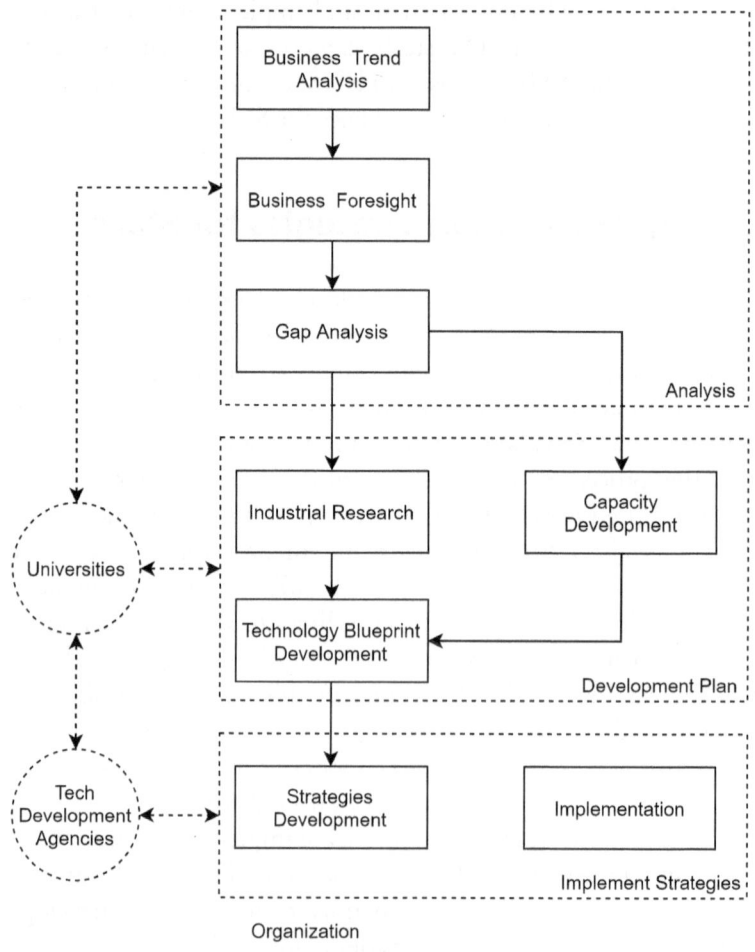

Fig. 12.1 Implementation strategies for SME 4.0 meta-model

12.2.1 Phase 1—Analysis

The first analysis phase comprises three steps, i.e., (1) Business Trend Analysis, (2) Business Foresight, and (3) Gap Analysis.

"*Business Trend Analysis*" is the first step of the meta-model. It is the process of comparing business over time to identify any consistent trends. The developed strategies must correspond with these trends and the business goals. The trend analysis comprises of three sub-steps, i.e., review of KPIs, trend analysis, and business benchmarking (see Fig. 12.2). Firstly, the review of KPIs must include financial and non-financial KPIs (Tippayawong et al. 2019) both in the well-known Balance Scorecard (BSC) approach (Kaplan and Norton 1998) and sustainable concepts (Gabriel and Pessl 2016; Prause 2015; Stubbs and Cocklin 2008). Then, the trend analysis can be Time Series Analysis or Multivariate Analysis. This is to assist Business Decision Making. Multiple Criteria Decision Making (MCDM) is often used for such applications. Finally, it is necessary to do business benchmarking.

Once the business trend is analyzed, it is necessary to conduct "*Business Foresight*" to conceptualize practices, capabilities, and ability of firms. The foresight enables firms to detect changes, understand the

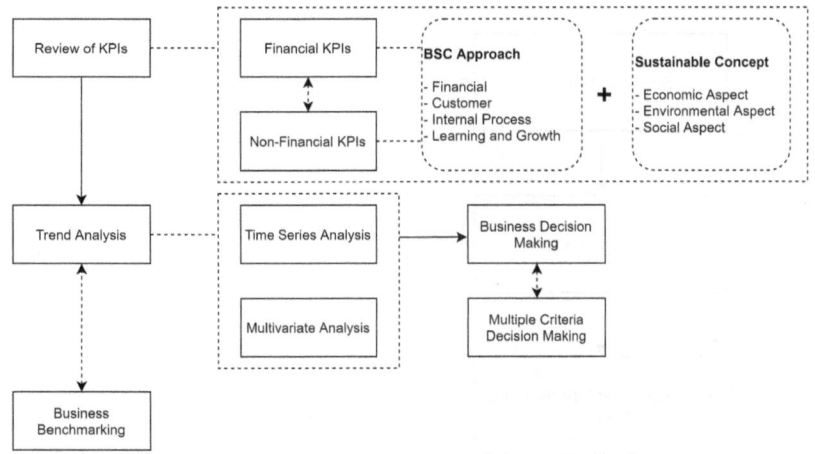

Fig. 12.2 Business trend analysis concept

consequences, and address appropriate responding actions (Rohrbeck 2010; Rohrbeck et al. 2015). The business foresight comprises six sub-steps, i.e., framework development, environment analysis, scanning signal analysis, scenario building, scenario analysis, and strategic development and planning (see Fig. 12.3).

The final step of Phase 1 is the "*Gap Analysis.*" The gap analysis involves the comparison of the actual performance with the desired performance or the foresighted goals. Three gaps must be identified before setting up the development plan in Phase 2. The gaps are from product/service delivery, perceived service, expected product/service, and

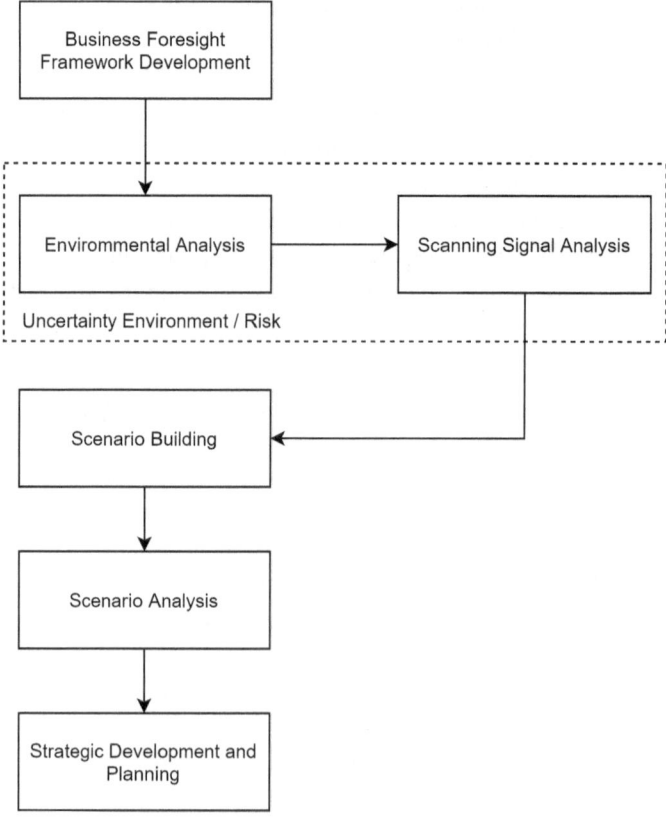

Fig. 12.3 Business foresight concept

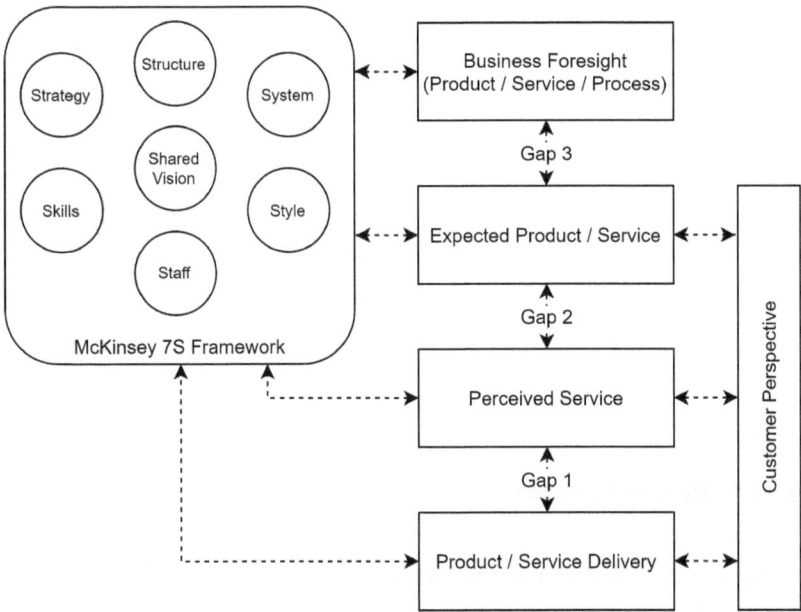

Fig. 12.4 Gap analysis concept

business foresight (see Fig. 12.4). Here, they shall be addressed in corresponding to the McKinsey 7S Framework (Hanafizadeh and Ravasan 2011; Singh 2013) and the customer perspective.

12.2.2 Phase 2—Development Plan

Phase 2 (Development Plan) comprises of three steps, i.e., Industrial Research, Capacity Development, and Technology Blueprint Development. They are as follows.

According to Commission Regulation (EU) No 651/2014 of 17 June 2014, Industrial Research is defined as "the planned research or critical investigation aimed at the acquisition of new knowledge and skills for developing new products, processes or services or for bringing about a significant improvement in existing products, processes or services. It comprises the creation of components parts of complex systems, and may

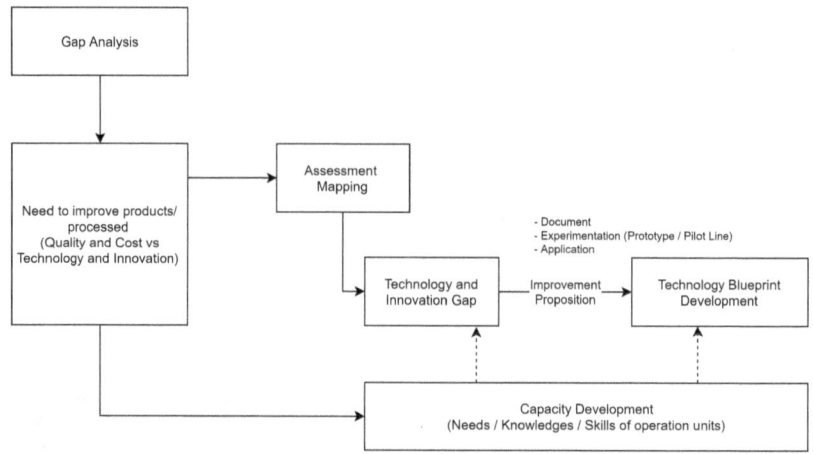

Fig. 12.5 Development plan concept

include the construction of prototypes in a laboratory environment or in an environment with simulated interfaces to existing systems as well as of pilot lines, when necessary for the industrial research and notably for generic technology validation" (add reference).

Therefore, the development plan phase requires input from the gap analysis in the previous step to identify the need to improve products/processed in terms of quality and cost with technology and innovation. Then, the assessment mapping is to yield technology and innovation gap. This must be aligned with the capacity development, i.e., needs, knowledge, and skills of operation units. This is to address the improvement proposition and thus to develop a technology blueprint (see Fig. 12.5).

While the first two phases involve universities and think tank, Phase 3 is mostly supported by tech-development agencies.

12.2.3 Phase 3—Implementation Strategies

This phase involves the development of strategies and implementation. Where the strategic development and planning concerns business strategy, business process, and business organization/function, these yield

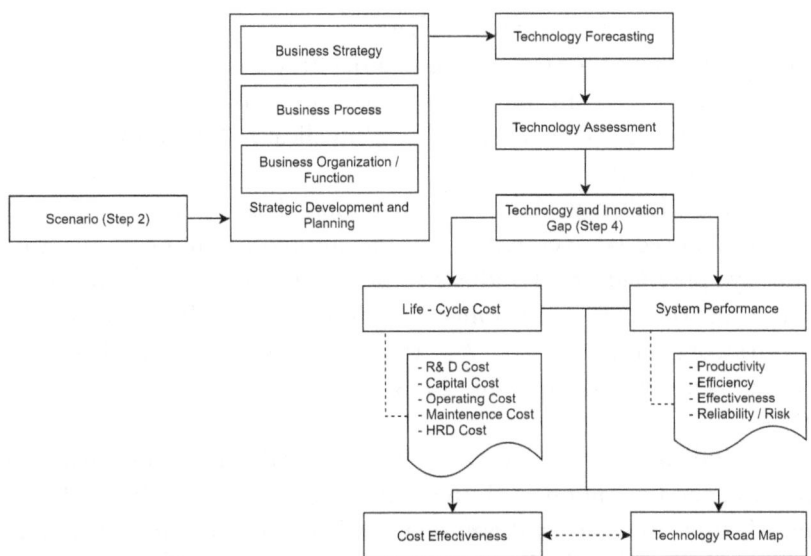

Fig. 12.6 Implementation strategies concept

the technology forecasting, assessment, and hence technology and inno-vation gap. The consideration is then on life-cycle cost and system performance. Here, the life-cycle cost can be research and development (R&D) cost, capital cost, operation cost, maintenance cost, or human resource development cost. The system performance can be produc-tivity, efficiency, effectiveness, and reliability/ risk. Finally, the technology roadmap can be reached with cost-effectiveness (see Fig. 12.6).

12.3 Industry 4.0 Implementation in Thailand

Thai industry has started to incorporate Industry 4.0 to its production system (Hotrawaisaya et al. 2019). Thailand enjoyed of several Industry 4.0 supporting policies from the Thai government, especially industrial transformation, ICT adoption, re-skilling, and e-government (Kohpai-boon 2020; Chinachoti 2018). However, most technological investment, expected at 1 billion USD in 2020, is on 10 targeted industries according to Thailand 4.0 development scheme, i.e., next-generation automotive, intelligent electronics, advanced agriculture and biotechnology, food processing and tourism, digital, robotics and automation, aviation and logistics, biofuels and biochemicals and medical hub (Kumpirarusk and Rohitratana 2018).

To date, Industry 4.0 implementation has become evident in several advanced large enterprises in the automotive industry, electronic industry, pharmaceutical industry, smart farming (Chetthamrongchai and Jermsittiparsert 2020; Jones and Pimdee 2017; Phungphol 2018; Tippayawong et al. 2016). Yet, the campaign is highly challenging (Korkueasuebsai and Pornsing 2018; Laosiritaworn and Chattinnawat 2019).

Thailand defines SMEs as companies with no more than 200 employees and 2 million THB in assets. According to the Office of Small and Medium Enterprises Promotion (OSMEP) of Thailand, there are 3 million SMEs. This SME sector contributes up to 43% of Thailand's GDP in 2019. SMEs make up 99.6% of total enterprises in the country, creating more than 10.5 million jobs.

However, according to Cisco APAC SMB digital maturity index, Thai SMEs digital readiness is low, ranked 11th out of 14 countries in Asia-Pacific. Thai SMEs are identified at the stage of digitally indifferent, i.e., reactive to market changes, digital efforts do not exist, no automation (the majority of processes are manual), digital technologies are not used, and not using cloud resources. Besides, Thai SMEs are lack of customer data, lack of digital skills and talent, and lack of a digital mindset.

Investment in information technology or even automation systems alone can be difficult. Financial risk along with managerial risk can be absolute. Therefore, the implementation of SME 4.0 for Thai SMEs has been scarce (Dallasega et al. 2019; Munkongsujarit 2016).

The Thai government has foreseen the opportunity and therefore assigned OSMEP to support SMEs by providing an online platform for B2B sales, training updates, and activities that will boost up the combined revenue of Thai SMEs to 2.3 trillion THB within the next five years. Moreover, there are financial supports and promotion from the board of Investment of Thailand (BOI), National Board of SMEs Promotion, and also financial providers, especially, Small and Medium

Enterprise Development Bank of Thailand (SME Bank). This expects to stimulate SMEs to become "smarter" and supportive to the SME 4.0 journey.

12.4 Case Study—Thai Agritech SME

The case study SME is a plant factory start-up in Thailand. The company is named "Wangree Health Factory Co., Ltd.". The company is inspired and initiated under an innovation ecosystem of Thailand, by which the private sector has been groomed by the university and government agencies. The project was called "STI Policy Management Program (PMP)," which is a series of training, networking, and industrial visit, hosted by the Ministry of Science and Technology in 2015. The case study company was established as a result of the project, leading to the business model of Thailand's first Agritech SME.

The company firstly aims at advancing the agriculture industry, being a Tech start-up. The idea is to upgrade the traditional agriculture industry, which is low-value-added to advanced innovative industry. Today, the agriculture industry contributes only 10% of Thailand's GDP, despite involving with nearly half of the population of Thailand from downstream to upstream. Most players in the industry are SMEs and low-tech. Cultivation and production are mostly traditional and labor-intensive. Productivity is low. It is the goal of the established company to overcome these hurdles using the Industry 4.0 concept.

The first establishment of the company is a plant factory testing facility in Chiang Mai. The pilot plant was supported by Science and Technology Park of Chiang Mai University and funded by the National Science Technology and Innovation Policy Office (STI) and National Science and Technology Development Agency (NSTDA), Ministry of Science and Technology of Thailand.

12.4.1 Business Idea of Agritech

Following the business idea development, the marketing survey is conducted first to investigate the expected demand, if it is aligned with the possible planned supply. Demand-side survey suggests that there is a considerable volume of the segmented customers who need clean, fresh, high-quality vegetables at a low price and at their convenience (Sukkarat and Athinuwat 2020). Expected sales volume, profit, and Return on Investment (ROI) are feasible if mass-produced in the economy of scale. Moreover, the business idea also addresses the sustainable, non-financial key performance indicators (KPI) as the food safety and quality as well as the environmental impact of the vegetable to be produced.

Looking at today agriculture industry, organic and hydroponic farms are globally flourishing and technologically saturated. Both agricultural techniques can address the needs of the customer with further benefit to the environmental and social perspectives. The cost of organic and hydroponic are comparable. However, organic is seasonal and the productivity is relatively low. It is pesticide-free but the size of the vegetable is normally smaller.

Here, the production resources are considered as the benchmarking of these two alternative technologies. To produce a vegetable of 5 tons/month, an organic farm may require 10 rais (1.6 hectares) of land, a 3 million liter of water, and a labor of 20. Maximum production is normally 6 crops per year. On the other hand, a hydroponic farm may require 6 rais (0.96 hectares) of land, 1.8 million liters of water, and a labor of 20. Possible productivity can reach up to 8–12 crops per year. Therefore, the Business Trend Analysis suggests that the company shall focus on the more superlative hydroponic option (see Fig. 12.7).

According to the study by Kasikorn Research Center, the domestic organic market size has been expanded to 2700–2900 million THB in 2019, with an annual growth of 10%. Demand has been driven mostly by the millennials and aging society. These customer segments account for almost 40% of the Thai population.

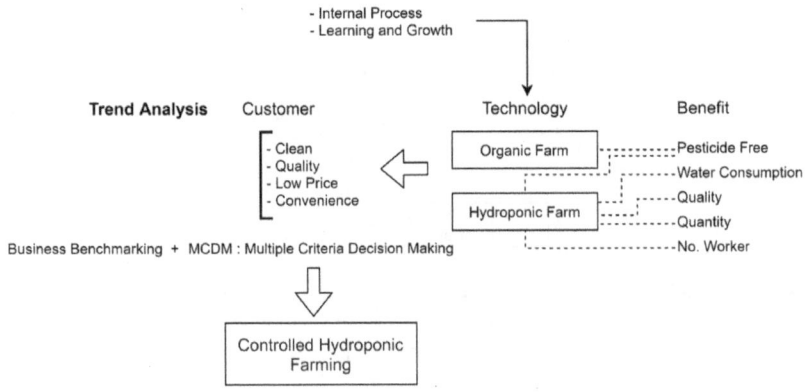

Fig. 12.7 Trend analysis of Agritech

12.4.2 Plant Factory—The Foresight of Agritech Business

The Business Foresight (see Fig. 12.8) suggests the investigation of the environmental analysis as there can be uncertainty and risk in the business foresight framework. In this case, the demand must meet the mass production capacity. The competitor and price also affect the competitiveness of the proposed business model. Moreover, there are risks, i.e., technological risk and operational risks. After scanning signal analysis, traditional Hydroponic planting techniques can be found limited in terms of productivity. A new scenario arises if the plant factory is more feasible.

The plant factory is categorized as one of the advanced agriculture systems which have been in the spotlight as a global prospect, owing to Industry 4.0 (Antonopoulos et al. 2019; Griffin et al. 2018; Katyal and Pandian 2020). This smart/precision farming allows fully autonomous planning, plowing, seed mapping, seeding, reseeding, and monitoring, using farming robots, sensors, IoT, and artificial intelligence (AI) (Zanwar and Kokate 2012). The technology is demanding due to the growing number of the world population but the shrinkage of arable land (Benke and Tomkins 2017). Not to mention the independence of

Environmental Analysis

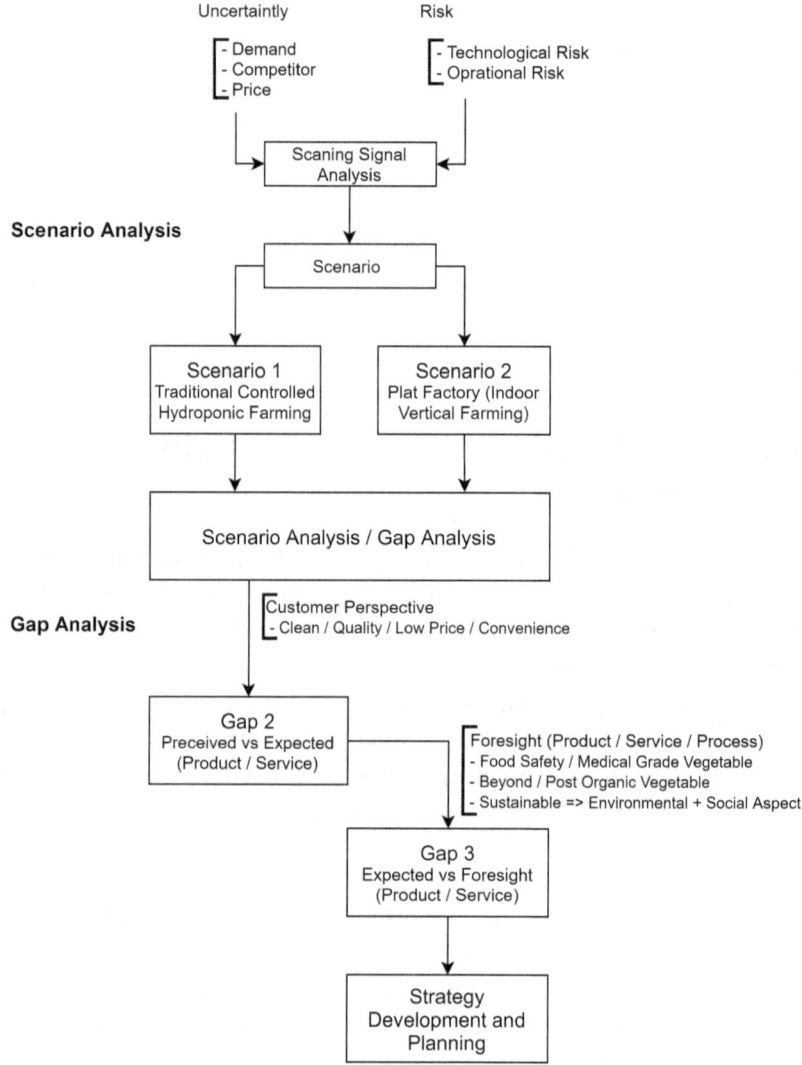

Fig. 12.8 Business foresight of Agritech

the climate change that directly affects the productivity of the traditional outdoor cultivation.

In terminology, the plant factory is referred to the facility with an artificial cultivation environment, including light, wind, temperature, moisture, and carbon dioxide concentrations. The control production parameters give the independency of the crop to the outdoor climate, which favors the steady production of high-quality vegetables (Goto 2012, Kim 2010; Kozai et al. 2019). The plant factory can produce vegetables 4 times faster than by typical outdoor cultivation. The productivity and quality, in terms of size, taste, texture, can be maximized and controllable. The factory requires only a small site owing to the vertical farming technology (multiple cultivation shelves system) which allows mass production for the economy of scales. The plant factory leverages the concepts of hydroponics which can grow plants without soil (Resh 1995) and the organic agriculture (Badgley et al. 2007; Willer and Lernoud 2019) with Industry 4.0 technologies such as data-driven and IoT-based agriculture (Gondchawar and Kawitkar 2016; Leksakul et al. 2015; Ramingwong et al. 2011; Suma et al. 2017). Thus, the vegetable from the plant factory can be cost-effective, clean, pesticide-free, and sustainable (Benke and Tomkins 2017; Santiteerakul et al. 2020). The idea is aligned with Thailand 4.0 targeted industry and the newly promoted bio-economy, circular economy, and green economy (BCG) model.

On the global level, the plant factory has become very popular and has been well received by many countries. For example, in Japan, currently, there are more than 200 plant factories in the operation. Where the biggest plant factory in Japan is in Miyagi prefecture, this farm is 2300 sqm., equipped with 18 cultivation racks reaching 15 levels high, producing 10,000 heads of lettuce per day. In Taiwan, the plant factory related to Foxconn can produce a vegetable of 2.5 tons per day. The size of the factory is 5000 sqm. with 14 plantation shelves. The product is supplied to Foxconn's staff kitchen.

While considering gap analysis, the customer perspectives are set as the cleanness, quality, price, and convenience are the bottom lines. The plant factory can address those issues. In fact, the quality of the vegetable and the production cost are beyond the expectation. The vegetable from the plant factory can be classified as medical-grade or beyond/post organic. The water used in the factory is reversed osmosis. It is so clean, as to wash with tap water, it will be dirtier. The plant factory also addresses the sustainability issues as it uses much less water. It requires only 1% of water, which is normally used in an organic farm.

With the advancement of technology, skilled workers are the plant factory requirement. Advanced agricultural and engineering skills are demanded. These complex Industry 4.0 systems of CPS, IoT, AI, and Big Data are beyond traditional (Rauch 2020).

12.4.3 Technology Blueprint Development—Plant Factory

After the business model is firmly analyzed, the development plan proceeds. The second facility of the company is set up in Nakorn-nayok, 113 km north-east of Bangkok (see Fig. 12.9). The project was further financially supported by the National Innovation Agency (NIA). This plant factory is 160 sqm. (0.018 hectares) with 6-m high multiple shelves. It can accommodate 50,000 plants or 5 tons of vegetables per crop. It initially requires 21–30 days per crop. The facility needs only 3 labor in operation. Moreover, the designed Standard Operating Procedure (SOP) allows individuals on the autism spectrum or elderly people to be able to work in this environment.

The facility has proven that the plant factory can be financial feasibility. Interestingly, the overall production cost can be low due to the capability to produce all-year-round. The plant factory can produce up

Fig. 12.9 Wangree plant factory in Nakornnayok, Thailand

to 20 crops per year. The equipment is today efficient and assessable with a considerably low cost such as LED technology, sensors, controllers. In this facility, the technology assessment focusing on the plant factory was conducted. This so-called inside-out and outside-in approach industrial research was assisted by Chiang Mai University, Maejo University, Ministry of Digital Economy and Society and Ministry of Industry of Thailand, Delta Electronics (Thailand) Public Co., Ltd, ASEAN Coordinating Committee on Micro, Small, and Medium Enterprise (ACCMSME), Price water house Coopers Thailand (PwC Thailand), Japan External Trade Organization (JETRO), Ministry of Foreign Affairs and Ministry of SME, and Startups of the Korean government. After years of project pitching and experiments, different production parameters can be optimized using Big Data, including a close system, artificial lighting system, environment controlled (temperature and moisture), carbon dioxide concentration, wind speed and direction controlled, PH controlled system, and Percent oxygen concentration in water. The recipe is designed for each vegetable if desired.

The designed plant factory is fully automated, multi-shelves, and smart (see Fig. 12.10). The plant is a CPS where the physical layer comprises of Farm Gate Way (FGW) Unit that connects with application unit, i.e., automated multi-shelf system, cultivation robot, light control, water control, temperature and humidity sensor, carbon dioxide concentration sensor, wind speed, and direction sensor (see Fig. 12.11). Then the data is collected and analyzed using the Big Data engine. AI is used to plan cultivation, determine production parameters, and control crop management. Figure 12.12 illustrates the process of developing technology blueprint after technology assessment and gap analysis.

Current productivity is comparatively superior to the organic farm. It requires a much smaller area, less water, less labor. The overall unit cost is competitive. Quality is also more desirable. The vegetable size can be 2 times bigger than those of organic farms. Taste and texture can also be controlled.

This pilot 160 sq.m. plant factory with the production capacity is 160 kg per day. The infrastructure investment and technology acquiring are estimated as much as 6-m THB. However, the plant factory can

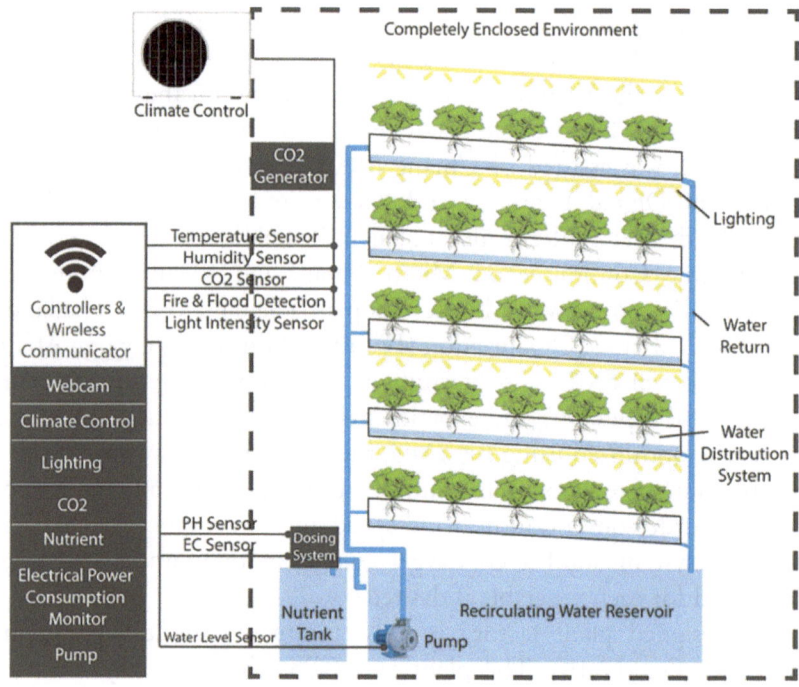

Fig. 12.10 Acquired technology in the plant factory of Agritech

generate a revenue of up to 3 m THB per month. The payback period can be as quick as 2 months.

12.4.4 Requirement of New Skills—Addressing SME 4.0

While the process can be fully automated, the number of operational workers can be reduced to minimal (Zsifkovits 2020). However, to design these sophisticating systems and advise the AI to the utmost efficiency (Woschank et al. 2020), multi-skill set workers are needed (Karacay 2018; Motyl et al. 2017). This includes professional technical production hard skills such as production management, logistics and supply chain engineering and management, robotics and automation

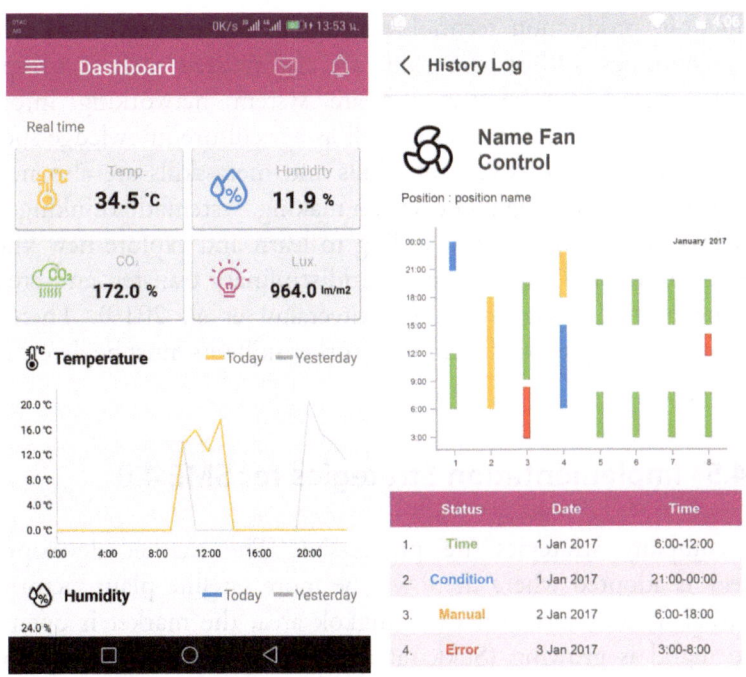

Fig. 12.11 Information systems developed for the plant factory

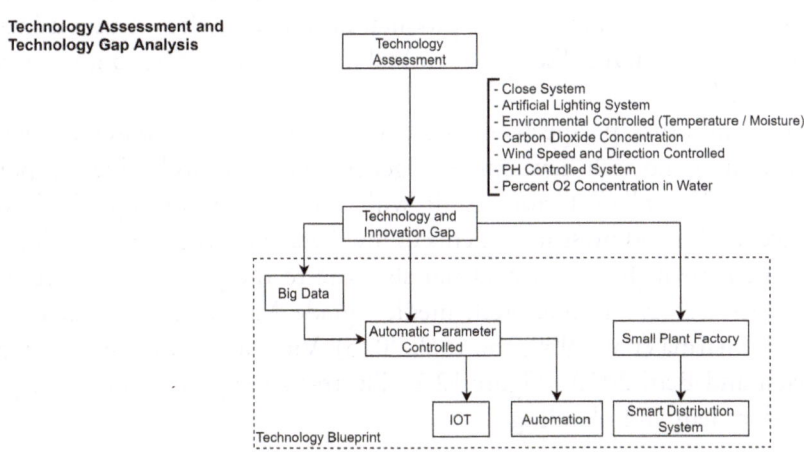

Fig. 12.12 Technology blueprint development of Agritech

production, production technology, engineering material, work-study, and ergonomics. Other hard skills are also needed such as computer engineer including hardware, software, system, networking, information technology, data analytics as well as agriculture knowledge such as plant physiology. Moreover, soft skills and meta-skills are also needed such as problems solving and decision making, systematic thinking, data analytics, and reasoning skill, willing to learn and explore new knowledge, creativity and innovation, multidisciplinary transfer, and creative thinking and idea generation (Santiteerakul et al. 2019). Therefore, appropriate strategies for developing worker skill sets must be determined accordingly.

12.4.5   Implementation Strategies for SME 4.0

The company strategies are progressive. The market development strategy is adopted where there will be more satellite plant factories as the urban indoor farm. In the Bangkok area, the market is open and the demand is growing (Sukkarat and Athinuwat 2020). This comes with a distribution and logistics improvement to address the need of the customer. The packaging development is also an issue to maintain the freshness and the quality of the product. Further complete chain business model, e.g., Amazon Fresh model, YesHealth iFarm of Taiwan, must be examined.

On the other perspectives, the company has been researching with universities and tech-development agencies with various business opportunities. The plant factory is fit with the concept of superfood for cancer and blood pressure patients or low Potassium vegetable for kidney disease patient. Further R&D can also lead to the production of edible vaccines (rabies vaccine, etc.), medical-grade Marijuana (Kumar et al. 2013; Sharma et al. 1999; Yao et al. 2015), Vitro meat (Bhat et al. 2015; Datar and Betti 2010). Figure 12.13 illustrates the processes of strategic development and planning.

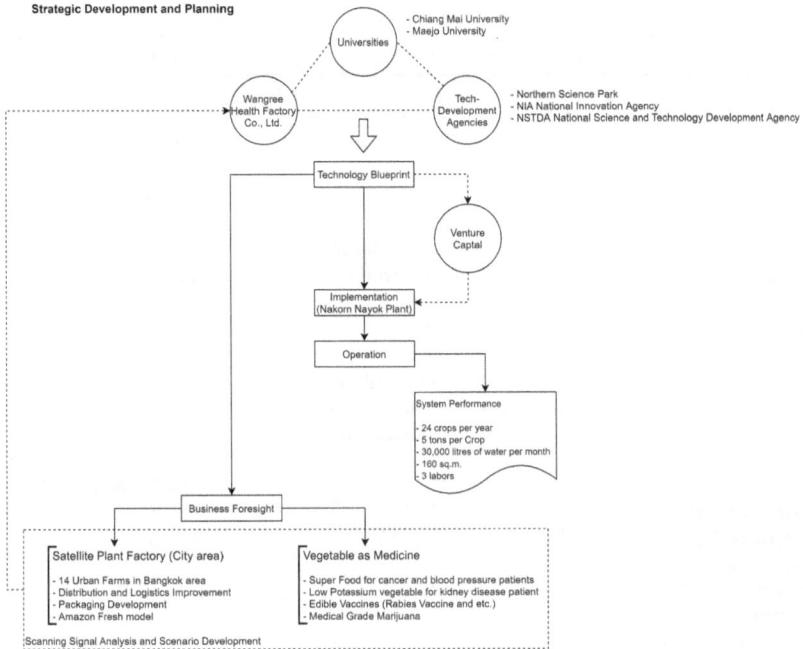

Fig. 12.13 Strategic development and planning of Agritech

12.5 Discussion

To review the targeted KPIs, the performance of the plant factory is benchmarked with organic and hydroponic farms. Table 12.1 summarizes the requirement with three different techniques of interest for producing 5 tons of vegetables per month.

Here, it can be seen that the plant factory is more desirable financially and more sustainable. However, skilled workers and advanced technology management can be critical as discussed.

Moreover, the business model is aligned with United Nations' Sustainable Development Goals (SDGs) (Stafford-Smith et al. 2017), i.e., zero hunger, good health and well-being, clean water and sanitation, affordable and clean energy, decent work and economic growth, industry, innovation and infrastructure, sustainable cities and communities, responsible consumption and production, climate action and life of land.

Table 12.1 Performance of each advance agriculture techniques

	Organic farm	Hydroponic farm	Plant factory
Specific Requirement	soil and nutrient	hard-infrastructure, water, and nutrient	hard-infrastructure, water and nutrient, multiple-shelf system, and automation system
Land Use	1.6 hectares	0.96 hectares	0.018 hectares
Water (Liter per month)	3 million	1.8 million	30,000
Number of workers	20	20	3
Product Characteristic	pesticide-free, smaller size	controllable size, and R&D potential	medical grade, pesticide-free, controllable size, taste, and texture, R&D potential
Crops per years	6–8	8–12	12–20
Average Cost per kg (THB)	80–120	50–70	42.25
Investment Cost (million THB)	12	10	6
Payback period (months)	4	1.5	2

The study demonstrates the use of the implementation strategies for SME 4.0 meta-model to the case study Thai Agritech SME. The company seeks a feasible business solution to address the demand for clean and quality fresh vegetables in Thailand. Following the first phases of analysis, which comprises business trend analysis, business foresight, and gap analysis, the case study company developed the business model of Plant Factory, which can utilize the benefit of Industry 4.0. New technology and investment are needed. Then to develop the plan, the company has been working in the triple-helix ecosystem. The company can then construct their technology blueprint to close the technology and innovation gap. Big Data as well as the Industry 4.0 concept leverages the production capability of the plant factory to be competitive to those other advanced agriculture systems. The new skill sets of labor are required to addressing the sophistication of this SME 4.0 including hard, soft, and meta-skills. Finally, the strategies are developed and planned as of further market expansion and evolution to Social start-up 4.0.

Acknowledgements This project has received funding from the European Union's Horizon 2020 research and innovation program under the Marie Skłodowska-Curie grant agreement No 734713. This research work was partially supported by Chiang Mai University—Thailand.

References

Antonopoulos, K., C. Panagiotou, C.P. Antonopoulos, and N.S. Voros. 2019. A-FARM Precision Farming CPS Platform. In *2019 10th International Conference on Information, Intelligence, Systems and Applications (IISA)*, pp. 1–3. https://doi.org/10.1109/IISA.2019.8900717.

Badgley, C., J. Moghtader, E. Quintero, E. Zakem, M.J. Chappell, K. Aviles-Vazquez, et al. 2007. Organic agriculture and the global food supply. *Renewable agriculture and food systems*, 86–108. https://doi.org/10.1017/S17 42170507001640.

Bär, K., Z.N.L. Herbert-Hansen, and W. Khalid. 2018. Considering Industry 4.0 aspects in the supply chain for an SME. *Production Engineering* 12 (6): 747–758. https://doi.org/10.1007/s11740-018-0851-y.

Benke, K., and B. Tomkins. 2017. Future food-production systems: Vertical farming and controlled-environment agriculture. *Sustainability: Science, Practice and Policy* 13 (1): 13–26. https://doi.org/10.1080/15487733.2017. 1394054.

Bhat, Z.F., S. Kumar, and H. Fayaz. 2015. In vitro meat production: Challenges and benefits over conventional meat production. *Journal of Integrative Agriculture* 14 (2): 241–248. https://doi.org/10.1016/S2095-3119(14)608 87-X.

Brozzi, R., D. Forti, E. Rauch, and D. Matt. 2020. The advantages of industry 4.0 Applications for sustainability: Results from a sample of manufacturing companies. *Sustainability* 12: 3647. https://doi.org/10.3390/su12093647.

Chetthamrongchai, P., and K. Jermsittiparsert. 2020. Ensuring environmental performance of pharmaceutical companies of Thailand: Role of robotics and AI awareness and technical content knowledge in industry 4.0 era. *Systematic Reviews in Pharmacy* 11 (1): 129–138. https://doi.org/10.5530/srp.2020.1.18.

Chinachoti, P. 2018. The readiness of human resource management for industrial business sector towards industrial 4.0 in Thailand. *Asian Administration & Management Review* 1 (2). https://ssrn.com/abstract=3269079.

Chonsawat, N., and A. Sopadang. 2019. The development of the maturity model to evaluate the smart SMEs 4.0 readiness. In *Proceedings of the International Conference on Industrial Engineering and Operations Management*. Bangkok, Thailand.

Dallasega, P., M. Woschank, S. Ramingwong, K.Y. Tippayawong, and N. Chonsawat. 2019. Field study to identify requirements for smart logistics of European, US and Asian SMEs. In *Proceedings of the International Conference on Industrial Engineering and Operations Management*. Bangkok, Thailand.

Datar, I., and M. Betti. 2010. Possibilities for an in vitro meat production system. *Innovative Food Science & Emerging Technologies* 11 (1): 13–22. https://doi.org/10.1016/J.IFSET.2009.10.007.

Gabriel, M., and E. Pessl. 2016. Industry 4.0 and sustainability impacts: Critical discussion of sustainability aspects with a special focus on future of work and ecological consequences. *Annals of the Faculty of Engineering Hunedoara—International Journal of Engineering* 14 (2): 131–136.

Galvao, A., C. Mascarenhas, C. Marques, J. Ferreira, and V. Ratten. 2019. Triple helix and its evolution: A systematic literature review. *Journal of Science and Technology Policy Management* 10 (3): 812–833. https://doi.org/10.1108/JSTPM-10-2018-0103.

Ganzarain, J., and N. Errasti. 2016. Three stage maturity model in SME's toward industry 4.0. *Journal of Industrial Engineering and Management* 9 (5): 1119–1128. http://dx.doi.org/10.3926/jiem.2073.

Gilchrist, A. 2016. Introducing industry 4.0. In *Industry 4.0*. Berkeley, CA: Apress. https://doi.org/10.1007/978-1-4842-2047-4_13.

Gondchawar, N., and R.S. Kawitkar. 2016. IoT based smart agriculture. *International Journal of Advanced Research in Computer and Communication Engineering* 5 (6): 838–842. https://doi.org/10.1109/ACCESS.2019.2932609.

Goto, E. 2012. Plant production in a closed plant factory with artificial lighting. In *VII International Symposium on Light in Horticultural Systems*, 956. https://doi.org/10.17660/ActaHortic.2012.956.2.

Griffin, T.W., J.M. Shockley, and T.B. Mark. 2018. Economics of precision farming. *Precision Agriculture Basics*, 221–230. https://doi.org/10.2134/precisionagbasics.2016.0098.

Hanafizadeh, P., and A.Z. Ravasan. 2011. A McKinsey 7S model-based framework for ERP readiness assessment. *International Journal of Enterprise Information Systems* 7 (4): 23–63. https://doi.org/10.4018/jeis.2011100103.

Hotrawaisaya, C., V. Pakvichai, and T. Sriyakul. 2019. Lean production determinants and performance consequences of implementation of industry 4.0 in Thailand: Evidence from manufacturing sector. *International Journal of Supply Chain Management* 8 (5): 559. https://doi.org/10.13140/RG.2.2.16491.69929.

Jazdi, N. 2014. Cyber physical systems in the context of Industry 4.0. In *IEEE International Conference on Automation, Quality and Testing, Robotics*. https://doi.org/10.1109/AQTR.2014.6857843.

Jones, C., and P. Pimdee. 2017. Innovative ideas: Thailand 4.0 and the fourth industrial revolution. *Asian International Journal of Social Sciences* 17 (1): 4–35. https://doi.org/10.29139/aijss.20170101.

Kaplan, R.S., and D.P. Norton. 1998. Putting the balanced scorecard to work. *The Economic Impact of Knowledge* 27 (4): 315–324.

Karacay, G. 2018. Talent development for industry 4.0. In *Industry 4.0: Managing the digital transformation*. Cham: Springer Series in Advanced Manufacturing, Springer. https://doi.org/10.1007/978-3-319-57870-5_7.

Katyal, N., and B.J. Pandian. 2020. A comparative study of conventional and smart farming. In *Emerging technologies for agriculture and environment*. Singapore: Springer. https://doi.org/10.1007/978-981-13-7968-0_1.

Kim, J.W. 2010. Trend and direction for plant factory system. *Journal of Plant Biotechnology* 37 (4): 442–455. https://doi.org/10.5010/JPB.2010.37.4.442.

Kohpaiboon, A. 2020. Industry 4.0 policies in Thailand. *Economic Working Paper* 2020–02.

Korkueasuebsai, O., and C. Pornsing. 2018. A study of factors and effects of Industry 4.0 policy on Thai electronics industry (Doctoral dissertation, Silpakorn University). http://ithesis-ir.su.ac.th/dspace/handle/123456789/1783.

Kozai, T., G. Niu, and M. Takagaki. 2019. *Plant factory: An indoor vertical farming system for efficient quality food production*. Academic Press.

Kumar, B.V., T.K. Raja, M.R. Wani, S.A. Sheikh, M.A. Lone, G. Nabi, et al. 2013. Transgenic plants as green factories for vaccine production. *African Journal of Biotechnology* 12 (43): 6147–6158. https://doi.org/10.5897/AJB 2012.2925.

Kumpirarusk, P. and K. Rohitratana. 2018. Industry 4.0: Future industries of Thailand. *WMS Journal of Management* 7 (3): 52–64. https://so06.tci-tha ijo.org/index.php/wms/article/view/147021.

Laosiritaworn, W., and W. Chattinnawat. 2019. Industry 4.0 gap analysis for Thai industries with association rules mining. In *Proceedings of the International Conference on Industrial Engineering and Operations Management*. Bangkok, Thailand.

Lasi, H., P. Fettke, H.G. Kemper, T. Feld, and M. Hoffmann. 2014. Industry 4.0. *Business & Information Systems Engineering* 6 (4): 239–242. https://doi. org/10.1007/s12599-014-0334-4.

Lee, J., B. Bagheri, and H.A. Kao. 2015. A cyber-physical systems architecture for industry 4.0-based manufacturing systems. *Manufacturing Letters* 3: 18–23. https://doi.org/10.1016/j.mfglet.2014.12.001.

Leksakul, K., P. Holimchayachotikul, and A. Sopadang. 2015. Forecast of off-season longan supply using fuzzy support vector regression and fuzzy artificial neural network. *Computers and Electronics in Agriculture* 118: 259–269. https://doi.org/10.1016/j.compag.2015.09.002.

Leydesdorff, L. 2010. The knowledge-based economy and the triple helix model. *Annual Review of Information Science and Technology* 44 (1): 365–417. https://doi.org/10.1002/aris.2010.1440440116.

Lu, Y. 2017. Industry 4.0: A survey on technologies, applications and open research issues. *Journal of Industrial Information Integration* 6: 1–10. https://doi.org/10.1016/j.jii.2017.04.005.

Matt, D.T., and E. Rauch. 2020. SME 4.0: The role of small-and medium-sized enterprises in the digital transformation. In *Industry 4.0 for SMEs: Challenges, opportunities and requirements*. Cham: Palgrave Macmillan. https://doi.org/10.1007/978-3-030-25425-4_1.

Mihiotis, A. 2014. Management of supply chain: X-to-order concepts vs make-to-stock model. *International Journal of Business Administration* 5 (3): 30. https://doi.org/10.5430/ijba.v5n3p30.

Modrak, V., Z. Soltysova, and R. Poklemba. 2019. Mapping requirements and roadmap definition for introducing I 4.0 in SME environment. *In Advances in manufacturing engineering and materials*. Cham: Springer. https://doi.org/10.1007/978-3-319-99353-9_20.

Motyl, B., G. Baronio, S. Uberti, D. Speranza, and S. Filippi. 2017. How will change the future engineers' skills in the Industry 4.0 framework? A questionnaire survey. *Procedia Manufacturing* 11: 1501–1509. https://doi.org/10.1016/j.promfg.2017.07.282.

Munkongsujarit, S. 2016. Business incubation model for startup company and SME in developing economy: a case of Thailand. In *2016 Portland International Conference on Management of Engineering and Technology*. IEEE. https://doi.org/10.1109/PICMET.2016.7806786.

Nakwa, K., and G. Zawdie. 2016. The 'third mission' and 'triple helix mission' of universities as evolutionary processes in the development of the network of knowledge production: Reflections on SME experiences in Thailand. *Science and Public Policy* 43 (5): 622–629. https://doi.org/10.1093/scipol/scw030.

Pérez, J.D.C., R.E.C. Buitrón, and J.I.G. Melo. 2018. Methodology for the retrofitting of manufacturing resources for migration of SME towards industry 4.0. In *International Conference on Applied Informatics*. Springer, Cham.

Phungphol, W., S. Tumad, K. Sangnin, and S. Pooripakdee. 2018. Creating passion for preparedness of automotive industry entrepreneurs for industry 4.0 era in the Southern part of Thailand. *International Journal of Business and Economic Affairs* 3 (1): 1–12. https://doi.org/10.24088/IJBEA-2018-31001.

Prause, G. 2015. Sustainable business models and structures for Industry 4.0. *Journal of Security & Sustainability Issues* 5 (2). https://doi.org/10.9770/jssi.2015.5.2(3).

Ramingwong, S., and W. Manopiniwes. 2019. Supportment for organization and management competences of ASEAN community and European Union toward Industry 4.0. *International Journal of Advanced and Applied Sciences* 6 (3): 96–101. https://doi.org/10.21833/ijaas.2019.03.014.

Ramingwong, S., W. Manopiniwes, and V. Jangkrajarng. 2019. Human factors of Thailand toward industry 4.0. *Management Research and Practice* 11 (1): 15–25.

Ramingwong, S., K.Y. Tippayawong, and A. Sopadang. 2011. On the development of I-community to improve production of off-season longan. *Australian Journal of Basic and Applied Sciences* 5 (10): 649–654.

Rauch, E. 2020. Industry 4.0+: The next level of intelligent and self-optimizing factories. *Book Advances in Design, Simulation and Manufacturing* 3: 176–186. https://doi.org/10.1007/978-3-030-50794-7_18.

Rauch, E., M. Unterhofer, R. Rojas, L. Gualtieri, M. Woschank, and D. Matt. 2020. a maturity level-based assessment tool to enhance the implementation of industry 4.0 in small and medium-sized enterprises. *Sustainability* 12: 1–18. https://doi.org/10.3390/su12093559.

Resh, H. M. 1995. *Hydroponic food production. A definitive guidebook of soilless food-growing methods* (No. Ed. 5). Woodbridge Press Publishing Company.

Rohrbeck, R. 2010. Corporate foresight: Towards a maturity model for the future orientation of a firm. In *Springer series: Contributions to management science.* Heidelberg and New York.

Rohrbeck, R., C. Battistella, and E. Huizingh. 2015. Corporate foresight: An emerging field with a rich tradition. *Technological Forecasting and Social Change* 101: 1–9. https://doi.org/10.1016/j.techfore.2015.11.002.

Rüßmann, M., M. Lorenz, P. Gerbert, M. Waldner, J. Justus, P. Engel, and M. Harnisch. 2015. Industry 4.0: The future of productivity and growth in manufacturing industries. *Boston Consulting Group* 9 (1): 54–89.

Safar, L., J. Sopko, S. Bednar, and R. Poklemba. 2018. Concept of SME business model for industry 4.0 environment. *TEM Journal* 7 (3): 626. https://doi.org/10.18421/TEM73-20.

Santiteerakul, S., A. Sopadang, and A. Sekhari. 2019. Skill Development for Industrial Engineer in Industry 4.0. In *Proceedings of IEEE—15th China-Europe International Symposium on Software Engineering Education.* Lisbon-Caparica, Portugal.

Santiteerakul, S., A. Sopadang, K.Y. Tippayawong, and K. Tamvimol. 2020. The role of smart technology in sustainable agriculture: A case study of Wangree plant factory. *Sustainability* 12: 4640. https://doi.org/10.3390/su12114640.

Sharma, A.K., A. Mohanty, Y. Singh, and A.K. Tyagi. 1999. Transgenic plants for the production of edible vaccines and antibodies for immunotherapy. *Current Science* 77 (4): 524–529.

Singh, A. 2013. A study of role of McKinsey's 7S framework in achieving organizational excellence. *Organization Development Journal* 31 (3): 39.

SME4.0. 2020. SME4.0 Project Objectives. http://www.sme40.eu/. Accessed on 10 June 2020.

Sopadang, A., N. Chonsawat, and S. Ramingwong. 2020. Smart SME 4.0 implementation toolkit. In *Industry 4.0 for SMEs: Challenges, opportunities and requirements.* Cham: Palgrave Macmillan. https://doi.org/10.1007/978-3-030-25425-4_10.

Stafford-Smith, M., D. Griggs, O. Gaffney, F. Ullah, B. Reyers, N. Kanie, N., et al. 2017. Integration: The key to implementing the Sustainable Development Goals. *Sustainability Science* 12 (6): 911–919. https://doi.org/10.1007/s11625-016-0383-3.

Stubbs, W., and C. Cocklin. 2008. Conceptualizing a "sustainability business model". *Organization & Environment* 21 (2): 103–127. https://doi.org/10.1177/1086026608318042.

Sukkarat, K., and D. Athinuwat. 2020. Study of consumption behavior and attitude of organic product consumer (in Thai). *Thai Journal of Science and Technology* 9 (1). https://doi.org/10.14456/tjst.2020.6.

Suma, N., S.R. Samson, S. Saranya, G. Shanmugapriya, and R. Subhashri. 2017. IOT based smart agriculture monitoring system. *International Journal on Recent and Innovation Trends in Computing and Communication* 5 (2): 177–181. https://doi.org/10.35940/ijitee.I7142.079920.

Tippayawong, K.Y., N. Niyomyat, A. Sopadang, and S. Ramingwong. 2016. Factors affecting green supply chain operational performance of the Thai auto parts industry. *Sustainability* 8 (11): 1161. https://doi.org/10.3390/su8111161.

Tippayawong, K.Y., S. Santiteerakul, S. Ramingwong, and N. Tippayawong. 2019. Cost analysis of community scale smokeless charcoal briquette production from agricultural and forest residues. *Energy Procedia* 160: 310–316. https://doi.org/10.1016/j.egypro.2019.02.162.

Willer, H., and J. Lernoud. 2019. *The world of organic agriculture. Statistics and emerging trends 2019.* Research Institute of Organic Agriculture FiBL and IFOAM Organics International.

Woschank, M., E. Rauch, and H. Zsifkovits. 2020. A review of further directions for artificial intelligence, machine learning, and deep learning in smart logistics. *Sustainability* 19: 1–23. https://doi.org/10.3390/su12093760.

Yao, J., Y. Weng, A. Dickey, and K.Y. Wang. 2015. Plants as factories for human pharmaceuticals: Applications and challenges. *International Journal of Molecular Sciences* 16 (12): 28549–28565. https://doi.org/10.3390/ijms161226122.

Zanwar, S.R., and R.D. Kokate. 2012. Advanced agriculture system. *International Journal of Robotics and Automation* 1 (2): 107–112. https://doi.org/10.11591/ijra.v1i2.382.

Zsifkovits, H., M. Woschank, S. Ramingwong, and W. Wisittipanich. 2020. State-of-the-art analysis of the usage and potential of automation in logistics.

In *Industry 4.0 for SMEs: Challenges, opportunities and requirements.* Cham: Palgrave Macmillan. https://doi.org/10.1007/978-3-030-25425-4_7.

Index

© The Editor(s) (if applicable) and The Author(s) 2021
D. T. Matt et al. (eds.), *Implementing Industry 4.0 in SMEs*,
https://doi.org/10.1007/978-3-030-70516-9